AF518363

Advances in Biochemistry and Biotechnology

Advances in Biochemistry and Biotechnology

— *Volume 1* —

Edited by

Chiranjib Chakraborty

Head, Department of Biotechnology, Institute of Applied Medicines & Research
9th Mile Stone, Duhai, Delhi-Meerut Road, Ghaziabad.
(Previously Manager, Biotechnology (R&D), Glenmark Laboratories, Mumbai;
Research Scientist, Biotechnology, Macleods Pharmaceuticals Ltd., Mumbai;
Lecturer, Department of Industrial Fish and Fishery, Burdwan Raj College, Burdwan, West Bengal, India)

2005

DAYA PUBLISHING HOUSE

Delhi - 110 035

© CHIRANJIB CHAKRABORTY (b. 1973 –)
First Edition, 2005

ISBN 81-7035-362-9

All rights reserved. Including the right to translate or to reproduce this book or parts thereof except for brief quotations in critical reviews.

Published by : **Daya Publishing House**
1123/74, Deva Ram Park
Tri Nagar, Delhi - 110 035
Phone: 27383999
Fax: (011) 23244987
e-mail: dayabooks@vsnl.com
website: www.dayabooks.com

Showroom : 4762-63/23, Ansari Road, Darya Ganj
New Delhi - 110 002
Phone: 23245578, 23244987

Laser Typesetting : **Classic Computer Services**
Delhi - 110 035

Printed at : **Chawla Offset Printers**
Delhi - 110 052

PRINTED IN INDIA

Preface

With all optimistic view, we admit that we know very little of the whole nature and its governing. We are still trying to understand nature. We are still trying to understand. We are trying to decipher the codes used for storing various information with the advanced technology of Biological sciences i.e. Biotechnology and Biochemistry, which have developed since last 50 years.

This book has been designed to incorporate contributors of several Indian scientists of admirable skill and standing in India who are participated into the advancement of this subject.

I would like to thank all the contributors for their contributions into this volume.

Views expressed in the contributions are exclusively those of the authors and responsibilities do not lie on the editors and the publisher.

Last but not the least; I take the privilege to admiring the Managing Director of Daya Publishing House, Mr. Anil. K. Mittal thankfully for completing this publication in a hot hurry.

Dr. Chiranjib Chakraborty

Contents

List of Contributors

A. Ghosh

Mosquito Research Unit, Department of Zoology, Burdwan University, West Bengal, India

Akash Verma

School of Biotechnology, Guru Ghasidas University, Bilaspur, Koni, Chhattisgarh

Anirban Banerjee

Department of Pharmaceutical Technology, National Institute of Pharmaceutical Education and Research, Sector 67. S.A.S. Nagar–160 062, Punjab

Arun Sharma

Head, Food Technology Division, B.A.R.C., Mumbai–400 085, Tel: 91-22-25505180/25505314, Fax: 91-22-25505151/25519613, Email: ksarun@apsara.barc.ernet.in

Arup Tewary

Aquaculture Research Unit, Department of Zoology, Vidyasagar University, Midnapore–721 102, West Bengal, India

Bidhan C. Patra

Aquaculture Research Unit, Department of Zoology, Vidyasagar University, Midnapore–721 102, West Bengal, India, Telephone: Office: (91) 03222-62297, Residence: (91) 03222-66035, Fax No.: (91) 03222-62329, e-mail: bcpatra@yahoo.com

Biplab Sarkar

Central Institute of Fresh Water Aquaculture, Kausalyaganga, Bhubaneswar, Orissa

Brajesh Barse

Department of Pharmaceutical Technology, National Institute of Pharmaceutical Education and Research, Sector 67. S.A.S. Nagar–160 062, Punjab

Chiranjib Chakraborty

Head, Department of Biotechnology, Institute of Applied Medicines & Research, 9th Mile Stone, Duhai, Delhi-Meerut Road, Ghaziabad

Debabrata Mukherjee

Lab. of National Fellow, Central Inland Fisheries Research Institute, C.G.O. Complex, Salt Lake, Kolkata- 64

G. Chandra

Mosquito Research Unit, Department of Zoology, Burdwan University, West Bengal, India

G.C. Mishra

Director, National Centre for Cell Science, NCCS Complex, Ganeshkhind, Pune – 411 007, Maharashtra, Tel: 020-5491065, Fax: 020-5697756, E-mail: gcmishra@nccs.res.in

Garima Sharma

Centre for Biomedical Engineering, Indian Institute of Technology Delhi, Hauz Khas, New Delhi–110 016

J.K. Deb

Department of Biochemical Engineering and Biotechnology, I.I.T., Delhi, New Delhi–110 016

M.A. Haniffa

Centre for Aquaculture Research and Extension, St. Xavier's College, Palayamkottai, Tirunelveli–627 002, Tel: +91 462 2560670: Fax: +91 462 2560670; E-mail: haniffacare@hotmail.com

Mani Shankar Bhattacharyya

Department of Pharmaceutical Technology, National Institute of Pharmaceutical Education and Research, Sector 67. S.A.S. Nagar–160 062, Punjab

Pankaj Soni

Department of Pharmaceutical Technology, National Institute of Pharmaceutical Education and Research, Sector 67. S.A.S. Nagar–160 062, Punjab

Prashant Mishra

Department of Biochemical Engineering and Biotechnology, Indian Institute of Technology Delhi, Hauz Khas, New Delhi–110 016

Prveen Kaul

Department of Pharmaceutical Technology, National Institute of Pharmaceutical Education and Research, Sector 67. S.A.S. Nagar – 160 062, Punjab

R.K. Saxena

Centre for Biomedical Engineering, Indian Institute of Technology Delhi, Hauz Khas, New Delhi–110 016

Ragini Gothalwal

Head of the Department, School of Biotechnology, Guru Ghasidas University, Bilaspur, Koni, Chhattisgarh, E-mail: ragini_gothalwal@hotmail.com

Rohit Sharma

Department of Pharmaceutical Technology, National Institute of Pharmaceutical Education and Research, Sector 67. S.A.S. Nagar–160 062, Punjab

S. Adhikari

Central Institute of Fresh Water Aquaculture, Kausalyaganga, Bhubaneswar, Orissa

S. Ayyappan

Deputy Director General, (Fisheries), ICAR, Krishi Anusandhan Bhaban, PUSA, New Delhi

S.A. Bapat

National Centre for Cell Science, NCCS Complex, Ganeshkhind, Pune 411 007, Maharashtra

Satyendra Gautam

Food Technology Division, Bhabha Atomic Research Centre, Mumbai–400 085

Sawraj Singh

Department of Pharmaceutical Technology, National Institute of Pharmaceutical Education and Research, Sector 67. S.A.S. Nagar–160 062, Punjab

Shyam Sundar Nandi

School of Biotechnology, Guru Ghasidas University, Bilaspur, Koni, Chhattisgarh, E-mail: nandibiotech@yahoo.co.in

Surajit Sinha

National Centre for Cell Science, NCCS Complex, Ganeshkhind, Pune 411 007, Maharashtra

T.K.Maiti

Department of Biotechnology, I.I.T.-Kharagpur (W.B), E.Mail- tkmaiti@hijli.iitkgp.ernet.in

U.C. Banerjee

Department of Pharmaceutical Technology, National Institute of Pharmaceutical Education and Research, Sector 67. S.A.S. Nagar–160 062, Punjab, Phone: 91-172-2214682-85 Extn. 2142, Fax 91-172-22114692, E-mail: ucbanerjee@niper.ac.in

Advances in Biochemistry and Biotechnology Volume 1 *Pages 1-9*
Edited by: **Dr. Chiranjib Chakraborty**
Published by: **DAYA PUBLISHING HOUSE**

From Editor Desk

Current Changing Scenario of Biotechnology and Biochemistry

Chiranjib Chakraborty
Head, Department of Biotechnology, Institute of Applied Medicines & Research, 9th Mile Stone, Duhai, Delhi-Meerut Road, Ghaziabad

Biotechnology and biochemistry has involved with many parts of the chemical industry, agriculture, and medicine. These two subjects have demarcation between basic and applied research, and new discoveries and innovations, in most cases, can find direct application. Innovations, techniques, and tools that have emerged and revolutionized modern biotechnology include genetic engineering, cell fusion technology, bioprocess technologies, and structure-based molecular designs including drug development, drug targeting, and drug delivery systems.

After 20 years of growing pains, the biotechnology sector is coming into its own as a mature industry, as highlighted in the article on Nature Reviews Drug Discovery (1). This is reflected in the number of companies that have made lucrative licensing deals with toptier pharmaceutical companies, as well as in the growing number of biotech companies that are on their way to becoming fully integrated drug development enterprises. With several leading biotech companies posting billion dollar earnings in 2001, there is little doubt that the major players in this industry have discovered a formula for success.

Leading biotech companies can now successfully compete with big pharmaceutical companies for the rights to late-stage compounds, and mid-size biotechs are using their considerable cash reserves to strike deals with smaller companies in attempts to become more integrated, product-focused organizations (2). Smaller biotechs have benefited by having the option of negotiating more equitable deals with other biotechs, who might have more at stake than big pharma companies and therefore be more committed to the success of a particular product. In 2001, there were more than 1,500 biotech companies in Europe and approximately 1,300 in the United States, but 2001 revenues for European biotechs were less than one-third that of US revenues (US $8 versus US $25 billion, respectively)(3). Most of the profitability within this sector came from those companies who have expanded their discovery and development capabilities over the years–in part, by making deals with or acquiring smaller, specialized biotech companies. Looking ahead, further intra-industry deal making and eventual consolidation within the biotech industry is inevitable if biotech companies are to survive to profitability.

Indian Biotechnology and Biochemistry Advancement

In the 1980s the Government of India considered the need for creating a separate institutional framework to strengthen biology and biotechnology research in the country. Scientific agencies supporting research in modern biology included: Council of Scientific and Industrial Research (CSIR), Indian Council of Agricultural Research (ICAR), Indian Council of Medical Research (ICMR), Department of Science and Technology, and University Grants Commission. Biotechnology was given an important boost in 1982 with the establishment of the National Biotechnology Board. Its priorities were human resource development, creation of infrastructure facilities, and supporting research and development (R and D) in specific areas. The success and impact of the National Biotechnology Board prompted the Government to establish a separate Department of Biotechnology (DBT) in February 1986. There have been major accomplishments in areas of basic research in agriculture, health, environment, human resource development, industry, safety, and ethical issues (4).

India led through the work of G.N. Ramachandran, in which he elucidated the triple helical structure of collagen. The Ramachandran plot has proven to be fundamental in solving the protein structure. Areas of biosystematics using molecular approaches, mathematical modeling, and genetics including genome sequencing for human beings, animals, and plants, will continue to have priority as we move into the next century. The tremendous impact of genome sequencing is increasingly evident in many fields. As an increasing number of new genes are discovered, short, unique, expressed sequenced tags segments are used as signatures for gene identification. The power of high throughput sequencing, together with rapidly accumulating sequenced data, are opening new avenues in biosciences. In the plant genome area, the sequencing of *Arabidopsis* and rice genome will soon be completed and cataloging and mapping of all the genes will be done.

There have been major achievements in basic bioscience in the last decade or so in India, where we have expertise in practically all areas of modern biology. The institutions under the CSIR, ICMR, ICAR, DST, and DBT have established a large number of facilities where most advanced research work in biosciences is being done. In the identification of new genes, development of new drug delivery systems, diagnostics, recombinant vaccines, computational biology, and many other related areas, considerable success has been achieved. Breakthroughs include studies on the three-dimensional structure of a novel amino acid, a long protein of mosquito (University of Poona), and demonstration of the potential of the reconstituted Sendai viral envelops containing only the F protein of the virus, as an efficient and site-specific vehicle for the delivery of reporter genes into hepatocytes (Delhi University).

In India we have achieved the cloning and sequencing of at least six genes, developed regeneration protocols for citrus, coffee, mangrove species, and new types of biofertilizer and biopesticide formulations, including mycorrhizal fertilizers. Research to develop new genetically improved (transgenic) plants for brassicas, mung bean, cotton, and potato is well advanced. Industries have also shown a keen interest in the options of biotechnology and are participating in field trials and pilot level productions. The successful tissue culture pilot plants in the country, one at TERI in New Delhi and the other at NCL in Pune. The world's first IVF buffalo calf (PRATHAM) was born through embryo transfer technology at the National Dairy Research Institute, Karnal. Multiple ovulation and embryo transfer, in vitro embryo production, embryo sexing, vaccines and diagnostic kits for animal health have also been developed.

With a coastline of more than 8,000 kilometers, and two island territories of Andaman and Nicobar and Lakshadweep, there is great potential for marine resource development and aquaculture.

To achieve an annual target production of 10 million metric tons of fish, scientific aquaculture offers great possibilities. In fact, aquaculture products are among the fastest moving commodities in the world. We have to continuously improve seed production, feed, health products, cryopreservation, genetic studies, and related environmental factors.

It is scientifically well established that an environmentally benign way of ensuring food security is through bioengineering of crops. For the billion people in developing countries, one billion do not get enough to eat and live in poverty. Is there any other strategy or alternative? Biotechnology will provide the new tools to breeders to enhance plant capacity. Since we know that 12 per cent of the world land is under agricultural crops, it is projected that the per capita availability may be reduced from 2.06 hectares to 0.15 hectare by 2050.

Our National Plant Genome Research Centre at Jawaharlal Nehru University dealing with plant molecular biology in different parts of the country were initially responsible for training significant numbers in crop biotechnology. There are innumerable possibilities of producing more proteins, vitamins, and pharmaceuticals from plant. Biotechnology is becoming a major tool in conservation biology. Twelve per cent of the vascular plants are threatened with extinction. Over 5,000 animal species are threatened worldwide, including 563 Indian species. India also has about 2000 species of vascular plants that are threatened.

Globally, there are about 35–40 biotechnology-derived therapeutics and vaccines in use and more than 500 drugs and vaccines in different stages of clinical trials. Every year about 12 million people die of infectious diseases. The main killers according to WHO are acute respiratory infection, diarrheal diseases, tuberculosis, malaria, hepatitis, and HIV-AIDS. There are vaccines being developed for many diseases, and diagnostic kits for HIV, pregnancy detection, and hepatitis are being developed. The technologies have been transferred to industry. National Brain Research Centre is being established to improve knowledge of the human brain and the brain diseases. The age-old system of Ayurveda practiced in India needs to be popularized and made an integral part of health care. The global market for herbal products may be around US$5 trillion by 2050. The coming together of biotechnology and informatics is paying rich dividends. Genome projects, drug design, and molecular taxonomy are all becoming increasingly dependent on information technology. Information on nucleotides and protein sequences is accumulating rapidly.

The number of genes characterized from a variety of organisms and the number of evolved protein structures are doubling every two years. In India it has established a national Bioinformatics Network with different Distributed Information Centres which help to establishment of genomic databases. It is also helping to the Indian scientists. There are about several approved MS, postdoctoral, and MD training programs in biotechnology in progress or just about to start, in different institutions and universities covering most Indian States. Short-term training programs, technician training courses, fellowships for students to go abroad, training courses in Indian institutions, popular lecture series, awards, and incentives form an integral part of the human resource development activities in India.

States are taking a keen interest in developing biotechnology-based activities. The States of Delhi,Uttar Pradesh, Arunachal Pradesh, Madhya Pradesh, Kerala, West Bengal, Jammu and Kashmir, Haryana, Mizoram, Punjab, Gujarat, Meghalaya, Sikkim and Bihar have already started large-scale demonstration activities and training programs.

India will require a strong educational and scientific base, clear public understanding of the value of new biotechnologies, and involvement of society in many of these biological ventures.

India has a large research and educational infrastructure comprising 29 agriculture universities, 204 central and state universities, and more than 500 national laboratories and research institutions.

It should therefore be possible to develop capabilities and programs so that these institutions act as regional hubs for the farming community, where they can get direct feedback about new technological interventions. It will be equally important to establish strong partnerships and linkages with industry, from the time a research lead has emerged until the packaging of the technology and commercialization are achieved (4-6).

Advancement in Drug Development, Pharmaceutical Industry and Medical Sector Throughout Globe

The biotech industry has produced several enabling technologies over the years to support faster, cheaper and more targeted drug discovery and development. Ultimately, the goal of these technologies is to reduce (potentially halve) discovery time and increase (potentially triple) the number of new chemical entities (NCEs) entering the development stage. The question is, how will these technologies contribute to the development of better and/or safer therapies for the future? Simply supplying more targets or more drug leads does not, in itself, assure success. What will increase the chances of success–and what some biotech-derived tools will facilitate–is the ability to define, with a high level of certainty, a population that will respond to a particular therapy, with minimal risk for failure. This emerging discipline has been termed 'Pharmacogenomics'(7).

At the heart of Pharmacogenomics are technologies designed to identify single-nucleotide polymorphisms (SNPs), which are small genetic variations that might be linked to disease susceptibility or to drug response. The most prominent use of SNP technology in the future is likely to involve defining populations of potential drug responders compared with potential non-responders for clinical trials, identifying individuals who are likely to experience adverse events in clinical trials or carrying out post-marketing analysis of drug safety and efficacy (8). In fact, the US Food and Drug Administration (FDA) has already voiced support for this approach (9), providing validation for Pharmacogenomics analysis as a valuable tool in the drug-approval process. So far, SNP analysis has confirmed the previous discovery(10-14) of the E4 allele of apolipoprotein (*APOE4*) as the susceptibility gene variant that is linked to common, alzheimer's disease (12), and allowed the discovery *SLC12A84* (16) and the insulin receptor (*INSR*) gene on chromosome 19 (17) as susceptibility gene variants in psoriasis and aurapositive migraine, respectively.

Although not yet adopted by most companies carrying out clinical trials, Pharmacogenomics has yielded valuable information for big pharma companies, such as Glaxo SmithKline (GSK), which has implemented this type of analysis in clinical trials (8). One example of the successful use of Pharmacogenomics analysis–not in the initial clinical trial process, but in post-launch safety analyses–was in a drug-rescue programme for GSK's irritable bowel syndrome (IBS) drug Lotronex (alosetron). The company voluntarily took Lotronex off the US market in 2001 after reports of serious adverse events–that is, ischaemic colitis and other severe complications–some of which resulted in death.

Subsequent Pharmacogenomics studies completed by GSK allowed the company to submit a supplemental new drug application (sNDA) to the FDA for a new and more narrowly defined subset patients in whom the risk/benefit ratio was favourable. The FDA has now approved the drug with the condition that it be prescribed according to a strict risk-management programme, and Lotronex was relaunched mid-2002. This decision makes Lotronex the first example of a drug that is being successfully reintroduced to market with a risk-management programme.

A further evolution in this technological approach will be its use in preventive medicine. It is quite conceivable at this time that pharmacogenetics will allow the proactive identification of populations at risk or predisposed towards particular diseases–for example, type 2 diabetes, hyperlipidaemia and high blood pressure. This will give rise to a whole new class of drugs that are designed to stave off the onset of acute disease. The pharmacoeconomic benefit and impact to society would be enormous, as the end-stage co-morbidities that are associated with many of these diseases (18).

Disciplines such as functional genomics and proteomics will serve to expand greatly our understanding of the biological roles of different genes and proteins, both in healthy and disease states.With the map of the human genome completed, functional-genomic studies are rapidly expanding the list of potentially therapeutic genes. Proteomics, which is still a fairly undeveloped discipline in terms of its potential to contribute to drug development, is rapidly maturing, and is widely anticipated to contribute to major advances in drug discovery(19). Understanding the proteome in all its complexity is the new frontier in disease biology, with important clinical ramifications, as it will increasingly refine our molecular understanding of specific diseases (20). As important as genomics is to drug discovery today, proteomics will one day supersede it in its importance in both drug discovery and the practice of clinical medicine.

The Future Revolutions Throughout the Globe

To meet the broader needs of the health-care market in the future, biotech companies are taking novel approaches to the treatment of disease outside the realm of traditional drug discovery and development. Many are pursuing stem-cell biology for use in a wide range of therapeutic applications, whereas others are refining tissue-engineering technologies to aid in the development of whole-organ replacement. Although these approaches will not become a clinical reality in the near future, we predict that they will substantially affect the market within the next 20 years. Stem-cell therapies, in particular, could potentially address a broad range of disease states (21), as the technologies that are being developed in this area could allow differentiation of stem cells into numerous cell types. Recent findings of adult stem cells are enabling tissue and organ engineering (22) that will one day allow wholesale transplantation of new tissues and organs. Discoveries about the growth and differentiation factors and conditions that drive tissue development could, in the future, allow the *in situ* regeneration of cells and tissues within patients.Although this field remains in its infancy, new technologies will

Figure: Evolutionary Tree and Advancement of Biotechnology and Biochemistry

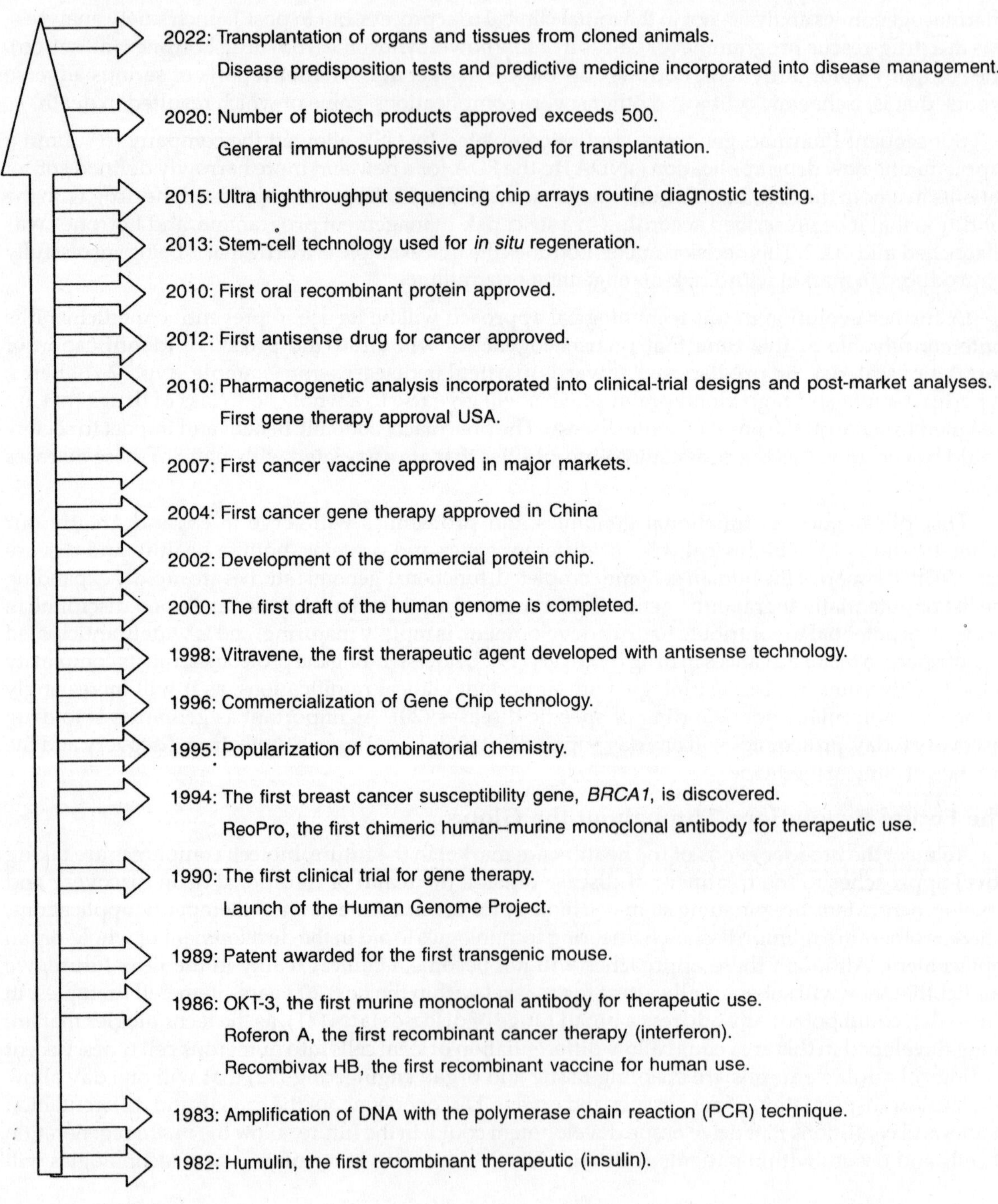

no doubt emerge to allow the isolation, expansion and differentiation of stem cells from various sources. This could, ultimately, lead to widespread therapeutic applications in areas including cardiovascular disease, neurodegenerative disease, metabolic disease, bone disease and repair, and haematological disorders. A similarly novel approach of nearer-term benefit, and representative of how biotechnology has overcome tremendous hurdles to bring truly novel therapies to late-stage clinical development, is gene therapy (23). Since its inception in the 1970s, the field of gene therapy has been fraught with safety concerns, primarily stemming from gene delivery–which historically has necessitated the use of viral vectors–but also from the need to control gene expression.

Despite these obstacles–and an abundance of negative press throughout the years when novel therapies did not live up to their potential or when safety concerns were a cause for halting clinical trials–a number of companies have stayed the course and survived to see their technologies reach late-stage clinical development. In fact, there are several gene therapies in Phase III trials at present, most of them targeting cancer indications. We predict that at least one of these will be approved by 2006. At present, cancer and cardiovascular disease–both leading causes of death and two of the biggest potential markets in the western world–are the main targets for gene therapies. Gene therapies that have reached Phase III clinical development for cancer include Introgen's head and neck cancer therapy Advexin–for which a BLA could be filed as early as 2004; Onyx Pharmaceutical's ONYX-015 for head and neck cancer; and Vical's Allovectin-7 (HLA-B7 gene therapy) for melanoma.Other active areas of development for gene therapies include infectious diseases (such as HIV/AIDS, hepatitis and human papillomavirus) and central nervous system disorders; in fact, the first gene-therapy trial for Parkinson's disease was approved in October 2002, amidst much excitement within the Parkinson's community, with plans to begin the Phase I trial by 2003. In the fourth quarter of 2002, there were more than 600 completed, ongoing or pending gene-therapy trials worldwide; in the United States alone, the FDA had 196 active gene-therapy Investigational New Drugs (INDs) (106 for *in vivo* gene transfer and 90 for *ex vivo* gene transfer). However, in light of some highly publicized safety concerns from clinical trials in recent years–specifically, the two deaths that led the FDA to halt Vascular Genetics' angiogenesis trial using vascular endothelial growth factor 2 (VEGF2), and the more recent reported case of leukaemia in a child involved in the X-SCID (X-linked severe combined immunodeficiency) gene trial–regulatory agencies in many countries are urging extra caution for gene-therapy trials. That said, many in the scientific community, and several biotech companies with expertise in this arena, hold unwavering support for this approach to treating disease. Despite significant obstacles, the reward potential for a successful gene therapeutic is huge, given the diseases that most companies are targeting (such as cancer and cardiovascular diseases). Similarly, the reward potential for stem-cell therapies or other tissue-replacement technologies, which might encounter obstacles much like those encountered by gene therapies, could be huge if these approaches offer effective treatments for degenerative disorders. In essence, novel biotechnology products, such as those previously mentioned as well as others such as cancer vaccines and antisense therapeutics, hold great potential because they might fill major areas of unmet need that are not currently met by traditional pharmaceuticals.

To achieve the goal of self-reliance in this field, India will require a strong educational and scientific base, clear public understanding of the value of new biotechnologies, and involvement of society in many of these biological ventures.

Arther Kornberg, Nobel Laureate, stated: *"Much has been said about the future impact of biotechnology on industrial development, but this does not yet apply to the less developed countries that lack this infrastructure and industrial strength. In view of the current power of biotechnology and its even brighter future, there is no question that the less developed countries must now position and strengthen their status in biotechnology."*

REFERENCES

1. Berg, C., Nassr, R. and Pang, K. (2002) The evolution of biotech. **Nature Rev. Drug Discov.** (1) 845–846.
2. Decision Resources, Inc. *Biotech Alliances* (2001): *Feeding and Riding the Tiger* (Decision Resources, Inc., Spectrum, Pharmaceutical Industry Dynamics, Issue 6, 22 May 2002).
3. Decision Resources, Inc. (2002) Outlook for the European Biotechnology Industry: Industry and Government Cooperation Key to Continued Growth (Decision Resources, Inc., Spectrum, Pharmaceutical Industry Dynamics, Issue 9.
4. Sharma, M.(2000) India: Biotechnology Research and Development.51-57
5. French, C.E., S.J. Rosser, G.J. Davies, S. Nicklin, and N.C. Bruce. (1999) Biodegradation of explosives by transgenic plants expressing pentaerythritol tetranitrate reductase. **Nature Biotechnology** 17(5) 491-494.
6. James, C. 1998. Global Review of Transgenic Crops: 1998. ISAAA Brief No. 8. Ithaca, N.Y.: International Service for the Acquisition of Agri-biotech Applications.
7. Lindpaintner, K. (2002) The impact of pharmacogenetics pharmacogenomics on drug discovery. **Nature Rev. Discov**. (1), 463–469.
8. Roses, A. D. (2002) Genome-based pharmacogenetics and pharmaceutical industry. **Nature Rev. Drug Discov.** (2) 541–549.
9. Lesko, L. J. and Woodcock, J. (2002) Pharmacogenomic-guided drug development: regulatory perspective. **Pharmacogenomics J.**(2) 20–24.
10. Roses, A. D. (2002) Apolipoprotein E alleles as risk factors Alzheimer's disease. **Annu. Rev. Med**. (47) 387–400
11. Farrer, L. A. (1997) Effects of age, sex, and ethnicity association between apolipoprotein E genotype and Alzheimer disease. A meta-analysis. APOE and Alzheimer Disease Meta Analysis Consortium. **JAMA** (278) 1349–1356.
12. Tang, M. X. (1998) The *APOE4* allele and the risk of Alzheimer disease among African Americans, whites, Hispanics. **JAMA** (279) 751–755.
13. Kardaun, J. W. (2000) Genotypes and phenotypes for apolipoprotein E and Alzheimer disease in the Honolulu-Asia aging study. **Clin. Chem.** (46) 1548–1554.
14. Graff-Radford, N. R. (2002) Association between apolipoprotein E genotype and Alzheimer disease in African American subjects. **Arch. Neurol**. (59) 594–600.
15. Lai, E., Riley, J., Purvis, I. and Roses, A. D. (1998) Ȧ 4-Mb high density single nucleotide polymorphism-based map around human *APOE*. **Genomics** (54) 31–38.

16. Hewett, D. (2002) Identification of a psoriasis susceptibility candidate gene by linkage disequilibrium mapping a localized single nucleotide polymorphism map. **Genomics** (79) 305–314.

17. McCarthy, L. C. (2001) Single-nucleotide polymorphism alleles in the insulin receptor gene are associated with typical migraine. **Genomics** (78) 135–149.

18. Tricia N., Carilee B., Rena Nassr and Kevin P.(2003) The further evolution of biotech. **Nature Reviews** (2): 75-80.

19. Burbaum, J. and Tobal, G. M. (2002) Proteomics in drug discovery. ***Curr. Opin. Chem. Biol.*** (6) 427–433.

20. Petricoin, E. F., Zoon, K. C., Kohn, E. C., Barrett, J. and Liotta, L. A. (2002) Clinical proteomics: translating benchside promise into bedside reality. **Nature Rev. Drug Discov.** (1) 683–695.

21. Lovell-Badge, R. (2001) The future for stem cell research. **Nature** (414) 88–91.

22. Bianco, P. and Gehron Robey, P. (2001) Stem cells in tissue engineering. **Nature** (414) 118–121.

23. Chiranjib Chakraborty and Atanu Bhatcharyya (2004): Pharmacogenomics: Approaches to new drug development. Biotech Books, New Delhi, India, pp 140-143.

Advances in Biochemistry and Biotechnology Volume 1 *Pages 10-52*
Edited by: **Dr. Chiranjib Chakraborty**
Published by: **DAYA PUBLISHING HOUSE**

1

Recent Developments in Chondrocyte Cultures and Cartilage Engineering

*R.K. Saxena, Garima Sharma & Prashant Mishra**
*Centre for Biomedical Engineering and *Department of Biochemical Engineering and Biotechnology, Indian Institute of Technology Delhi, Hauz Khas, New Delhi – 110016*

ABSTRACT

Chondrocytes play a major role in physiological and load bearing properties of articular cartilage. Articular cartilage tissue is important for normal and friction free movements of synovial joints. Chondrocytes are embedded in an abundant extracellular hydrated matrix comprising predominantly of type II collagen and proteoglycans. The collagens are organized into an insoluble fibrous network of high tensile strength, which contain the soluble deformable PGs that give the tissue its ability to undergo reversible deformation. The biomechanical properties of articular cartilage rely on the biochemical composition and integrity of its extra cellular matrix. The metabolism of articular cartilage represents a fine balance between synthesis and degradation of extracellular matrix. It is a feature of normal growth development and maintenance of articular cartilage in adult. In degenerative joint disease or osteoarthritis, cartilage shows a distinct loss of proteoglycan, marked collagen network fibrillation, increased tissue hydration and loss of

* Correspondent Author

compressive stiffness of the tissue. These changes are followed by enhanced level of tissue degrading enzymes like matrix metalloproteases-1 (MMP-1), matrix metalloproteases-3, (MMP-3), matrix mettaloproreases-7 (MMP-7), Collagenase, Cathepsin D, hyaluronidase, aggrecanase and various other serine proteases. The progressive irreversible loss of matrix proteoglycans result in diminished tissue function and it likely to exacerbate the disease process by exposing underlying bone to greater mechanical forces. Hence chondrocyte tissue culture systems can be employed as a model to study the cartilage metabolism and to elucidate the complex mechanism regulating the synthesis of proteoglycan. The organoid or explant culture is very useful for studying synthesis, turnover and drug responses with respect to the metabolism of matrix molecules. In monolayer culture chondrocytes are grown at low density with addition of serum. After a short culture period chondrocytes dedifferentiate and are fragmented to fibroblast like cells, change the pattern of their proteoglycan synthesis and overgrown by fibroblast like cells.

Another means of growing chondrocytes is cultivation in suspension or semisolid media using agar, alginate and other gel forming substances. The most recent *in vitro* model of chondrocyte culture employs encapsulating cells in alginate beads. This culture system offers multifold advantages over other models. The chondrocytes remain viable and phenotypically stable for months after encapsulating them in alginate beads. A great advantage of alginate bead culture is that it can be dissolved by chelating substances and allows further investigations of isolated chondrocytes in culture medium. Many studies have suggested that cultivation of chondrocytes in alginate stimulate synthesis of type II collagen and large aggrecans. The alginate system has extensively been used to examine the effect of different catabolic/anabolic agents on synthesis and turn over of extracellular matrix of cartilage. The cartilage tissue engineering approaches to cartilage repair are also focused on developing efficient methods of growing chondrocytes in culture medium. There is growing interest in the development of cartilage repair systems, which involve the implantation of autologous chondrocytes into defect sites for the treatment of localized full, or partial thickness lesions in articular cartilage. The chondrocytes isolated from small tissue biopsy are cultured *in vitro* to increase their number. The chondrocytes are expanded in specially designed biomaterials scaffolds (alginate, collagen fibrin etc), which provides them mechanical stability and supports even cell distribution. The cartilage constructs can be implanted at the site of articular cartilage injuries for maintenance and management of synovial joints. Recent developments in chondrocyte culture and its implications in tissue engineering and rehabilitation have been reviewed.

INTRODUCTION

Chondrocytes are unique type of cellular configurations, responsible for the maintenance of interior milieu of weight bearing synovial joint articular cartilage (AC). Articular cartilage covers the entire condylar mating surfaces of movable joints of the body. It is a highly specialized and uniquely designed connective tissue material that forms the smooth, glistening and friction free lining on the subchondral bones. Chondrocytes are embedded in the extracellular ground matrix of AC and responsible for the providing load bearing and load shearing properties during dynamic joint activities. It provides a wonderful friction free environment for complete maneuvers and flawless locomotion. In addition, articular cartilage is also capable of dissipating the mechanical energy so that the joint structure is protected from any harmful damage by an impact force. Physiologically, articular cartilage is virtually an isolated tissue, devoid of all blood vessels, lymph channels, and nerves. It receives nutrition from the synovial fluid and has two primary functions:

1. To distribute joint loads over a wide area, thus decreasing the stresses sustained by the contacting joint surfaces (shock absorption).
2. To allow relative movement of the opposing joint surfaces with minimal friction and wear (Joint lubrication).

It is well known that degeneration/wearing of AC can lead to several disabling problems and some times complete failure of joints. Maintenance of good health of load resistant cartilage by virtue of well-being of chondrocytes, provides an excellent, dynamic and better quality of life for human populations. In view of this vision currently several research studies are in progress for understanding the properties of articular chondrocytes *in vivo* and *in vitro*, in order to provide a suitable therapeutic method for the prevention of onset and control of progression of degenerative joint disease.

Chondrocyte are metabolically dynamic cells (1) and secrets extracellular matrix largely made up of proteoglycans aggrecans (load shearing material, provides compressive strength), and collagens (load baring material, provides tensile strength and holds proteoglycans aggrecans in mass). Also chondrocytes are responsible for assembly, incorporation, organization and turnover of these macromolecules in cartilage (2,3). Thus chondrocytes form the structural and functional units of articular cartilage. Type II collagen is the major form of collagen in articular cartilage together with smaller amount of other collagen like type I, IV, IX and X (4). The triple helix of type II collagen forms the basic architecture of the tissue. The three dimensional fibrilar network of collagen provides tensile stiffness and strength to the cartilage (5). Large molecular weight aggrecans and other proteoglycans are embedded within this fibrous network and provide compressive stiffness to the tissue (6). These interactions are mediated by electrostatic and covalent cross-linking between charged groups on both the species. In normal physiological conditions, chondrocytes maintain fine balance between synthesis and degradation of these components. However, in certain pathological conditions such as osteoarthritis, this equilibrium gets disturbed in response to various systemic and local etiological factors like aging, trauma, mechanical stresses, obesity and genetics. Pathological cartilage shows a loss of proteoglycan (7), marked collagen network fibrillation (8), increased tissue hydration and loss of compressive stiffness. These changes are followed by altered level of tissue degrading enzymes like matrix metalloproteases-1 (MMP-1), matrix metalloproteases-3 (MMP-3), (9), matrix mettaloproreases-7 (MMP-7), (10) collagenase, cathepsin D, hyaluronidase, aggrecanase (11) and various other serine proteases. These disturbances lead to gradual cartilage degeneration (12). An understanding of the factors mediating cartilage breakdown and their regulation, is of therefore great importance, since from them an approach can be made to design therapies for the reduction of pathological cartilage destruction and enhancement of repair in this tissue. Progress to this goal is dependent upon identifying the degradative events and agents, potentially active in articular cartilage tissue. A number of *in vitro* studies on cartilage have investigated various aspect of regulation of biosynthesis of cartilage matrix, its processing and turnover. Investigators have tried to develop the *in vitro* models of cartilage employing isolated human chondrocytes maintained in culture medium and also with intact articular cartilage tissue culture (13). The aim of these studies is to define the unique metabolic properties of chondrocytes and to develop culture conditions that accurately maintain the cartilage phenotype (14-16). Among the various *in vitro* model systems available for chondrocyte and cartilage culture, includes, cartilage organ culture (17), isolated chondrocytes culture in agarose (18-19), alginate (20-22) and collagen gel (23,24), and as high-density monolayers (25).

In this review we have presented an overview of various chondrocyte culture techniques, and their advantages and disadvantages with respect to the desired application. Also the focus has been given on discussing the current strategies of chondrogenesis and recent advances of *in vitro* cartilage regeneration in direction of providing functional cartilaginous constructs that can be used for controlled *in vitro* studies of chondrogenesis and potentially for *in vivo* cartilage repair, and restoration of joint function in degenerative joint disease like osteoarthritis.

Chondrocytes: The Architect of Articular Cartilage

Chondrocytes are specialized cells, which produce and maintain the extracellular matrix of cartilage, a tissue that is resilient and pliant. Although the number of cells in unit volume of articular cartilage is small in comparison to other tissues (1-10% of the tissue volume) (26), the chondrocytes show considerable metabolic activity (1). The activity is associated with a continuous physiological turnover of the extracellular ground substance in adult articular cartilage, involving its synthesis and degradation.

Articular cartilage is devoid of nerve supply and is generally considered to be an avascular tissue. The cartilage cell thus provide the body with a metabolically economical load bearing material, and the investigators, with interesting problems in understanding life in an avascular environment and modes of synthesis and degradation of matrix molecules such as collagen and proteoglycan. The unique nature of the chondrocytes lies in its ability to form and regulate an organized tissue at sites remote from blood vessels, nerves and lymphatics.

Chondrocytes are one of the few cell types that can survive without being close to a blood supply. However, once the cartilage is damaged, the chondrocytes' distance from a blood supply means that they lack the ability to regenerate themselves, as this process requires an increased amount of nutrients and access through the blood stream and secretion of regenerative proteins. Chondrocyte is surrounded by a specialized matrix distinctly different from the bulk of the intercellular matrix, which makes up more than 90% of the tissue volume. The chondrocytes are surrounded by a specialized matrix to form a structural unit named as *chondron*. Each chondron consist of a 'chondrocyte', its 'lacunar space' and a 'pericellular rim', (27) which represent the primary structural, functional and metabolic unit responsible for cartilage matrix homoeostasis.

The Physical Environment of the Chondrocytes

The chondrocytes of adult cartilage lie isolated from each other and are embedded in a matrix consisting mainly of collagen network cross linked within high concentration of polyanionic proteoglycans. The total ionic composition of extracellular fluid in cartilage is higher due to presence of high concentration of free cations and low concentration of anions relative to external medium like synovial fluid (28). Also the presence of fixed negative charges on proteoglycan ensures that the pH in loaded cartilage is around 0.5 units lower than the synovial fluid pH (29). In addition the chondrocytes obtain their energy by anaerobic respiration therefore the concentration gradient of lactic acid and other metabolites also exist at the articular cartilage joint surface varying from 6% O_2 to less than 1% in the deep layers (30).

In addition to control the ionic concentration and pH of the cartilage extracellular medium, proteoglycan also play their role as selective permeability barrier for entry of large molecules like immunoglobulins and serum albumin and allow very restricted entry of small molecules like cytokines and interleukins in cartilage (31). Thus an intact cartilage chondrocytes are exposed to very low concentration of circulating factors than most of the other cells.

The Chondrocyte: A Cell Under Pressure

Articular cartilage has unique material properties that enable it to perform its physiological functions over a lifetime and under a wide range of loading conditions. Articular cartilage is subjected to high compressive and shears stresses, and during normal activity pressure may rise to 100-200 atm (10-20 Megapascal, MPa) within milisec (32). A biomechanical study performed on cadaveric limbs loaded in stimulated gait have shown that human hip experiences stress in the range of 5-10 Mpa that are 300-800% of normal body weight (33-35). During normal walking the load may range form 2.5 times bodyweight (36) to 5 times bodyweight (37). Thus *in vivo* chondrocytes are constantly exposed to high mechanical loads, which depend on body weight and vary with posture and joint activity. Duration of loading cycle is another very critical factor, as it alters the extracellular environment of the cartilage in a complex manner.

Thus it is clear that resilience and endurance capacity of articular cartilage in physiological condition are determined by the loading pattern and depends critically upon the structure, composition and integrity of its extracellular matrix. The cyclic stress and strain generated during the repetitive load in joint motion may cause fatigue failure within the tissue. The mechanical load not only affect the material properties of the cartilage matrix, but it has been known that chondrocytes also respond to mechanical forces and within limits are able to remodel cartilage in line with the prevailing level of stresses (38). Thus chondrocytes always function in a demanding environment and their metabolic activities get altered in response to mechanical loads. Over the past 40 years, there has been ample demonstration of a surprisingly active level of metabolism in articular cartilage, and it is now clear that the chondrocytes, despite their sparse appearance in adult cartilage, synthesize and assemble the matrix components and direct their distribution within the tissue.

Various Models for the Study of the Articular Cartilage

Since a long time animal models are widely used to investigate the interaction between mechanical and biological factors, to study the metabolic response of cartilage (39-41), and to develop and evaluate tissue-engineering techniques for reconstruction of damaged articular cartilage. The aim of these studies is to engineer articular cartilage for the functional restoration of a joint surface that appears anatomically, histologically, biochemically and mechanically similar to the original joint surface. There are various critical issues in the selection of an appropriate animal model, such as the problems related to availability of the animal, age of the animal, the benefits and limitations of the model systems, scaffold selection with emphasis on polymers, and the most important one is, evaluation of the tissue-engineered articular cartilage in such models. As these models have uncontrollable biological processes so inference of the results is another critical aspect of these studies. Moreover, *in vivo* models have proven valuable only for studying the overall gross tissue response but they are limited in addressing specific questions concerning cellular response to mechanical stimuli and the process of chondrogenesis during the cartilage repair and regeneration. These concerns are now being addressed through the use of new and innovative *in vitro* model systems. Much of the information currently available on cartilage mechanobiology and biochemistry has been realized from tissue culture experiments. There are three basic approaches for this: (1) the intact synovial joint, with all the biomechanical forces placed on it (2) the cells in their native matrix in tissue culture; and (3) culture of chondrocytes after releasing them from the extracellular matrix.

The culture techniques, employing cartilage slices or isolated chondrocytes have proven to be very valuable in studies of metabolic activity and regulation of proteoglycan, secretion, its assembly in aggregates and their turnover (23,24), determining the effect of mechanical, hormonal and autocrine factors on proteoglycan homeostasis. Also, the regeneration of functional articular cartilage by formation of neocartilage is another most tantalizing prospect and intimidating challenge for bioengineers today. The tissue engineering approaches to cartilage repair are focusing on developing efficient methods to form functional cartilage via expansion of chondrocytes obtained from various sources, in synthetic medium and implantation at the defect sites for treatment of localized full or partial thickness lesions in young adults (42-45). As the cartilage is a sparsely cellular tissue so obtaining sufficient number of healthy chondrocytes to grow cartilage *in vitro*, remains one of the greatest challenges in healing injured or degenerated cartilage. The cells may be obtained through culturing primary autologous/herterologous chondrocytes (43), mesenchymal stem cells (46) and embryonic stem cells (47). There are various ways of expansion of chondrocyte population as: monolayer culture, during this process chondrocyte are known to dedifferentiate and adopt a fibroblastic phenotype and lose their characteristic pattern of matrix protein expression, and the other one is chondrocyte culture technique using scaffold material. In the later approach chondrocytes are seeded into a biomaterial, which can be derived from natural (collagen, fibril, hyaluronic acid, alginate) or synthetic sources (poly ethylene glycol, poly lactic acid) (48) Also the scaffold free approach have been used where chondrocytes are cultured at high densities in the absence of any supporting material (49).

There is a hierarchy of culture systems that can be employed as models in studies of cartilage metabolism:

Organ/Explant Culture System

This is the most extensively studied model system in which cartilage explants from calves, steers, pigs and humans have been used for culturing of cartilage and have provided a wealth of information relating to the mechanical and biochemical regulation of cartilage metabolism (50-54). These studies are invaluable because they provide the window through which the behavior of chondrocytes in their native extracellular matrix can be studied.

In organ or explant culture, cartilage together with its extracellular matrix is grown *in vitro* and the chondrocytes function in maintenance mode *i.e.* they do not proliferate, rate of collagen metabolism is very low whereas proteoglycan metabolism is in homeostasis. As expected the articular cartilage explants can be maintained in steady state for several weeks under appropriate nutritional conditions (55).

The aided advantage of explant culture system is that it preserves the immediate pericellular microenvironment of the chondrocytes with its macromolecules including chondrons, cell-binding proteins like fibronectin and laminin, and organization of the collagen fibrilar network in the territorial matrix surrounding the chondrocytes. Thus, it permits a more realistic picture for evaluation of metabolic signals transmitted from and to the chondrocytes, through the complex architecture of the tissue. The organ culture system has been also used to examine incorporation of newly synthesized labelled molecules into cartilage matrix and their turn over (56).

As in cartilage explant culture; chondrocytes are maintained in homeostasis similar to *in vivo* condition, this has inspired various researchers to choose this model system to investigate different aspects of proteoglycan metabolism in further detail. The studies concerning the regulation of proteoglycan synthesis, its turnover (56-58), and response of cartilage to damage by proteinases (59) lipopolysaccharide (58) and interleukin-1 (56, 60, 61) have been extensively done. Also it has been studied that organ culture maintain constant level of collagen type II for several weeks (58,62).

The cartilage organ culture model is ideal for investigating the effects of various agents on cartilage metabolism. It is of particular interest to assess the ability of therapeutic agents to protect cartilage from damage, and to enhance the repair process in damaged cartilage. Morales and Hascall (63) have presented various studies of the effect of fetal calf serum (57), Insulin like growth factors IGF-1 (64), IGF-2 (65), TGF-β (66), IL-1 (67) and hyaluronic acid (68), on proteoglycan homeostasis in cartilage explants of bovines. Cartilage explant studies have also been performed to investigate the interaction between mechanical and biological factors including the effect of unloading, static loading and overloading (41, 69-71). The major challenge in organ culture is to assure sufficient gas and nutrient exchange in order to maintain viability of cells throughout the tissue mass.

A disadvantage of the organ culture system is that extraction of newly synthesized aggregating proteoglycan from cartilage is a difficult process and requires the use of nondenaturing solvents (72). The presence of matrix containing large anionic molecules makes the process of extraction and analysis of m RNA and protein, exceedingly difficult. In addition the explant system is not suitable for studies related to the cell surface receptors etc.

Isolated Chondron Culture System

Another alternative for studying chondrocyte metabolism is to isolate and culture chondrons. One or more chondrocytes in articular cartilage are surrounded by a pericellular matrix (PCM), which consist of type VI collagen and proteoglycan and together with the chondrocyte have been termed as the 'chondron' (27,73). The pericellular matrix has been hypothesized to play important roles in regulating the biomechanical, biophysical, and biochemical signals that the chondrocyte perceives during normal joint activity. In a recent study, Graff (74) reported that the presence of native pericellular matrix in chondron enhanced the development of mechanically functional neocartilage in porcine pellet cultures. Histological and contemporary arguments suggest that the chondrons act as a fluid filled compression resistant bladder, which seems to protect chondrocytes function by dampening the mechanical, physicochemical and osmotic changes, induced in the matrix by dynamic loading of joint.

Chondron culture has received very limited attention because of the limited yield and viability after mechanical isolation of chondrons from cartilage tissue. Studies to date have focused on developing a variety of techniques to exploit the isolated chondrons as a model of cell matrix interactions in articular cartilage. The isolated chondrons model offers the potential to study dynamic chondron function independently of the bulk extracellular matrix, and development of techniques in this area is vital for understanding articular cartilage function.

Monolayer or Two Dimensional Chondrocyte Culture Systems

The cartilage engineering and autologous chondrocyte implantation approaches are based on expansion and culturing of cartilage cells removed from their native extracellular matrix. The treatment of traumatic or congenital cartilage defects is limited by the fact that cartilage has no intrinsic repair mechanism and the amount of cartilage available for autologous transplantation is always limited. Currently, chondrocyte expansion using monolayer culture is most widely used technique in cartilage tissue engineering. For isolation of chondrocytes from cartilage, the tissue is enzymatically digested with various matrix-degrading enzymes like hyaluronidase, pronase and collagenase. The chondrocytes, thus released from the matrix are maintained in cell culture medium, where they proliferate and lay down new matrix. Studies have confirmed that under appropriate culture conditions the chondrocytes display their *in vivo* phenotypic properties like spherical shape, synthesis of type II collagen and cartilage specific proteoglycan (75,76).

There are two methods of cultivation of chondrocytes isolated from articular cartilage:

Low-density Monolayer Culture

In low-density monolayer culture, after separating the chondrocytes from cartilage, they are grown at low density (3500-5000cells/cm^2) in presence of serum. However various studies have suggested that *in vitro* multiplication of chondrocytes in low-density monolayers result in dedifferentiation of these cells. They lose their specific phenotype and become more fibroblast-like (77-79). Production of the specific cartilage matrix components collagen type II and aggrecans is gradually shifted toward collagen type I (80-82), and small proteoglycans and also alteration occurs in pattern of chondrocytes surface receptors proteins (83). Finally the chondrocyte are overgrown by fibroblast-like cells (84).

High-density/Organoid Culture

The chondrocytes are seeded in high density (<30,000 cells/cm^2) in presence of fetal bovine serum. Various studies have shown that in high-density culture chondrocytes remain stable for 2-3-weeks period and then either transformed to fibroblast like cells (85-87) or to hypertrophic cartilage cells with mineralization of matrix (88,89). One disadvantage of high-density culture is that it does not yield pure cartilage specific cells but a mixed population of cells with perichondrium and nonspecific connective tissue material between the cartilage nodules (83, 86, 87, 89).

In a recent study (90), attempts have been made to regulate the cultivation of chondrocytes from mouse embryo cartilage in monolayer culture and slow down their transformation or their overgrowth by fibroblast-like cells, employing b1 subfamily of integrin receptors, which mediate chondrocyte-matrix interactions. This interaction can be regulated or synergized by growing chondrocytes on collagen type II coated surface in presence of growth factors, such as IGF-I, which stimulates many chondrocyte functions. The results of this study are very promising as cells maintained their round phenotype till the end of cultivation and a higher cell density in comparison to control. The cartilage-specific surface receptors (integrins of the b1-group) could also be demonstrated on the membrane of these cells. Hence, collagen type II and IGF-I prevent chondrocyte dedifferentiation to fibroblast like cells and this model allows a pure chondrocyte culture.

Monolayer chondrocytes culture has been useful in several studies concerning the effect of soluble factors like cytokines, on cartilage metabolism. The technical procedures of culturing and analysis are simple and many variables are relatively well controlled. The disadvantage of phenotypically different population of cells is sometimes less considered. Monolayer, or two-dimensional, culture of disaggregated, anchorage-dependent cells *in vitro,* has allowed progress in understanding regulation of cell differentiation, growth, and function as well as cell-to-cell interactions that modulate these processes.

Monolayer chondrocytes culture with high cell density are useful for studying the intracellular events involved in proteoglycan biosynthesis (91) and in some cases, the process of proteoglycan aggregation (92). Only very little cell associated matrix get accumulated in these cultures, and the chondrocytes secret newly synthesized macromolecules more or less directly into the medium. As a consequence, proteoglycan and other matrix components synthesized in the presence of radiolabelled precursors are easily recovered from the medium in relatively large amounts with a high specific radioactivity (93). For the same reason monolayer cultures are not useful for examining proteoglycan turnover or for studies of mechanism and molecules involved in matrix assembly.

Suspension or Three Dimensional Culture System

A most popular means of growing chondrocytes is cultivation in semisolid media. Chondrocytes grown in solid materials or gels like agarose, alginate and collagen, maintain their spherical shape, reestablish an abundant cell associated matrix and permit the maintenance of their differentiated state. These culture systems are most relevant models as they reflect how chondrocytes are seeded onto various scaffolds; the approach of using scaffolds is a very important aspect of cartilage engineering.

A variety of matrixes have been used to culture chondrocytes three dimensionally, mainly employing natural materials like agarose, alginate, collagen and fibrin; and also synthetic materials like Poly glycolic acid (PGA), Poly lactic acid (PLA) and polyurethanes etc. In particular, agarose and alginate represent the two most widely employed gel systems, and have been extensively used to examine the effect of different agents on synthesis and turn over of proteoglycan by articular chondrocytes of several species (14,16,19,21,72,94). These *in vitro* models are also useful in comparing the properties of specific cell populations from different regions of articular cartilage from different joints (95). In addition, gel suspension culture can be used as a source of matrix-embedded chondrocytes to be tested in biomechanical experiments. Gel containing suspended chondrocytes with their newly formed and laid down matrix can be loaded by varying loads in order to determine the stress-strain relationship (19) of chondrocytes and provide a useful tool in studying the etiology of osteoarthritic like disorders.

Agarose-Chondrocytes Suspension Culture

The entrapment of chondrocytes within agarose (14,18,96) or on agarose (97) has become a popular culture system. Most of the proteoglycans, which are synthesized by the cells and are retained in the vicinity and accumulate with time, forms a matrix that appears to share many of the physicochemical properties of normal cartilage matrix (19). Studies have confirmed that chondrocyte culture in agarose gel maintain the expression of aggrecans (cartilage specific large aggregating proteoglycan) and collagen types characteristic of articular cartilage (predominantly type II, also type IX and XI) (18,72,98). The maintenance of chondrocyte phenotype during long-term culture in agarose has shown the development of a mechanically functional cartilage like matrix (19) Also chondrocytes in agarose gel respond biosynthetically to static and dynamic mechanical load in a manner similar to that of intact organ culture (99). A significant advantage of this culture system is that the cartilage matrix as collagen type II and keratan sulfate, produced by chondrocytes, can be separately characterized using techniques of immunohistochemistry (100).

Alginate-Chondrocyte Suspension Culture

Alginate bead system was firstly used by Jourdian and colleagues (20), to study the metabolism of proteoglycan in human articular chondrocytes. Since then alginate gels have been extensively studied for cartilage tissue engineering applications as a matrix for cell encapsulation and culture. Suspension of chondrocytes in alginate gel promotes the retention of the chondrocytes phenotype and studies have demonstrated that bovine articular chondrocytes suspended in alginate beads remain phenotypically stable for as long as 8 months (16,21), and synthesize cartilage matrix components, including type II collagen and aggrecan (101).

Alginates are naturally derived polysaccharides block copolymers, composed of regions of sequential β-mannuronic acid (M-blocks), and α-L-glucoronic acid (G-blocks) monomers. The length of the M- and G-blocks and their sequential distribution along the polymer chain, varies depending on the source of the alginate (102). Alginates undergo reversible gelation in aqueous solution under mild conditions through interaction with divalent cations such as Ca^{2+} that can cooperatively bind between the G-blocks of adjacent alginate chains creating ionic interchain bridges (103).

Interestingly, the alginate bead culture system has unique property that makes it possible to study the accumulation and turnover of macromolecules in two distinct compartments of cartilage matrix. The matrix that rapidly forms around the cells has been termed as cell-associated matrix (CM), and believed to correspond to the pericellular and territorial matrix compartments, and further removed matrix (FRM), which corresponds to the interterritorial areas of the cartilage matrix, forms further away from the cells (13,104). Using this unique property of alginate culture systems, recently a study has been done on chondrocytes of different developmental stages, to visualize the age related differences in the accumulation and rate of synthesis of proteoglycan, hyaluronic acid and collagen in the two matrix compartments (105). These studies performed on chondrocytes isolated from articular cartilage of young and adult bovine animals, suggested that alginate system could be an effective model in determining the biochemistry of age related cartilage changes like osteoarthritis.

Another very important feature of alginate system is that the beads can be quickly solubilized by chelating substances like ethylene diaminetetraccetic acid (EDTA), and phosphates, and cells can be released after a certain time in culture. The aggrecans and collagen molecules can be readily extracted from each compartment and analyzed further. Thus alginate culture system facilitates the use of experimental approaches not accessible while culturing chondrocytes in other systems like monolayers, explants or agarose gel.

Although alginate culture keeps chondrocytes in a spherical confirmation, supports differentiated phenotype and synthesis of cartilage specific markers, yet chondrocytes suspended in alginate do not adhere to the matrix as they would *in vivo*. For this an important modification in the protocol of using alginate matrix for chondrocyte culture has been to encourage the attachment of cells to the alginate beads via integrins and other cell surface receptors. A most recent study, which has been carried out by Genes (106) deals with designing a modified alginate surface by covalent attachment of GGGG**RGD**Y peptide sequence to low viscosity alginate. The hypothesis is that RGD modification of alginate enables chondrocytes to attach to them via integrins, which are mechanoreceptors and can sense changes in the mechanical environment of the matrix to which they are attached.

The histological and electron microscopical studies of osteoarthritic chondrocytes in alginate culture indicated that after three passages in monolayer, chondrocytes display typical round morphology (107). Study has also demonstrated that human osteoarthritic chondrocytes embedded in alginate beads after their culture in monolayer for three passages developed a pericellular matrix enriched in keratan sulfate and chondrotin sulfate glycosaminoglycans. Consequently, the alginate system appears to be an ideal model for the study of cartilage matrix components in a three dimensional system.

Despite these interesting findings the morphometric and biochemical characterization of the different matrix compartments of chondrocytes cultured in alginate needs to be established further; and it has yet to be shown that how these two compartments can be readily separated at various time points in culture.

Collagen-Chondrocyte Sponge Culture

A relatively recent approach in three-dimensional culturing of chondrocytes, is to seed them with porous collagen sponges. Histochemical and immunohistochemical analyses have confirmed the presence of cartilage specific matrix in cells (108). Analysis of the cartilage-specific genes, aggrecan and collagen type II; have demonstrated the preservation of the chondrocyte molecular signature compared to dedifferentiated monolayer culture (109). Preliminary investigations indicated that articular and endochondral chondrocytes retain their differences in these collagen sponges (110). Also the collagen sponge has been used to determine the effects of insoluble, soluble, and mechanical factors on chondrogenesis. In this direction attempts have been made to study the effect of hyaluronan by incorporating it in collagen sponge. It was found that small amount of hyaluronan (2% w/w) in 3D collagen scaffolds enhanced chondrogenesis, but a greater amount was found to be inhibitory (111).

This finding lends support to the idea that biomimetic, or "smart" scaffolds can be developed with optimal signals for specific histogenesis.

Micro Carrier Suspension Culture

Attempts to maintain the original phenotype of chondrocytes is the most challenging aspect of cartilage engineering and its regeneration. For this attempts have been made in monolayer culture by seeding chondrocytes at high density and to entrap cells in a three-dimensional matrix or scaffold such as agarose and alginate beads, as already been discussed. Despite these modifications in culture conditions, not all chondrocyte population reverted to the spherical, type II collagen synthesizing cells and therefore these methods are not able to provide sufficient numbers of chondrocytes with original phenotype needed for engineering the neocartilage. There is need for a culture system that would allow chondrocytes to proliferate as well as maintain their original characteristics. For this the most recent technique of culturing cells that has been shown to facilitate cell proliferation as well as maintain synthesis of cellular products is the 'microcarrier suspension culture'. This culture system is based on the original 'bioreactor' design to grow bacteria in large quantities. The same principle is used to expand anchorage-dependent mammalian cells in microcarriers and thus mimic other cells types that grow in suspension culture (112). Under controlled levels of pH, oxygen, nutrient supply and mechanical agitation more than 80 anchorage-dependent cell types have been propagated using various types of microcarriers like glass beads, charged molecules such as dextran, sepharose and cross-linked collagens (113-116). Frondoza (117) studied the proliferation and production of cartilage matrix components in human articular chondrocytes in microcarrier suspension culture. Various types of microcarriers, which were used in the study are: cytodex 1 (180pm dextran beads); cytodex 2 (155 pm dextran beads); cytodex 3 (175 pm denatured pig skin type I collagen-coated dextran beads); and cellagen (collagen type I derived beads). From the study it was concluded that out of the four microcarriers, cellagen beads, which are made out of collagen type I, were the most suitable, promotes cell attachment, cell proliferation and allows easy retrieval of chondrocytes with mild trypsin treatment. The retrieved chondrocytes were metabolically active and can be utilized for further experiments.

Recent studies indicated that microcarrier suspension culture system enhanced the growth of bone cells and chondrocytes from rat calvaria proliferating in cytodex 1 microcarriers and continued producing alkaline phosphatase and other cellular products (118,119). These studies have confirmed earlier observation that microcarrier suspension culture has the advantage of increasing division rates of chondrocytes. Thus, limited number of chondrocytes obtained from cartilage could be expanded to greater numbers over a shorter period of time. Studies have also documented success of repairing cartilage defects with chondrocytes grown in microcarrier suspension culture (43). Mass cultures of chondrocytes retrieved from microcarriers may be made available for cartilage regeneration.

It is apparent from the above discussion that interest in the application of gel suspension culture system for *in vitro* proliferation of cartilage has grown tremendously. The potential advantages of these cultures are not restricted to the fact that the cell environment is physically stabilized and that the cells kept well in place after transplantation. These three dimensional systems, as outlined in the sections above, offer possibilities to obtain controlled cell multiplication in a phenotypically stable fashion. It is noteworthy that growth factors like interleukins, cytokines have been shown to stimulate the proliferation of chondrocytes, induce the differentiation of mesenchymal cells into chondrocytes, and stimulate the synthesis of cartilage specific molecules. Improved protocols for the culture of human chondrocytes in gel suspension systems are likely to offer novel opportunities in the treatment of articular cartilage lesions.

Table 1.1: Comparative Study of Different Chondrocyte Culture Systems

Characterstics	*Monolayer*	*Explant*	*Suspension*	*References*
Chondrocyte phenotype	Poor (at low density type I collagen)	Excellent (90% type II collagen)	Very good (in agarose/alginate, 70% type II collagen)	16, 21, 62, 81, 82, 98
Cellularity	High	Low	Medium	16, 21,55, 85,
Expansion Capacity (to increase cell no)	Able to expand (10 X)	No expansion (1X)	Able to expand (10X- 25X)	16, 21,55
Difference in cell population	Has not been shown	Yes	Yes at least to some extent	13,72, 87, 89, 95, 99
Matrix dependent control of cell behavior	No	Yes	Disrupted initially reestablished later	13, 16,51, 86, 91
Penetration of soluble factors	Very good	Poor (uneven)	Good	63, 64, 111
Metabolic activity with respect to matrix constituents	Strongly stimulated anabolism	Steady state with slow turn over	Enhanced metabolic activity, turnover rate similar to *in vivo*	55, 104, 105
Technical handling	Easy	Difficult	Less difficult	93
Quantification of gene products, isolation of macromolecules	Easy to isolate m-RNA	Difficult to isolate m-RNA	Easy to isolate m-RNA (in alginate)	72, 88, 93, 109
Matrix supramolecular organization study	Limited possibilities, not very useful	Possible, but difficult	Possible	56, 57, 104
Nutrition to cells	Good and constant	Poor	Good and constant in many cases	–

Chondrocyte Culture and Cartilage Engineering: Role in *in vitro* Regeneration of Articular Cartilage

In 1743, Hunter stated, *"From Hippocrates to the present age, it is universally allowed that ulcerated cartilage is a troublesome thing and that once destroyed in not repaired"* (120). Several factors may explain why cartilage demonstrates very little capability for self-repair after various injuries. The most obvious one is that chondrocytes are not required to proliferate as in some other tissues (*i.e.* skin), and mature adult chondrocytes have a relatively low metabolic activity (121). Also the articular cartilage is avascular, alymphatic and aneural tissue; and inherent healing of cartilage defect can be achieved only when the defect penetrates the subchondral bone and triggers response from the mesenchymal cells present in bone marrow. Except in the case of full-thickness defects, there is no direct access to progenitor cells and also the resident chondrocytes may be obstructed by the extracellular matrix from migrating to fill the defect site (122,123).

A substantial amount of research has been conducted on articular cartilage, with significant advances in our understanding of the biological repair process being made over the past four decades. While many cartilage repair techniques have been proposed, none have successfully regenerated long-lasting and functional hyaline cartilage to replace the damaged tissue (124). Most of the surgical interventions to repair damaged cartilage have been aimed towards the treatment of clinical symptoms rather than the regeneration of hyaline cartilage, such as pain relief and functional restoration of joint

(125). These approaches include tissue adhesives (126,127), enzymatic treatments (128), laser solder welding (129), autograft cell/tissue transfer via periosteal grafts (130), autologous osteochondral grafting such as mosaicplasty (131) and the carticel method (43). Now the rapidly emerging field of tissue engineering holds great promise for the regeneration of functional cartilage tissue substitutes *in vitro*, for subsequent implantation *in vivo*. Tissue engineering has a primary goal of restoring, maintaining or improving tissue function by employing the principles of engineering and life sciences (132). Tissue engineering of articular cartilage involves the isolation of articular chondrocytes or their precursor cells that may be expanded *in vitro* and then seeded into a biocompatible matrix, or scaffold, for cultivation and subsequent implantation into the joint. The concept of tissue engineering approach in cartilage repair was first proposed by Green (133), clinical applications of this tissue engineering approach was first attempted by a Swedish group (43) with fair-to-excellent results. The success of a neo-cartilage depends upon various factors, which include cell type, scaffold characteristics and culture conditions of isolated chondrocytes in these scaffolds. Stabilization of the differentiated phenotype of the cells and the optimization of neomatrix deposition, composition and organization are crucial factors that have to be considered in cartilage engineering.

The strategy for tissue engineering of cartilage can be classified into six phases:

1. Seeding of bioresorbable polymer scaffold, characterization and toxicity study
2. Seeding of the chondrocytes populations into the polymeric scaffold in static culture (Petri plate)
3. Growth of premature tissue in a dynamic environment (spinner flask)
4. Growth of mature tissue in a physiological environment (bioreactor)
5. Surgical implantation and evaluation *in vivo*
6. Tissue engineered transplant remodeling

Some of the most important aspects of cartilage engineering are described as following:

Identification of Cell Type

For any tissue engineering based method, the critical factors while selecting the cell type are proliferative capacity of cells, phenotypic stability and immunogenicity. However in fully differentiated autologous cells there is no concern of immune response but the issues related to proliferative capacity, donor site morbidity and limited life span, have brought the usage of other cell populations like mesenchymal stem cells and progenitor cells into the forefront of such applications. Various cell types which are being used for expansion and production of neocartilage are described as following:

Differentiated Chondrocytes

Currently the autologous chondrocytes transplantation is the only cell-based method for articular cartilage repair approved by FDA. The procedure involves harvesting of cartilage from healthy regions of joints, their isolation and expansion and subsequent transplantation at the affected site. However the mature chondrocytes have only limited proliferative potential and the number of cell divisions they undergo *in vitro*, decreases with age (134). Differentiated chondrocytes grown as a monolayer loose their native phenotype and dedifferentiate with passaging. However the chondrocyte phenotype may be rescued by use of scaffolds (18,22), reducing oxygen tension (135) and use of growth factors such as basic fibroblast growth factor (FGF-2), transforming growth factor β1 (TGFβ-1) (136) and high concentrations of fetal calf serum (43). The method of autologous chondrocyte implantation is effective in patients with healthy cartilage at other than the focal site of injury, but in the patients with symptoms of arthritis and other joint diseases, own chondrocytes may not be useful and hence an alternative source is required.

Differentiated chondrocytes can also be obtained from non-articular, hyaline cartilage tissues such as nasal or rib cartilage. However, additional work will be necessary to assess whether cells of non-articular origin can be successfully used for articular cartilage repair.

Mesenchymal Stem/Progenitor Cells (MSCs/MPCs)

A stem cell is a non-committed cell from young and adults, and can give rise to specialized cells, tissues and organs. Adult bone marrow is a major source of undifferentiated stem cells (bone marrow stromal cells BMSCs), which contribute to the regeneration of mescenchymal tissues such as bone, cartilage, ligament, tendons and stroma. (46). The multipotent nature of MPCs makes them theoretically ideal candidates for cartilage repair (137). Gao and coworkers (138) tested this hypothesis by attempting repair of osteochondral defects using MPCs loaded in injectable calcium phosphate and hyaluronate derivative. After that Noth (139) also tried *in vitro* engineering of cartilage by using MPCs on PLA scaffolds. This approach provides the added benefits of minimizing handling and maximizing biocompatibility. However, the mechanism of stem cell chondrogenesis is a complicated one and regulated by highly complicated consortium of growth factors, signaling molecules, cell-cell and cell–matrix adhesion molecules. Thus designing the microenvironment of the progenitor cell to obtain the desired functional stage is the most challenging part.

The cells with chondro-osteoprogenitor properties have been isolated from several tissues, including periosteum, bone marrow, spleen, thymus, skeletal muscle, adipose tissue, skin and retina (140-144) and recently in low efficiency, from peripheral blood (145). The use of progenitor cells has not always been straight forward in cartilage engineering for various reasons like hardly accessible tissue source, low frequency, limited information on the particular cell system and lack of understanding about growth requirements of the cells.

Embryonic Stem Cells (ES)

Embryonic stem cells are isolated from the preimplantation embryos, human blastocysts and fetal tissues. Introduction of ES cells in 1998 gave the scientific community an exciting new prospect of cell-based therapies in tissue engineering. Human ES cells can be maintained in culture for more than two years, thus have ceaseless proliferation and pluoripotent nature. However, human ES cells have not yet been reported to be used specifically for musculoskeleton repair due to issues related to ethics, therefore researchers have directed the use of murine ES cells toward the chondrogenic (146) and osteogenic lineages (147) with some success. The culturing of ES cells requires use of feeder layers and technically a cumbersome process. A lot of research is required in this direction to fully understand and optimize the potential of these cells in cartilage tissue engineering.

Coculturing of Chondrocytes/Mesenchymal Cells

Engineering cartilage using multiple cell types is most recent approach and provides the opportunity for cells to provide regulated and synergistic effect on each other for functional tissue development. Culturing of chondrocytes with MSCs induces the chondrocyte expansion by providing paracrine effect of growth factors for chondrogenesis and differentiation of passage cultured MSCs in chondrocytes. The approach also deals with the problem of scarcity of cell source. In a very recent study carried out by Tsuchiya *et al.* (148), human MSCs and bovine mature articular chondrocytes were mixed and cultured under appropriate conditions in pellet culture. The culture showed upregulation of cell proliferation, cartilaginous matrix production and type II collagen gene expression.

The coculture strategy appears to be useful for the *in vitro* expansion of mature chondrocytes without decreasing their differentiation ability. Further trials are in progress to clarify the effect using primary MSCs and autologous chondrocytes cells.

Identification of Appropriate Scaffold

A scaffold provides the cell with a three dimensional matrix upon which it adhere, proliferate and produce matrix components and may be tailored to serve as a vehicle for cell delivery at the site of injury. Scaffold also performs tissue inductive effect by controlling release of growth factors and thus forms functional tissue structure and fills the lesion.

Scaffolds designed for use in cell-based therapies to repair damaged articular cartilage should ideally provide the following characteristics:

1. A three-dimensional (3D) and highly porous structure to support cell attachment, proliferation and extra-cellular matrix (ECM) production;
2. An interconnected/permeable pore network to promote nutrient and waste exchange;
3. A biocompatible and bioresorbable substrate with controllable degradation rates by hydrolytic or enzymatic mechanisms;
4. Minimization of leaching of toxic products like monomers, crosslinkers and solvents;
5. A suitable surface chemistry for cell attachment, proliferation, and differentiation;
6. Mechanical properties to support, or match, those of the tissues at the site of implantation;
7. An architecture which promotes formation of the native anisotropic tissue structure; and
8. A reproducible architecture of clinically relevant size and shape

Selection of scaffold is a considerable challenge in musculoskeleton tissue repair keeping in view the highly demanding environment of load bearing joint. Articular cartilage imposes some special requirement on scaffold material as evident from its unique functional environment, biochemical composition, lack of vascularity and concomitant inability to injury and disease. A variety of biomaterials, autologous and heterologous, naturally occurring and synthetic, biodegradable and non-degradable and bioceramics, have been introduced as potential cell-carrier substances for cartilage repair.

Among the naturally occurring biomaterials are various forms of types I and II collagen-based material, as scaffold matrices (149, 150) gels (151,152) or collagen–alginate composite gels (152), alginate, fibrin, chitosan and glycosmainoglycan based scaffolds. The synthetic degradable biomaterials include polyglycolic acid (PGA) (153,154) and poly-L-lactic acid PLLA (155,156) and PGA–PLLA copolymers (157, 158) etc. Several investigators have also found that some non-biodegradable polymer substances, such as polytetrafluoroethylene (PTFE), (159) polyethylmethacrylate (178), and hydroxyapatite/Dacron composites, (160) also facilitate the restoration of an articular surface in musculoskeleton disorders.

Natural and synthetic biomaterials have their respective advantages and disadvantages. Natural polymers have good cytocompatibility, biaoactivity, cell interaction and hydrophilicity, but they are mechanically too weak to maintain the desired shape and to withstand physiological loading. Synthetic polymers have relatively good mechanical strength and enable precise control over the rate of degradation, porosity and microstructure but their surfaces are hydrophobic and lack cell-recognition signals. Although results of preliminary clinical studies have indicated towards the potential of naturally derived polymers, but concerns about the feasibility of source of these materials for clinical applications, as well as the fear of infection at the site have prompted researchers to investigate the use of synthetic polymers. The synthetic biomaterials can be easily mass-produced and their properties can be tailored for specific applications. There is also the possibility that growth factors or other pharmacologics could be included in the scaffold to encourage certain cellular responses, such as proliferation or differentiation. However only few of these polymers have been FDA approved for use in humans (162).

Some of the most commonly used natural and synthetic polymers, their applications and evaluation in cartilage engineering has been discussed as following:

Natural Scaffold Materials

Collagen

Collagen is the naturally occurring and most abundant protein of the tissue with very high mechanical and shear resistance properties. These properties obviously make it a suitable candidate for delivering chondrocytes at the site of injury. It can be prepared in various forms like solution or shaped into membrane films, threads, sponges and, more recently, into acidic hydrogels. Collagen scaffolds have been already used in *in vitro* characterization of chondrocytes and stem cell behavior study (163, 164). Wakitani *et al.* (165) reported that MSCs in collagen gel brought about the repair of full thickness cartilage defects in rabbit. Collagen matrices containing glycosaminoglycans are also under investigations for inducing hyaline cartilage repair (166).

A most recent study conducted by Ohno (167) has shown that both type I and type II collagen sponge had equivalent effect in cell-seeding efficiency and chondro specific gene expressions. Another study evident that redifferentiation of dedifferentiated bovine chondrocytes occurs when cultured in vitro in a type I collagen- PLGA hybrid mesh (168). These results indicate that the type I collagen sponge is as useful as type II collagen sponge for constructing hyaline cartilage like tissue. Van Susante 1995 (169) demonstrated that type I collagen gels enhance the proliferative capacity of chondrocytes better than alginate beads. Human chondrocytes from non-fibrillated areas of osteoarthritic articular cartilage cultured in a collagen type I microcarrier suspension system, multiplied at least 20 fold within 2 weeks (170). It has also been shown that chondrocytes cultured at high density in collagen type I gels can maintain their phenotype by synthesizing type II collagen and cartilage specific proteoglycans (171).

Collagen matrices are usually derived from xenogeneic sources (mostly porcine and cow skin), the immunological concern can be eliminated by purification of immunogenic telopeptides. Still the possibility of xenozoonoses, the microbial transmission from the animal tissue to the human recipient, has to be carefully considered and remains one of the primary obstacles in cartilage transplantation using collagen (172).

Fibrin

It is a natural polymer that forms during blood coagulation and plays an important role in homeostasis and wound healing. Fibrin glue has been considered for use in a variety of orthopaedic procedures, including fixation of chondral fragments, chondral chips, and osteochondral fragments (173). Fibrin preparations have also been considered as a material for delivering chondrocytes or growing cartilaginous tissue for cartilage repair (174, 175). Chondrocytes interact with fibrin glue in the autologous chondrocyte transplantation procedure, when injected chondrocytes are sealed within cartilage defects by application of fibrin glue to overlying periosteum (176). However, chondrocyte growth and matrix production vary between fibrin preparations (177). In an equine study the incorporation of IGF within polymerized fibrin improved the histology, amount of type II collagen in chondro lesions; however the mechanical properties of the tissue and biochemical features were not consistent with normal articular cartilage (178). Recently Gelse *et al.* (179) implanted MSCs transfected with adenoviral vector containing bone morphogenetic protein–2 (BMP 2) and IGF-1 c-DNA suspended in fibrin glue in rabbit full thickness cartilage lesions and demonstrated that cells were able to attach to the wounded cartilage and cell growth is consistent. Fibrin gel is useful as a model because it is an FDA approved material, and can also be obtained from a patients' own blood (180).

Alginate and Agarose

These two polysaccharides are derived from seaweed and form hydrogels that have the advantage of allowing a uniform distribution of cells throughout implantable scaffolds. The hydrogels have potential to be injected at the defective site and *in situ* form a gel structure with cells encapsulated in them. Agarose has been used extensively in *in vitro* studies of chondrocyte behavior however it does not resorb well, and may elicit a foreign body giant cell response *in vivo* (181). Alginate has been widely used for delivering chondrocytes. Diduch *et al.* (182) reported that alginate beads containing bone marrow stem cells, remained within rabbit osteochondral defects and progressively filled the defects with regenerated tissue. Histological analysis confirmed the chondrocytic phenotype and good amount of proteoglycan synthesis, also binding between the regenerate and host tissue was excellent. A more recent study by Dausse *et al.* (183) however, showed that despite good histological results with cell-seeded alginate, biochemical and mechanical properties of the regenerated tissue were significantly inferior to those evaluated in native cartilage. An alteration of surface properties of scaffold using cell adhesion proteins as RGD has introduced further scope of research of alginate as a carrier in cartilage engineering (106).

Hyaluronic Acid

Hyaluranon is a major component of cartilage matrix. It has unique physico-chemical properties and distinctive biological functions. It binds specifically to proteins in the extracellular matrix and on the cell surface. Partial esterification of hyaluronic acid increases its stability and makes the molecule particularly suitable for cartilage engineering. (184). Butnariu-Ephrat, (185) implanted MSCs in a hyaluranon-based glue in goat cartilage defects with little success. Other studies also indicated at much less potential of hyaluranon for cartilage engineering but in a recent study Girotto, (186) evaluated the re-differentiation capacity of cultured chondrocytes on two bioresorbable non-woven biomaterials composed of hyaluronic acid benzyl esters: HYAFF-11, and its sulphated derivative, HYAFF-11S. These experiments have shown that HYAFF-11 and sulphated HYAFF-11 offer an excellent support for *in vitro* chondrocyte culture, generating a functional cartilage construct that expresses specific genes for collagen types II, IX, × and aggrecans. Sulphated HYAFF-11 specially promoted the rate of cell proliferation in both human and chicken cells. Chondrocytes grown in three-dimensional HYAFF-11 scaffolds demonstrated gene expression characteristic of differentiation/re-differentiation. This may prove to be an effective delivery system for chondrocyte cells for the treatment of articular cartilage defects, and an *in vitro* system for studying cartilage differentiation. Further research is on to study human mesenchymal stem cell differentiation towards cartilage generation using sulfated HYAFF-11 scaffolds.

Gelatin-Chondrotin-Hyaluronan Tri-copolymer

Recently a tri-copolymer scaffold produced from chondroitin-6-sulfate, hyaluronan, and gelatin has been used to study cartilage dedifferentiation and synthesis of extracellular matrix to fill the cartilage lesions (187).

Inclusion of chondrotin sulfate in scaffold may promote the secretion of proteoglycan and type II collagen (188). Hyaluronan is present in high amounts in the ECM during embryonic cartilage development, and also facilitates the integration of engineered cartilage (189). Since gelatin is basically denatured collagen and presumably retains informational signals for cell attachment, it was hypothesized that a tri-copolymer formed from gelatin, chondrotin, and hyaluronan would mimic cartilage matrix and provide the necessary information for cell attachment to meet the requirement for dynamic reciprocity (the mechanism by which cell synthesizes and secretes ECM and is then, in turn, regulated by the ECM) (190) for cartilage tissue engineering. However the results of the study have shown that although chondrocytes were evenly distributes in the scaffold, secrete ECM, retain their phenotype, and secrete type II collagen but the amount of extracellular matrix is less to synthesize cartilage. The less than satisfactory amount of matrix production may be due to inhibition of cell mitosis by chondrotin. Further research in this field is under progress to stimulate cell proliferation in engineered tissue. The gelatin/chondrotin sulfate/hyaluronan tri-copolymer deserves to be tested as a candidate for cartilage tissue engineering. More ECM secretion with lacuna formation and more GAG production might be expected after longer periods of culture of chondrocytes.

Chitosan

Chitosan is a partially deacetylated derivative of chitin, found in arthropod exoskeletons. Structurally, chitosan is a linear polysaccharide consisting of β(1-4) linked D-glucosamine residues with a variable number of randomly located *N*-acetyl-glucosamine groups. It shares some characteristics with various glycosaminoglycans (GAGs) and hyaluronic acid present in articular cartilage. Depending on the source and preparation procedure, chitosan's average molecular weight may range from 50 to 1000 kDa.

Much of the potential of chitosan as a biomaterial stems from its cationic nature and high charge density in solution. Due to this it can form insoluble ionic complexes or complex coacervates with a wide variety of water-soluble anionic polymers like natural polysaccharides GAGs and alginate, as well as synthetic polyanions such as poly (acrylic acid) (191-193). Chitosan has been used as a biomaterial in drug delivery systems and it has recently drawn a considerable attention in its applications for repair of articular cartilage. Lu *et al.* (194) has demonstrated that the chitosan solution caused a significant increase in the density of chondrocyte in the knee articular cartilage, suggesting that the chitosan could be potentially beneficial to the wound healing of articular cartilage. Also Mattioli-Belmonte (195) has shown that the bone morphongenetic protein (BMP)-7, associated with *N*, *N*-dicarboxymethyl chitosan facilitates the repair of articular cartilage lesions. Chitosan forms insoluble ionic complexes with GAGs. A study has demonstrated that when associated with various polyelectrolytes GAGs chitosan has a protective effect against GAGs hydrolysis by their specific enzymes (193). These hydrogels have also been successfully used to deliver growth factors to chondrocytes (196).

Further studies are necessary to investigate the material and mechanical properties (such as porosity, compressive elastic modulus, permeability, and viscoelastic properties) of the chitosan scaffold. A development of a chitosan-based material that can support chondrogenesis may be significant not only in terms of the quality of neocartilage produced, but also in terms of the ability of that tissue to integrate with the host matrix.

Synthetic Scaffold Materials

Polylactic Acid (PLA)

The most widely used synthetic polymer to construct temporary scaffolds for tissue engineering includes poly (glycolic acid) (PGA), poly (lactic acid) (PLA), and their copolymer poly (D-L-lactic-co-glycolic acid) (PLGA) (118, 197-199). PLA, PGA and PLGA are biocompatible and are among few synthetic polymers approved by the Food and Drug Administration (FDA) for certain human clinical application such as surgical sutures and some implant devices.

Chu *et al.* (200) performed a study in rabbits in which allogenic perichondral cell/PLA composites were implanted for repair of cartilage and in same type of study, Dounchis *et al.* (201) used autogenous perichondrial cells. The results of both the studies showed grossly successful repairs followed by inconsistent subchondral bone reformation. Also the newly formed cartilaginous tissue had biochemical properties inferior to those of normal articular cartilage. More recently, Giurea *et al.* (202) modified the procedure by pre incubating the cells in matrices prior to implantation for improving their retention in scaffold. Caterson *et al.* (203) have shown that marrow MSCs displayed chondrocytic differentiation when cultured within PLA or PLA/alginate scaffolds in presence of growth factors TGFβ-1. Goomer *et al.* (204) used a PLA scaffold to implant perichondral cells transfected with parathyroid hormone receptor TGF β-1 gene in rabbits. In a composite scaffold study Frenkel *et al.* (205) used a D,D,L,L-polylactic acid and collagen device that has separate tissue-specific microenvironments for regeneration of bone and cartilage. The treatment of BMP-2 induced a high-quality, hyaline tissue that integrated well with the host tissue for 24 weeks.

Polyglycolic Acid (PGA) and Copolymers

Chondrocytes from human, bovina and many other sources have been cultured on the copolymers of poly glycolic acid with poly lactic acid (PLA), poly ethylene glycol and other viscous materials. These copolymers have been widely used for over 20 years. (206).

Freed (118) conducted a study to demonstrate that *in vitro* chondrocytes grown on PGA and PLLA copolymer for 1-6 months maintained the 3D shapes of the original polymer scaffolds, appeared glistening white macroscopically, contained S-GAG and type II collagen and histologically resembled to cartilage. The chondrocytes secreted cartilage matrix to fill the void spaces in polymer scaffolds that simultaneously biodegrade. This study demonstrated the feasibility of culturing isolated chondrocytes on biodegradable polymer scaffold to regenerate 3 D neocartilage.

Utilization of PLLA/PGA copolymer poly (DL-lactic-co-glycolic acid) (PLGA) allows one to control the rate of degradation of the scaffold. It is significant as residence time of the polymer at the defect site must be sufficient to serve its scaffolding purpose- without being so long as to impede tissue degeneration. Recently, Cohen *et al.* (207) used alginate beads to deliver chondrocytes homogeneously throughout PLA/PGA pads implanted in rabbits; at 12 weeks they demonstrated good histological and biochemical response. Sherwood *et al.* (208) designed a biphasic scaffold an upper, cartilage region composed of D, L-PLGA/L-PLA and a lower, bone portion consisting of an L-PLGA/tricalcium phosphate composite. Chondrocytes preferentially attached to the cartilage portion of the device, and biochemical and histological analyses showed that cartilage formed during a 6-week *in vitro* culture period. In another study, Kawanishi *et al.* (209) utilized only PLGA copolymer to develop a bilayered porous scaffold that combine freeze-drying and porogen leaching techniques. Thus pore size at both the layers can be precisely controlled. These bilayered porous scaffolds were revealed to have good potential in generating artificial cartilage.

Polyethylene Glycol (PEG) Hydrogel Scaffold

Hydrogels are attractive material for cartilage engineering because of their high water retaining properties and tissue like elastic properties. Hydrogels provide an environment for chondrocytes to retain their native form surrounded by lacunae. Examples of naturally occurring hydrogels are alginate and fibrin glue, which have been extensively used for encapsulation of chondrocytes and successfully regenerate cartilaginous tissue. However, limitations exist with naturally occurring hydrogels with respect to gel properties, gelation processes and also mechanical compressive properties. As an alternative to these natural hydrogels, synthetic and photocrosslinkable hydrogels based on polyethylene glycol (PEG) have been developed. (210). These hydrogls provide greater control over the hydrogel macroscopic properties. In addition, photo polymerization of hydrogles allow temporal and spatial control of the polymerization and gelation process, which increase the ease of handling for the surgeons as well as the ability to form complex architecture *in vivo*.

Further research is in progress to determine optimum crosslinking density in PEG hydrogels such that the gel mechanics is maximized, while a critical water content for cell viability and critical mesh size for extracellular matrix distribution is maintained.

Polyurethanes Scaffold

For years, biomedical segmented polyurethanes have extensively been used in various implantable devices like pacing leads insulation, indwelling catheters, intra-aortic balloons and mammary implants (211,212). While one of the primary features of these polyurethanes is their molecular stability *in vivo*, for the last several years there has been an increasing interest in designing biodegradable polyurethanes, which degrade *in vivo* to nontoxic by-products (213-216). Recently biodegradable polyurethanes have successfully been used as cancellous bone graft substitutes in animals. (217,218) However, very little attention has been paid to biodegradable polyurethanes as scaffolding materials for chondrocytes. To date, only hydroxybutyrate-co-valerate polyol-based polyurethane has been used to culture rat chondrocytes (219). A recent study by Grad (220) demonstrated that three-dimensional porous scaffolds prepared from linear biodegradable polyurethanes might serve as a carrier for cartilage tissue engineering, although several limitations need yet to be overcome. Firstly, the retention of matrix macromolecules in the scaffold has to be improved, and secondly, the scaffold structure should be modified to maintain the appropriate chondrocytic phenotype of the cells. A possible strategy to overcome these problems might be to decrease the scaffold pore size without impairing a homogeneous cell distribution within the scaffold. Also a too low pore size in the scaffolds may potentially limit the access to nutrients (111). Alternatively, the cells might be suspended in a gel matrix, such as agarose, alginate, chitosan, fibrin, or collagen, in order to improve extracellular matrix retention, before seeding them into the scaffold. The scaffold would have important function in helping to maintain the mechanical rigidity while the gel matrix would permit, in addition to improving the retention rate, a three-dimensional configuration of the chondrocytes that would facilitate the preservation of their differentiated phenotype (169).

Poly (hydroxybutyrate-co-hydroxyhexanoate) Scaffolds

Polyhydroxybutyrate (PHB) is finding some interesting potential applications as an implant material due to its biocompatibility and resorbability. PHB has already been used as temporary stents, bone plates, patches, nails and screws. The limitation associated with its use in soft tissue engineering is its brittle mechanical properties. To circumvent this, materials such as triethylcitrate have been used as plasticizers with PHB. Blend copolymer of PHA consisting of 3-hydroxybutyrate and 3-hydroxyvalerate (P (HB-co-HV)) were reported to be widely used with its improved mechanical properties over that of PHB (221). PHA with better elastomeric properties, such as poly (hydroxybutyrate-co-hydroxyhexanoate) (PHBHHx), is a potential candidate to meet special application as scaffolds in cartilage engineering (222) and research are being carried out in this direction.

Study by Deng (223) suggests the production of extracellular matrix of rabbit articular cartilage chondrocytes grown on scaffolds of PHB blended with PHBHHx. The m-RNA level of type II collagen of chondrocytes seeded on scaffolds consisting of PHBHHx were higher than that of PHB-only scaffold, throughout the culture period, suggesting the positive effect of PHBHHx on extracellular matrix production. Second-harmonic generation (SHG) imaging technique, combined with confocal fluorescence microscopy (CFM) revealed that PHBHHx in PHB scaffold provided better surface properties for anchoring type II collagen filaments and their penetration into internal layers of the scaffolds. Also glycosaminoglycan showed a sharp increase in construct of PHB/PHBHHx scaffold after 7-day cultivation, while only a small increase was observed in all other tested scaffolds. It appears that the presence of right proportion of PHBHHx in the composite system of PHB/PHBHHx, highly favors the production of extracellular matrix of articu

Titanium Alloy and Calcium Polyphosphate Scaffolds

One of the procedures for cartilage repair involves transplantation of osteochondral plugs (subchondral bone and overlying articular cartilage) obtained from donor sites within the same joint, into the cartilage defect (224, 225). This treatment is termed as 'mosaicplasty' and has been widely used over the past several years. But there are several major disadvantages with this technique like donor site morbidity, movement of implanted plugs relative to each other, the formation of repair tissue between plugs (*i.e.* spaces between juxtaposed round plugs), and the unpredictable integration of transplanted plugs with host tissues (226, 227). Thus misalignment of the grafts during placement can compromise graft viability because of formation of fibrous tissue at the perimeter of the grafts and/or abrasion of graft cartilage if protruding above the intended surface of articulation.

To circumvent some of these problems an approach has been developed that consists of generating a biphasic (*i.e.* cartilage and porous bone substitute) "osteochondral-type" plug in vitro (228, 229). This approach precludes the need to harvest donor plugs. The biphasic plug is composed of cartilaginous tissue that is formed by isolated chondrocytes placed on the surface of the porous substrate. The porous substrates, which were utilized, are porous calcium polyphosphate (CPP) and porous titanium alloy (Ti6Al4 V) (229, 230). Although these substrates have very different compositions and surface roughness, chondrocytes cultured on the top surface of these materials will form a continuous layer of cartilaginous tissue. The cells maintain their phenotype as characterized by the synthesis of large proteoglycans and type II collagen. Chondrocytes attached to 3D-porous substrates, Ti alloy and CPP, remained less spread than cells in monolayer culture and exhibited a ring-like, subcortical organization of actin filaments. The presence of these features may be involved in favoring tissue formation by the cells. Further studies are ongoing to evaluate how these porous materials induce these effects in chondrocytes. Knowing this it should be possible ultimately to synthesize *in vitro* the novel materials that would incorporate the characteristics of osteochondro graft and enhance cartilage tissue formation.

PLLA-Collagen Hybrid Scaffold

To integrate into the skeletal system, adequate mechanical strength is required of tissue engineered cartilage grafts to enable surgical handling and mechanical loading. Because biodegradable synthetic polymers can easily be processed into intended shapes and have good mechanical strength, resulting in easy handling and withstanding tissue collapse, they have been used as scaffolds for tissue engineering of cartilage. But lack of cell interaction and the hydrophobic property of these materials hinder uniform cell seeding. Many efforts have been made on seeding chondrocytes on synthetic polymer scaffolds but it is still difficult to achieve uniform cell distribution. The concept of hybrid scaffolds is a promising one where synthetic material is hybridized as skeletons with other more mechanically weak but bioactive natural materials. In few recent studies PLLA sponge of large pore size, hybridized with collagen sponge has been used as the scaffold for *in vivo* tissue engineering of bovine articular cartilage (231-234). The PLLA–collagen sponge retained the advantages of both PLLA and collagen, and facilitated the formation of cartilaginous tissue, suggesting that it would be a useful scaffold for cartilage tissue engineering. Further quantitative studies are required to address the precise seeding efficiency utilized by the hybrid scaffold. The effect on biomechanical composition and mechanical character of constructs should also be evaluated.

Table 1.2: Summary of Performance of Various Scaffolds in Cartilage Engineering

Scaffold	*Advantages*	*Disadvantages*	*References*
Heterologous	1. Native protein of body	1. Incomplete integration with host tissue	149, 150, 165, 172, 235
A. Natural Scaffolds	2. Good mechanical properties	2. Xerogenic source may provoke immune response	
(i) Collagen	3. Can form various shapes	3. Early repair may thin over time	
	4. Excellent histological results when carrying BMP-2 and growth factors		
(ii) Fibrin	1. Can be obtained from own blood	1. Poor mechanical properties	173, 178, 179, 180
	2. Provide good cell attachment and wound healing	2. Does not permit host cell in growth	
	3. Produce good result as carrier of c-DNA for growth factor and BMP-2		
(iii) Agarose/alginate	1. Uniform cell distribution within implant	1. Foreign body giant cell reaction (in agarose)	181, 182, 183, 236
	2. Injectible and easily degradable	2. Inferior mechanical properties	
	3. Good histological and biochemical properties	3. Does not resorb well	
(iv) Hyaluronan	1. Good binding with joints	1. Poor biochemical and mechanical assessment	138, 184, 185, 237
	2. Native to joints	2. May induce chondrolysis	
	3. Binds to many proteins		
(v) Chitosan	1. Binds to polyanionic compounds to form hydrogels	1. Poor mechanical properties	191, 192, 195
	2. Deliver growth factors and BMP-7 along with chondrocytes	2. Inferior biochemical properties	
		3. May induce foreign body response	

Contd...

Table 1.2–Contd...

Scaffold	Advantages	Disadvantages	References
B. Synthetic Scaffolds	1. Biocompatible and biodegradable	1. Inferior biochemical properties	200, 201, 202, 205
	2. Good interaction with host in presence of BMP-2	2. Gross repair with inconsistent sub-chondral bone formation	
(i) Poly lactic acid (PLA)	1. Excellent retention at implant site		
(ii) Poly glycolic acid (PGA)	1. Good short term results in some animals		118, 206
(iii) PLGA Copolymer	1. Controlled pore size		207, 208, 209
	2. Good histological and biochemical results		
	3. Maintain 3D shape		
(iv) Polyethylene glycol (PEG)	1. High water retaining properties	1. Poor Biomechanical properties	210
	2. Controllable pore size and elasticity		
Autologous Scaffolds			
(i) Perichondrium/ Periosteum	1. Applicable for young patients with part of healthy cartilage	1. Incomplete filling of defects	130,196
	2. Ease of surgical handling	2. Not satisfactory with people age 40 years	
	3. Carry undifferentiated progenitor cell	3. Graft detachment during movement	
(ii) Mosaicplasty	1. Pain relief and restoration of joint function	1. Creates defects in non-weight bearing regions	131, 225
(iii) Autologous chondrocytes transplants	1. Hyaline like repaired tissue	1. Requires double surgery (harvesting /implantation)	176, 238
	2. Graft matures with time	2. Technically difficult surgery	
	3. Pain relief and recovery of function	3. Cells are expanded in vitro	

The designing and evaluation of scaffolds is one of the major upcoming field of tissue engineering and several materials are in the infancy of investigation for appropriateness for cartilage repair, including poly (ethylene glycol)-terephthalate and poly (butylenes terephthalate) copolymer blocks, oligo (poly (ethylene glycol) fumarate, poly (*N*-isopropylacrylamide), and carbon fiber scaffolds. The versatility of many of the scaffolds discussed above is often counter balanced by a lack of one or more of the cited parameters of successful implant design. The vast effort being mounted to formulate scaffolds with multifarious role in tissue-regeneration will result in the highly efficient clinical devices for articular cartilage repair.

Critical Conditions in Articular Cartilage Tissue Engineering

Some of the critical aspects, which have to be monitored closely during *ex vivo* cartilage engineering are as following:

Role of Mechanical Stimulations

It is clear that mechanical stimuli have their role in synthesis and organization of components of articular cartilage that in turn affect the cellular microenvironment. *In vitro* studies have demonstrated that like most cell types, chondrocytes are also connected to their extracellular environment via focal adhesions, which mediate mechanotransduction and affect fundamental cellular activities. A scaffold coated with any of the adhesion molecule like fibronectin and laminin may serve as a substratum to promote cell attachment and tissue formation.

The influence of mechanical stimuli on the development of articular cartilage is especially important because other stimuli provided via vascular, lymphatic and neural systems are not necessarily available to chondrocytes due to avascular, aneural and alymphatic nature of cartilage. Application of mechanical stimuli like hydrostatic pressure, fluid-flow-induced shear, or interaction with a solid object generally promotes expression of cartilage molecules and synthesis of a cartilaginous material. (239,240). Although the precise signaling pathways that are involved in mediation of mechanostimuli are not completely understood, evidence suggests that some types of mechanical stimuli like turbulent flow, or suspended as freely floating composites in rotating vessels, are desirable for articular cartilage synthesis and modeling and clear differences were observed in the distribution of deposited matrix within the PGA scaffolds and in properties of the synthesized cartilaginous material (241). The metabolic activities of chondrocytes are significantly affected by varying load parameters, frequency and load duration. The role of mechanical stimuli would be very important while designing the bioreactor for cartilage genesis *in vitro*.

Role of Cell Seeding Density

In the literature, few studies have addressed the effects of the initial cell density on cartilage genesis. In a recent study Mauck (242) investigated the role of initial cell seeding density and nutrient supply on cartilage formation in agarose hydrogels and studied the effect of mechanical intermittent deformational loading. The young's modulus and dynamic modulus were calculated and showed that scaffolds seeded at high cell density (60 million cells/ml) show >2 fold increase in material properties in comparison to the constructs with low cell density (10 million-cells/ml). Also the chondrocytes seeded at density less than 10 million cells per ml at polylactide–polyglycolide scaffolds results in very little cartilaginous material synthesis (243). As the cell seeding density increases, the integrity of the cartilaginous product improves and its distribution becomes more homogeneous. Another systemic study investigated the effect of initial cell seeding density on cartilage formation and revealed that seeding density ranging from 20-100 million cells per ml resulted in formation of clinically appropriate cartilage when implanted subcutaneously into nude mice; whereas lower seeding density (*e.g.*, 10 million cells per ml) exhibited reduced cartilaginous formation as assessed by wet weight and thickness (244).

Role of Media Composition

To obtain the optimum number of chondrocytes and their secreted matrix products is the core of *in vitro* cartilage genesis. A number of different growth media with varying nutrient compositions are commercially available. The medium for chondrocyte culture should contain an energy source, inorganic salts, amino acids, vitamins and a buffering component, usually either a bicarbonate buffer or HEPES. Glucose, pyruvate and glutamate are included in growth medium as source of energy. Fetal bovine serum is a complex undefined media containing mitogenic peptide growth factors and many attachment factors to maintain differentiation and viability of chondrocytes. Other factors, which may be included, are those, which promote cell adhesion and spreading (*e.g.*, fibronectin and vitronectin), trace minerals, transport proteins such as transferrin and albumin, growth factors, and hormones. Typically 10% of the volume of growth media consists of fetal bovine serum or calf serum. For the synthesis of cartilage, supplements like ascorbic acid, proline, and some non-essential amino acids may be included. The chondrocyte culture is maintained at 37ºC in presence of 5% CO_2 and penicillin and streptomycin antibiotics. A combination of commercially available media components like Dulbecco's Modified Eagle Medium (DMEM) can easily supply all these components. Various growth factors like TGF-β, IGF-I, b FGF and BMPs may also be added in the medium to enhance the synthetic properties of cartilage.

Role of Scaffold Surface Properties and their Modifications

There are three mechanisms for attachment of cells to scaffolds:

1. Coating with physiological adhesion proteins such as fibronectin (245),
2. Cell contact with specific synthetic peptides derived from extracellular matrix molecules (243, 246, 247) and
3. Cell interaction with a suitably charged substratum such as a polylysine-coated surface.

The serum components like fibronectin and vitronectin are primary glycoproteins and promote adhesion of various synthetic peptides that are known to interact with cell surface adhesion molecules (249). Use of polylysine allows cells to spread on substratum via charge-mediated interactions (243). Also synthetic peptides, which interact with cell surface adhesion molecules like integrins (246) and proteoglycans (248) are likely to play a significant role in cartilage engineering.

Challenges for Tissue Engineered Cartilage

Cartilage tissue engineering has emerged as a potential alternative to joint replacement and represents an exciting direction in the efforts to solve the problem of cartilage regeneration. The aspect of tissue engineering that involves *ex vivo* fabrication of cartilage constructs occupies a central position in addressing this complex problem in a logical manner. Functional restoration of cartilage tissue is highly desirable yet remain medically challenging.

Mechanical Stability of Engineered Cartilage

The primary goal of *in vitro* engineering hyaline articular cartilage is to achieve the native biomechanical properties of the tissue. Although various studies have suggested as described in above sections that *in vivo* tissue-engineered cartilage achieves some pressure resistance and stiffness values which is comparable to those of native human cartilage. However, yet it is not clear that up to what extent of physiological loading the newly construct cartilage transplants could remain resilient. Also the regenerated cartilage reveals clear differences between native and transplanted cartilage as the distribution of cells lack typical zonal organization and appears somewhat random. Also the typical collagen architecture gets altered in more loosened and irregular helix formation. The water retaining property of the tissue is also under investigation, which is a very important feature of normal cartilage for its weight bearing compressive stiffness properties.

Fixation of Graft at the Site of Injury

Another major problem in cartilage engineering is the fixation of the cartilage transplant to the subchondral bone at the site of injury. In theory, the artificially grown cartilage layers could be attached directly to the defect joint surface using fibrin glue or collagen-fibrin layer. Also the use of resorbable pins has been mentioned in various studies (250, 251). Currently, clinical studies on patients with *in vitro* engineered cartilage flaps are ongoing. Another strategy that has gained attention is to develop an osseo-integrating interface for cartilage implant using either a calcium carbonate or hydroxyapatite–tricalcium phosphate (HA–TCP) scaffold or an engineered osteoblast layer (252). The ultimate aim of these studies is to achieve a permanent, solid connection between cartilage and bone tissue.

Immunological Aspects

Cartilage was long considered an immunologically privileged tissue. However, studies have shown that cartilage, and specifically chondrocytes, do possess antigens and are capable of eliciting an immune response. Two major immunological risks are linked to the use of engineered cartilage transplants. First, the invasion of scaffold by host giant cells or granulocytes. Second, the exposures of hidden cell surface and matrix protein epitopes, which are usually masked from the immune system and might therefore, generate immune response. The persons who have been operated for cartilage transplants, showed humoral reactions against type IX and XI collagens, which are associated with collagen, type II fibrils (253). In response to isolated chondrocyte allografts in full-thickness articular cartilage defects in rabbit knees, host lymphocytes accumulated around the allografts. According to the study by Kawabe (178) only 26% of the grafts contained viable cartilage cells after three weeks and both cell-mediated cytolytic reaction and a humoral immune response were present. Collagen and proteoglycan degrading enzymes were secreted in response to a factor produced by macrophases; and their further synthesis is inhibited by lymphocytes produced factors. Other studies have shown that antibodies may also be directed against other antigenic sites on chondrocytes or against specific sites in the surrounding matrix (254). Many other reports have confirmed the antigenicity of isolated chondrocytes; however, the immune response does not cross the intact matrix (255, 256). The matrix appears to prevent recognition of antigens on the chondrocytes, protecting them from the host immune response (257, 258)

Clearly, even though tissue engineering is usually regarded as an autologous therapy, a major clinical breakthrough would be, to solve the immunological problems associated with the technique.

Future Directions in Cartilage Engineering

The key to successful repair and regeneration of cartilage is to provide the repair site with sufficient chondrogenic cells in a suitable delivery vehicle to ensure maximal differentiation and deposition of right extracellular matrix. New optimized culture methodologies and bioreactors that provide appropriate mechanical and other guidance clues in order to ensure the successful and long-term functioning of engineered tissue. As we gain more and more information about the identity of morphogens for chondrocyte differentiation, it might be possible to orchestrate massive cartilage regeneration by clever combination of smart 3D scaffolds and such morphogenic factors. The management of cells both *in situ* and *ex vivo*, will be crucial to the success of such tissue engineering efforts. Also concentrated efforts are on to optimize the use of biochemical regulatory signals and growth factors like TGF-β, recombinant human BMP-2 (rhBMP-2) in chondrocytes differentiation which may help in understanding the complex phenomenon of cartilage degeneration. In the future, it can be envisioned that a combination of growth factors or cytokines may be incorporated into scaffolds to modify different cell functions, such as differentiation, metabolism, and synthesis of extracellular matrix. Recently, researchers have begun to combine the above chondrocyte culture models with external mechanical stimuli so as to improve the quality of generated cartilage tissue. Regulation of the biomechanical environment is a powerful modulator in tissue engineering of articular cartilage. The application of extrinsic mechanical forces can have a profound effect on chondrocyte function and cartilage regeneration. Effects of agitation/stirring, microgravity, and shear stress have a direct role on cell metabolism and extracellular matrix synthesis. It is essential to quantify the stress–strain fields experienced by individual cells attached on scaffolds and how these mechanical stimuli affect their mechanotransductional abilities.

Modification of scaffold surface material properties in order to enhance cell adhesion and to promote repair of partial and full thickness defects of articular cartilage are only now beginning to be considered. In the future, it may be desirable to modify scaffold external and internal surfaces in a preferential way such that preferred chemotaxis can be achieved. Another promising avenue is the identification and characterization of novel cell adhesive extracellular matrix molecules that may play real roles in cartilage development.

A very important aspect of research in chondrocyte culture and artificial cartilage engineering is development of clinical standards for assessments of neocartilage. As may be expected when assessing complex biological systems, the variables, which potentially affects the experimental outcomes of cartilage repair, needs thorough evaluations. For example evaluation of scaffolds with regards to mechanical properties, permeability, porosity, material properties, degradation rate, and biocompatibility. The experimental approach employed in the *in vivo* assessment of engineered cartilage also needs to be standardized with regards to the animal models used, surgical approach, defect size, location, defect creation, and duration of healing. Standard protocols should be allowed for multidisciplinary assessment of healing using well-defined tests.

It is essential and desirable that multiprong and multidimensional efforts should be made sincerely for the development of clinical standard protocol for chondrocyte cell tissue engineering. So the gigantic problem of degenerative joint disease like osteoarthritis should be tackled and diagnosed at early stages of development and remedial measures for the treatment and repair process may be initiated for the prevention of this dreaded disease.

REFERENCES

1. Mankin, H.J. and Lippiello, L. (1969) The turnover of adult rabbit articular cartilage **J. Bone Jt. Surg.** (51A) 1591.
2. Stockwell, R.A. (1979) Biology of cartilage cells, Cambridge, Cambridge University press.
3. Muir, H. (1995) The chondrocyte, architect of cartilage: Biomechanics, structure, function and molecular biology of cartilage matrix macromolecules. **Bioessays** (17) 1039-1048.
4. Mayre, R. Collagens: (1989) What is their function, are they involved in articular disease. **Arthritis Rheumatism** (32) 241-246.
5. Eyre, D.R., Wu, J.J. and Woods, P.E. (1991) The cartilage collagens: Structural and metabolic studies. **J. Rhematol.** (18) 49-51.
6. Roughley, P.J. and Lee, E.R. (1994) Cartilage proteoglycans: Structure and potential function. **Microsc. Res. Technol.** (28) 385-397.
7. Mankin, H.J., Dorfman, H., Lippiello, L. and Zarins, A. (1971) Biochemical and metabolic abnormalities in articular cartilage from osteoarthritic human hips. **J. Bone Joint Surg.** (53 A) 523-537.
8. Maroudas, A. (1976) Balance between swelling pressure and collagen tension in normal and degetenerative cartilage. **Nature** (260) 808-809.

9. Imai, K., Ohta, S., Mastumoto, T., Fujimoto, N., Sato, H. and Seiki, M. (1997). Expression of membrane type I matrix metalloproteases and activation of progelatinase A in human osteoarthritic cartilage. **Am. J. Pathol.** (151) 245-256.

10. Ohta, S., Imai, K., Yamashita, K., Mastumoto, T., Azumano, I. and Okada, Y. (1998) Expression of matrix metalloproteases-7 (matrilysin) in human osteoarthritic cartilage. **Lab. Investi.** (78) 79-87.

11. Lohmander, L.S., Neame, P.J. and Sandy, J.D. (1993) The structure of aggrecan fragments in human synovial fluid-evidence that aggrecanase mediate cartilage degeneration in inflammatory joint disease, joint injury and osteoarthritis. **Arthritis Rheum.** (36) 1214-1222.

12. Mynarik, V. and Trating, S. (2000) Physiochemical properties of normal articular cartilage and its MR appearance. **Invest. Radiol.** (35) 589-594.

13. Hauselmann, H.J., Masuda, K., Hunziker, E.B., Neidhart, M., Mok, S.S., Michel, B.A. and Thonar, E.J.M.A. (1996) Adult human articular chondrocytes cultured in alginate gel forms a matrix similar to native human articular cartilage. **Am. J. Physiol.** (Cell Physiol.) (271) 742-742.

14. Aydellotte, B.A., Schleyerbach, R., Zeck, B.J. and Kuettner, K.F. (1986) Articular chondrocytes cultured in agarose gel for study of chondrocytic chondrolysis. In *Articular cartilage biochemistry* (eds. Kuettner, K.E., Schleyerbach R. and Hascall V.) Ravan press, New York. pp235-254.

15. Adolphe, M. and Benya, P. (1992) Different types of cultured chondrocytes- the in vitro approach to the study of biological regulation. In *'Biological regulation of chondrocytes'* (ed M. Adolphe), CRC press Boca Raton, pp105-139.

16. Hauselmann, H.J., Fernandes, R.J., Mok, S.S., Schmid, T.M., Block, M.B., Kuettner, K.E. and Thonar, E.J. (1994) Phenotypic stability of bovine articular chondrocytes after long-term culture in alginate beads. **J. Cell. Sci.** (107) 17-27.

17. Handley, C.J., McQuillan, D.J., Campbell, M.A. and Bolis, S. (1986) Steady state metabolism in cartilage explants. In *'Articular cartilage biochemistry'* (eds Kuettner, K.E., Schleyerbach, R. and Hascall V.C.) Raven Press, New York, pp163-179.

18. Benya, P.D. and Shaffer, J.D. (1982) Dedifferentiated chondrocyte reexpress the differentiated collagen phenotype when cultured in agarose gel. **Cell** (30) 215-224.

19. Buschman, M.D., Gluzband, Y.A., Grodinzsky, A.J., Kimmura, J.H. and Hunziker, E. (1992) Chondrocyte in agarose culture synthesizes a mechanically functional extracellular matrix. **J. Orthop. Res.** (10) 745-758.

20. Guo, J., Jourdian, G.W. and Maccallumn, D.K. (1989) Culture and growth characteristics of chondrocytes encapsulated in alginate beads. **Connect. Tissue Res.** (19) 277-297.

21. Hauselmann, H.J., Aydellotte, M.B., Schumacher, B.I., Kuettner, K.E., Gitelis, S.H. and Thonar, E.J.M.A. (1992) Synthesis and turnover of proteoglycan by human and bovine articular chondrocytes cultured in alginate beads. **Matrix** (12) 116-129.

22. Bonaventure, J., Kadhom, N., Cohensolal, L., Ng., K.H., Bourguignon, J., Lasselin, C. and Frisinger, P. (1994) Reexpression of cartilage specific genes by dedifferentiated human articular chondrocytes cultured in alginate beads. **Exp. Cell Res.** (212) 99-104.

23. Kimmura, T., Yasui, N., Ohsawa, S. and Ono, K. (1984) Chondrocyte embedded in collagen gel maintains cartilage phenotype during long term cultures. **Clin Orhtop. Relat. Res.** (186) 231-239.

24. Maor, G., Vonder mark, K., Reddi, H., Heingard, D., Franzen, A. and Sibermann, M. (1987) Acceleration of cartilage and bone differentiation of collagenous substrata. **Collagen Relat. Res.** (7) 351-370.

25. Solursh, M. and Meir, S. (2001) Effects of cell density on the expression of differentiation by chick embryo chondrocytes. **J. Exp. Zool.** (187) 311-322.

26. Bywater, *E.*G.L. (1937) The metabolism of joint tissues. **J. Path. Bact.** (44) 237.

27. Poole, C.A., Flint, M.H. and Beaumont, B.W. (1987) Chondrons in cartilage: Ultrastructural analysis of the pericellular microenvironment in adult human articular cartilage. **J. Orthop. Res.** (5) 509-522.

28. Grodzinsky, A.J. (1983) Electromechanical and physicochemical regulation of cartilage strength and metabolism. **Crit. Rev. Bioengg.** (9) 133-199.

29. Maroudas, A. (1979) Physico-chemical properties of articular cartilage. In: *Adult articular cartilage*, ed. Freeman, M.A.R. (London: Pitman medical) pp215-290.

30. Holm, S., Maruodas, A., Urban, J., Selstam, G. and Nachemson, A. (1981) Nutrition of the intervertebral disc: Solute transport and metabolism. **Conn. Tiss. Res.** (8) 101-119.

31. Maroudas, A., Popper, O. and Grushko, G. (1992) Partition coefficient of IGF-1 between cartilage and external medium in the presence and absence of FCS. **Trans. Am. Orhtop. Res. Soc.** (16) 398.

32. Hodge, W.A., Fijan, R.S., Carlson, K.L., Burgess, R.G., Harris, W.H. and Mann, R.W. (1986) Contact pressure in the human hip joint measured *in vivo*. **Proc. Natl. Acad. Sci.** (83) 2879-2883.

33. Afoke, N.Y., Byers, P.D. and Hutton, W.C. (1987) Contact pressure in the human hip joint. **J. Bone Joint Surg. Br.** (69) 536-541.

34. Bergmann, G., Graichen, F. and Rohmann, A. (1993) Hip joint loading during walking and running measured in two patients. **J. Biomech.** (26) 969-990.

35. Eisenhart, R., Adam, C., Steinlechner, M., Muller-Gerbl, M. and Ecstein, F. (1999) Quantitative determination of joint in configurity and pressure distribution during stimulated gait and cartilage thickness in the human hip joint. **J. Orthop. Res.** (17) 532-539.

36. Rydell, N. (1965) Forces in the hip joint, Part II intravital measurement. In *Biomechanics and Related Bioengineering Topics.* ed. by Kennedy R.M. (Pergamom Press, Oxford) pp 351

37. Paul, J.P. (1965) Bioengineering studies of the force transmitted by joints. Part II: Engineering analysis. In *Biomechanics and Related Bioengineering Topics.* ed. by Kennedy, R.M. (Pergamom Press, Oxford) pp 369.

38. Helminen, H., Jurvelin, J., Kjiviranta, I., Paukkonen, K., Saamanen, A.M. and Tammi, M. (1987) Joint loading effects on articular cartilage: a historical review. In *Joint loading: biology and health of articular structures.* Eds. Helminen, H., Kjiviranta, I., Tammi, M., Saamanen, A.M. and Paukkonen, K., Jurvelin, J. (Bristol: John Wright) pp 1-46.

39. Pond, M.J. and Nuki, G. (1973) Experimentally induced osteoarthritis in dog. **Ann. Rheum. Dis.** (32), 387-388.

40. Altman, R.D., Tenenbaum, J., Latta, L., Riskin, W., Blanco, L.N. and Howell, D.S. (1984) Biomechanical and biochemical properties of dog cartilage in experimentally induced osteoarthritis. **Ann. Rheum. Dis.** (43) 83-90.

41. Wu, D.D., Burr, D.B., Boyd, R.D. and Radin, E.L. (1990) Bone and cartilage changes following experimental varus or vagus tibial angulation. **J. Orthop. Res.** (8) 572-585.

42. Bentley, G., Greer, R.B. (1971) Homotransplantation of isolated epiphyseal and articular cartilage chondrocytes into joint surfaces of rabbits. **Nature** (230) 184-197.

43. Brittberg, M., Lindhal, A., Nilsson, A., Ohisson, C., Isaksson, O. and Peterson, L. (1994) Treatment of deep cartilage defect in the knee with autologous chondrocyte transplantation. **New Eng. J. Med.** (331) 889-895.

44. Paige, K.T., Cima, L.G., Yaremchuk, M.J., Schloo, B., Vacanti, J.P. and Vacanti, C.A. (1996) De novo cartilage generation using calcium alginate-chondrocytes constructs. **Plastic Reconstruct Surg.** (97) 168-178.

45. Kawamura, S., Wakitani, S., Kimura, T., Maeda, A., Calpan, A.I., Shino, K. and Ochi, T. (1998) Articular cartilage. Rabbit experiments with a collagen gel-biomatrix and chondrocytes cultured in it. **Acta Orthop. Scand** (69) 56-62.

46. Pittenger, M.F., Mackay, A.M., Beck, S.C., Jaiswal, R.K., Douglas, R., Mosca, J.D., Moorman, M.A., Simonetti, D.W., Craig, S. and Marshak, D.R. (1999) Multilineage potential of adult human mesenchymal stem cells. **Science** (284) 143-147.

47. Thomson, J.A., Itskovitz-Eldor, J., Shapiro, S.S., Waknitz, M.A., Swiergiel, J.J., Marshall, V.S. and Jones, J.M. (1998) Embryonic stem cell lines derived from human blastocysts. **Science** (282) 1145-1147.

48. Wong, M. (2003) Chondrocyte in cartilage tissue engineering. **Bioforum Eur.** (7): 6, 324-325.

49. Grogan, S.P., Rieser, F., Winkelmann, V., Berardi, S. and Malini–Varlet, P. (2003) A static, closed and scaffold-free bioreactor system that permits chondrogenesis *in vitro*. **Osteoarthritis Cartilage** (11) 403-411.

50. Parkinnen, J.J., Ikonen, J., Lammi, M.J., Laakkonen, J., Tammi, M. and Helminen, H.J. (1993) Effect of cyclic hydrostatic pressure on proteoglycan synthesis in cultured chondrocytes and articular cartilage explants. **Arch. Biochem. Biophys** (300) 458-465.

51. Kim, Y.J., Sah, R.L., Grodinzsky, A.J., Plass, A.H. and Sandy, J.D. (1994) Mechanical regulation of cartilage biosynthetic behaviour: physical stimuli. **Arch. Biochem. Biophys** (311) 1-12.

52. Buschmann, M.D., Kim, Y.J., Wong, M., Frank, E., Hunziker, E.B. and Grodinzsky, A.J. (1999) Stimulation of aggrecan synthesis in cartilage explants by cyclic loading is localized to regions of high interstitial fluid flow. **Arch. Biochem. Biophys**. (366) 1-7.

53. Bonnassar, L.J., Grodinzsky, A.H., Frank, E.H., Davila, S.G., Bhaktav, N.R. and Trippel, S.B. (2001) The effect of dynamic compression on the response of articular cartilage to insulin like growth factor-1. **J. Orthop. Res.** (19) 11-17.

54. Jin, M., Frank, E.H., Quinn, T.M., Hunziker, E.B. and Grodinzsky, A.J. (2001) Tissue shear deformation stimulates proteoglycan and protein biosynthesis in bovine cartilage explants. **Arch. Biochem. Biophys** (395) 41-48.

55. Handley, C.J., Mcquillan, D.J., Campbell, M.A. and Bolis, S. (1986) Steady state metabolism in cartilage explants eds. Keuttner, K.E., Schleyerbach, R. and Hascall, V.C. (New York, Raven press), pp163-179.

56. Campbell, N.A., Handley, C.J., Hascall, V.C., Campbell, R.A. and Lowther, D.A. (1984) Turnover of proteoglycan in culture of bovine articular cartilage. **Arch. Biochem. Biophys.** (234) 275-281.

57. Hascall, V.C., Morales, T.I., Hascall, G.K., Handley, C.J. and McQuillan, D. (1983) Biosynthesis and turnover of proteoglycan in organ culture of bovine articular cartilage. **J. Rheumatol. Suppl.** (10) 45-52.

58. Morales, T., Wahl, L.M. and Hascall, V.C. (1984) The effect of lipolpolysaccharides on biosynthesis of proteoglycans from calf articular cartilage cultures. **J. Biol. Chem.** (259) 6720-6729.

59. Bartholomew, J.S., Handley, C.J. and Lowther, D.A. (1985) The effect of trypsin treatment on proteoglycan biosynthesis by bovine articular cartilage. **Biochem. J.** (227) 429-437.

60. Tyler, J.A. (1985) Chondrocytes mediated depletion of articular cartilage proteoglycan *in vitro*. **Biochem. J.** (225) 493-507.

61. Sandy, J.D., Neame, P.J., Boynton, R.E. and Flannery, C.R. (1991) Catabolism of aggrecan in cartilage explants. **J. Biol. Chem.** (266) 8683-8685.

62. Tian, X., Chen, S., Morales, T.I. and Hascall, V.C. (1989) Biochemical and morphological study of steady state and lipopolysaccharide treated bovine articular cartilage explant cultures. **Connect. Tissue. Res.** (19) 195-218.

63. Morales, T.I. and Hascall, V.C. ('1989) Factors involved in the regulation of proteoglycan metabolism in articular cartilage. **Arthritis Rheum.** (32): 10, 1197-1201.

64. McQuillan, D.J., Handley, C.J., Campbell, M.A., Bolis, S., Milway, V.E. and Herington, A.C. (1986) Stimulation of proteoglycan biosynthesis by serum and insulin like growth factor-1 in culture bovine articular cartilage. **Biochem. J.** (240) 423-430.

65. Luyten, F.P., Hascall, V.C., Nissley, S.P., Morales, T.I. and Reddi, A.H. (1988) Insulin like growth factors maintain steady state metabolism of proteoglycan in bovine articular cartilage explants. **Arch. Biochem. Biophys.** (267) 416-425.

66. Roberts, A.B., Flanders, K.C., Kondaiah, P., Thompson, N.L., Van Obbereghen- Schilling, E., Wakefield, L., Rossi, P., de Drombrugghe, B., Heline, U. and Sporn, M.B. (1988) Transforming growth factor beta; biochemistry and roles in embryogenesis, tissue repair and remodeling and carcinogenesis. **Recent Prog. Horm. Res.** (44) 157-197.

67. Morales, T.I. and Hascall, V.C. (1989) Specificity of action of lipopolysaccaride on bovine articular cartilage metabolism: a comparison with interleukin-1. **Connect. Tissue Res.** (19) 255-275.

68. Weibkin, O.W., Hardingham, T.E. and Muir, H. (1975) Hyaluronic acid proteoglycan interaction and the influence of hyaluronic acid on proteoglycan synthesis by chondrocytes from adult cartilage. Extracellular matrix influences on gene expression. Edited by Slavkin, H., Greulich, R. (New York, Academic press).

69. Jurvelin, J., Kiviranta, I., Tammi, M. and Helminen, J.H. (1986) Softening of canine articular cartilage after immobilization of the knee joint. **Clin. Orthop. Rel. Res.** (207) 246–252.

70. Saamanen, A.M., Tammi, M., Kiviranta, I., Jurvelin, J. and Helminen, H.J. (1987) Maturation of proteoglycan matrix in articular cartilage under increased and decreased joint loading. A study in young rabbits. **Conn. Tissue Res.** (16) 63–175.

71. Behrens, F., Kraft, E.L. and Oegema, T.R. (1989) Biochemical changes in articular cartilage after joint immobilization by casting or external fixation. **J. Orthop. Res.** 7: 335–343, 1989.

72. Aydelotte, M.B., Greenhill, R.R. and Kuettner, K.E. (1988) Differences between subpopulations of cultured bovine articular chondrocytes II Proteoglycan metabolism. **Connect. Tissue. Res.**(18): 223-234.

73. Poole, C.A., Flint, M.H. and Beaumont, B.W. (1988 a) Chondrons extracted from canine tibial cartilage: Preliminary report on their isolation and structure. **J. Orthop. Res**. 6: 1988a

74. Graff, R.D., Kelley, S.S. and Lee G.M. (2003) Role of pericellular matrix in development of a mechanically functional neocartilage. **Biotechnol. Bioengg**. (82) 4, 457–464.

75. Benya, P.D. and Brown, P.D. Modulation of chondrocyte phenotype *in vitro*. (1985) In *Articular cartilage biochemistry;* Kuettner, K.E., Schleyerbach, R., Hascall, V.C. eds. New York, Raven Press pp 219-230,

76. Vondermark, K. (1986) Differentiation, modulation and dedifferentiation of chondrocytes. **Rheumatology** (10) 272-315.

77. Grundmann, K., Zimmermann, B., Barrach, H.J. and Merker, H.J. (1990) Behaviour of epiphyseal mouse chondrocyte populations in monolayer culture. **Virchows Arch. A. Path. Anat. Histol** (389) 167–187.

78. Quarto, R., Dozin, B., Tacchetti, C., Campanile, C., Malfatto, C. and Cancedda, R. (1990) *In vitro* development of hypertrophic chondrocytes starting from selected clones of dedifferentiated cells. **J Cell Biol.** (110) 1379–1386.

79. Shakibaei, M. (1995) Integrin expression on epiphyseal mouse chondrocytes in monolayer culture. **Histol. Histopathol.** (10) 339–349.

80. Benya, P.D., Padilla, S.R. and Nimni, M.E. (1977) The progeny of rabbit articular chondrocytes synthesize collagen types I, III and type I trimer, but not type II. Verifications by cyanogens bromide peptide analysis. **Biochemistry** (16) 865–872.

81. Vondermark, K., Gauss, V., Vonder, H. and Mu¨Ller, P. (1977) Relationship between cell shape and type of collagen synthesized as chondrocytes lose their cartilage phenotype in culture. **Nature** (267) 531–532.

82. Vondermark, K. (1980) Immunological studies on collagen type transition in chondrogenesis. **Curr. Top. Dev. Biol**. (14) 199–225.

83. Shakibaei, M., Zimmermann, B. and Merker, H.J. (1995) Changes in integrin expression during chondrogenesis *in vitro* (An immunomorphological study). **J. Histochem. Cytochem**. (43) 1061–1069.

84. Norby, D.P., Malemud, C.J. and Sokoloff, L. (1977) Difference in the collagen types synthesized by lapine articular chondrocytes in spinner and monolayer culture. **Arthritis Rheum**. (20) 709–715.

85. Schröter-Kermani, C., Hinz, N., Risse, P., Zimmermann, B. and Merker, H.J. (1991) The extracellular matrix in cartilage organoid culture: Biochemical, immunomorphological and electron microscopic studies. **Matrix** (11) 428–440.

86. Shakibaei, M., Schröter-Kermani, C. and Merker, H.J. (1993a) Matrix changes during long-term cultivation of cartilage (organoid or high-density culture). **Histol. Histopathol**. (8) 463–470.

87. Shakibaei, M., Abou-Rebyeh, H. and Merker, H.J. (1993b) Integrins in ageing cartilage tissue *in vitro*. **Histol. Histopathol**. (8) 715–723.

88. Gerstenfeld, L.C. and Landis, W.J. (1991) Gene expression and extracellular matrix ultrastructure of a mineralizing chondrocyte cell culture system. **J Cell Biol** (112) 501–513.

89. Zimmermann, B., Schröter-Kermani. C., Shakibaei, M. and Merker. H.J. (1992) Chondrogenesis, cartilage maturation and transformation of chondrocytes in organoid culture of mouse limb bud mesodermal cells. **Eur. Arch. Biol.** (103) 93–111.

90. Shakibaei, M. (2000) Adhesion rates and chondrocyte phenotype in monolayer cultures under the influence of collagen type II and IGF-I. **Acta Medica Iranica** (3) 138-147.

91. Thonar, E.J.M.A., Buckwalter, J.A. and Kuettner, K.E. (1986) Maturation related differences in the structure and composition of proteoglycan synthesized by chondrocytes from bovine articular cartilage. **J. Biol. Chem.** (261) 2467-2474.

92. Kimura, J.H., Hardingham, T.E. and Hascall, V.C. (1980) Assembly of newly synthesized proteoglycan and link proteins into aggregates in culture of chondrosarcoma chondrocytes. **J. Biol. Chem.** (255) 7134-7143.

93. Lohmander, L.S. (1990) Biosynthesis of cartilage proteoglycan: An analysis of posttranslational events by different *in vitro* labeling protocols. In '*Methods in Cartilage Research*, eds Maruodas, A. and Kuettner, K.E. (Academic press, San Diego, CA) pp148-152.

94. Hauselmann, H.J., Stefanovic-Racie, M., Michel, B.A. and Evans, C.H. (1998) Differences in nitric oxide production by superficial and deep human articular chondrocytes. Implications for proteoglycan turnover in inflammatory joint diseases. **J. Immunol.** (160) 1444-1448.

95. Chubinskaya, S., Huch, K., Mikecz, K., Cs- Szabo, G., Hasty, K.A., Kuettner, K.E. and Cole, A. (1996). Chondrocytes MMP-8: Upregulation of neutrophil collagenase by interleukin- 1b in human cartilage from knee and ankle joints. **Lab Invest.** (74) 232-240.

96. Brukner, P., Horler, I., Mendler, M., Houze, Y., Winterhalter, K.H., Eich-Bender, S.G. and Spycher, M.A. (1989) Induction and prevention of chondrocyte hypertrophy in culture. **J. Cell. Biol.** (109) 2537-2545.

97. Archer, C.W., Mc Dowell, J., Bayliss, M.T., Stephens, M.D. and Bentley, G. (1990) Phenotypic modulation in sub-population of human articular chondrocytes *in vitro*. **J. Cell Sci.** (97) 361-371.

98. Sun, D., Aydelotte, M.B., Maldonado, B., Keittner, K.E. and Kim, H. (1986) Clonal analysis of population of chondrocytes from rat chondrosarcoma in agarose culture. **J. Orthop. Res.** (4) 427-436.

99. Buschmann, M.D., Gluzband, Y.A., Grodinzsky, A.J. and Hunziker, E.B. (1995) Mechanical compression modulates matrix biosynthesis in chondrocyte/agarose culture. **J. Cell. Sci.** (108) 1497-1508.

100. Aydelotte, M.B., Schumacher, B.L. and Kuettner, K.E. (1992) Heterogeneity of articular chondrocytes. In *Articular Cartilage and Osteoarthritis* eds. Kuettner, K.E., Schleyerbach, R. Peyron, J.G., Hascall, V.C. New York: Raven press, pp-237-249,

101. Petit, B., Masuda, K., D'Souza, A.L., Otten. L., Pietryla, D., Hartmann. D.J., Morris. N.P., Uebelhart. D., Schmid, T.M. and Thonar. E.J.M.A. (1996) Characterization of crosslinked collagen synthesized by mature articular chondrocytes cultured in alginate beads: comparison of two distinct matrix compartments. **Exp. Cell. Res.** (255) 151-161.

102. Martinsen, A., Skjaak-Braek, G. and Smidsroed, O. (1989) Alginate as immobilization material: I Correlation between chemical and physical properties of alginate gel beads. **Biotechnol. Bioengg.** (33) 79–89.

103. Rowley, J.A., Madlambayan, G. and Mooney, D.J. (1999) Alginate hydrogels as synthetic extracellular matrix materials. **Biomaterials** (20) 45–53.

104. Mok, S.S., Masuda, K., Häuselmann. H.J., Aydelotte, M.B. and Thonar, E.J. (1994) Aggrecan synthesized by mature bovine chondrocytes suspended in alginate. Identification of two distinct metabolic matrix pools. **J. Biol. Chem.** (269) 33021–33027.

105. Kamada, H., Masuda, K., D'souza, A.L., Lenz, M.E., Pietryla, D., Otten, L. and Thonar E.J.M.A. (2002) Age related differences in the accumulation and size of hyaluronan in alginate culture. **Arch. Biochem. Biophys** (408) 192-199.

106. Genes, N.G., Rowley, J.A., Mooney, D.J. and Bonassar, L.J. (2004) Effect of substrate mechanics on chondrocyte adhesion to modified alginate surfaces. **Arch. Biochem. Biophys.**

107. Cimpean, A., Zarnescu, O., Caloianu, M., Moldovan, L. and Efimov, N. (2000) Glycosaminoglycan Synthesis by Human Osteoarthritic Chondrocytes Cultured in Alginate Beads. **Roum. Biotech. Lett.** (5) 2, 127-134.

108. Mizuno, S. and Glowacki, J. (1996) Three-dimensional composite of demineralized bone powder and collagen for *in vitro* analysis of chondroinduction of human dermal fibroblasts. **Biomaterials** (17) 1819–1825.

109. Allemann, F., Yates, K.E. and Glowacki, J. (2000) Preservation of chondroblastic gene expression signature in 3D culture devices. **J. Bone Miner. Res.** (15) S468.

110. Eid, K., Mizuno, S. and Glowucki, J. (1998) Effect of PTH on matrix production by articular and endochondral chondrocytes in 3D collagen sponges. **Bone** (23) S349.

111. Allemann, F., Mizuno, S., Eid, K., Yates, K.E., Zaleske, D. and Glowacki, J. (2001) Effects of hyaluronan on engineered articular cartilage extracellular matrix gene expression in 3D collagen scaffolds. **J. Biomed. Mater. Res.** (55) 13–19.

112. Herbert, D., Elsworth, R. and Telling, R.C. (1956) The continuous culture of bacteria; a theoretical and experimental study. **J. Gen. Microbiol. (**14) 601-622.

113. Hirtenstein, M. and Clark, J. (1980) Attachment and proliferation of animal cells on microcarriers (CytodexTMl). In *Tissue Culture in Medical Research*, eds. Richards, R.J., Rayan, K.T., Oxford: Pergamon Press, pp 97-104.

114. Posillico, *E.G.* (1986) Microencapsulation technology for large-scale antibody production. **BioTechnology** (4)114-117.

115. Zuhlke, A., Roder, B., Widdecke, H. and Klein, J. (1993) Synthesis and application of new microcarriers for animal cell culture. Part I. Design of, polystyrene based microcarriers. **J. Biomater.** (5) 65-78.

116. Roder, B., Zuhlke, A., Widdecke, H. and Klein, J. (1993) Synthesis and application of new microcarriers for animal cell culture. Part II. Application of polystyrene microcarriers. **J. Biomater.** (5) 79-88.

117. Frondoza, C., Sotiabi, A. and Hungerford, D. (1996) Human chondrocytes proliferate and produce matrix components in microcarrier suspension culture. **Biomaterials** (17) 879-888.

118. Freed, L.E., Vunjak-Novakovic, G. and Langer, R. (1993) Cultivation of cell-polymer cartilage implants in bioreactors. **Cell Biochem.** (51) 257-264.

119. Tang, J.S., Chao, C.F. and Au, M.K. (1994) Growth and metabolism of cultured bone cells using microcarrier and monolayer techniques. **Clin. Orthop. Rel. Res.** (300) 254-258.

120. Hunter, W. (1743) Of the structure and diseases of articulating cartilages. **Philos. Trans.** 470:514–521.

121. Buckwalter, J.A. and Mankin, H.J. (1998) Articular cartilage: tissue design and chondrocyte–matrix interactions. **AAOS Inst Course Lect** (47) 477–486.

122. Coutts, R.D., Sah, R.L. and Amiel, D. (1997) Effect of growth factors on cartilage repair. **AAOS Inst Course Lect** (46) 487–494.

123. Hunziker, E.B. (1997) Articular cartilage repair: are the intrinsic biological constraints undermining this process insuperable? **Osteoarthritis Cartilage** (7) 15–28.

124. Buckwalter, J.A. and Mankin, H.J. (1997) Articular cartilage Part II: degeneration and osteoarthrosis, repair, regeneration, and transplantation. **J Bone Jt. Surg.** (79A) 612–632.

125. Johnson, L.L. (1989) Arthroscopic abrasion arthroplasty historical and pathological perspective: present status. **Arthroscopy** (2) 54-69.

126. Ahsan, T., Lottman, L.M., Harwood, F.L., Amiel, D. and Sah, R.L. (1999) Integrative cartilage repair by ß-aminoproprionitrile. **J Orthop. Res.** (17) 850-857.

127. Ahsan, T. and Sah, R.L. (1999) Biomechanics of integrative cartilage repair. **Osteoarthritis Cartilage** (7) 29-40.

128. Caplan, A.I., Elyaderani, M., Mochizuki, Y., Wakitani, S. and Goldberg, V.M. (1997) Principles of cartilage repair and regeneration. **Clin Orthop** (342) 254-269.

129. Zuger, B.J., Ott, B., Mainil-Varlet. P., Schaffner, T., Clemence. J.F., Weber, H.P. and Frenz, M. (2001) Laser solder welding of articular cartilage: tensile strength and chondrocyte viability. **Lasers Surg Med.** (28) 5, 427-34.

130. O'Driscoll, S.W., Keeley, F.W. and Salter, R.B. (1986) The chondrogenic potential of free autogenous periosteal grafts for biological resurfacing of major full-thickness defects in joint surfaces under the influence of continuous passive motion. An experimental investigation in the rabbit. **J Bone Joint Surg. Am.** Sep (68) 7, 1017-35.

131. Hangody, L., Kish, G., Karpati, Z. Udvarhelyi, I., Szigeti, I. and Bely, M. (1998) Mosaicplasty for the treatment of articular cartilage defects: application in clinical practice. **Orthopedics** (21) 751–756.

132. Goldberg, V.M. and Caplan, A.I. (1999) Biologic restoration of articular surfaces. **AAOS Instruction Course Lectures** (48) 623-627.

133. Green, W.T. (1977) Articular cartilage repair: Behavior of rabbit chondrocytes during tissue culture and subsequent allografting. **Clin. Orhtop. Relat. Res.** (79-A) 612-632.

134. Dozin, B., Malpeli, M., Camardella, L., Cancedda, R. and Pietrangelo, A. (2002) Response of young, aged and osteoarthritic human articular chondrocytes to inflammatory cytokines: molecular and cellular aspects. **Matrix Biol.** Aug; (21) 5, 449-59.

135. Domm, C., Schunke, M., Christesen, K. and Kurz, B. (2002) Redifferentiation of dedifferentiated bovine articular chondrocytes in alginate culture under low oxygen tension. **Osteoarthritis Cartilage** (10) 1, 13-22.

136. Jakob, M., Demarteau, O. and Schafer, D. (2001) Specific growth factors during the expansion and redifferentiation of adult human articular chondrocytes enhance chondrogenesis and cartilaginous tissue formation *in vitro*. **J. Cell Biochem**. (81) 368–377.

137. Wakitani, K., Kimura, T. and Hiruka, A. (1989) Repair of rabbit articular surface with allograft chondrocytes embedded in collagen gel. **J. Bone Joint Surg**. (71) 74-80.

138. Gao, J., Dennis, J.E., Solchaga, L.A., Goldberg, V.M. and Calpan, A.L. (2002) Repair of osteochondral defect with tissue engineered two-phase composite material of injectable calcium phosphate and hyaluranon sponge. **Tissue Engg.** (8) 827-837.

139. Noth, U., Tuli, R., Osyczka, A.M., Danielson, K.G. and Tuan, R.S. (2002) *In vitro* engineered cartilage constructs produced by press-coating biodegradable polymer with human mesenchymal stem cells. **Tissue Engg.** (8) 131-144.

140. Mizuno, S. and Glowacki, J. (1996) Chondroinduction of human dermal fibroblasts by demineralized bone in three-dimensional culture. **Exp Cell Res** (227) 89-97.

141. Bosch, P., Musgrave, D.S. and Lee, J.Y. (2000) Osteoprogenitor cells within skeletal muscle. **J.Orthop. Res**. (18) 933–944.

142. Levy, M.M., Joyner, C.J. and Virdi, A.S. (2001) Osteoprogenitor cells of mature human skeletal muscle tissue: an *in vitro* study. **Bone** (29) 317–322.

143. Zuk, P.A., Zhu, M. and Mizuno, H. (2001) Multilineage cells from human adipose tissue: implications for cell-based therapies. **Tissue Engg**. (7) 211–228.

144. Huang, J.I., Beanes, S.R., Zhu, M., Lorenz, H.P., Hedrick, M.H. and Benhaim, P. (2002) Rat extramedullary adipose tissue as a source of osteochondrogenic progenitor cells. **Plast. Reconstr. Surg.** (109) 1033–1043.

145. Kuznetsov, S.A., Mankani, M.H., Gronthos, S., Satomura, K., Bianco, P. and Robey, P.G. (2001) Circulating skeletal stem cells. **J. Cell Biol**. (153) 1133–1140.

146. Nakayama, N. (2003) Macroscopic cartilage formation with embryonic stem cell derived mesodermal progenitor cells. **J. Cell Sci.** (116) 10, 2015-2028.

147. Buttery, L.D. (2001) Differentiation of osteoblasts and *in vitro* bone formation from murine embryonic stem cells. **Tissue Engg**. (7) 89-99.

148. Tsuchiya, K., Chen, G., Ushida, T., Matstuno, T. and Tateishi, T. (2004) The effect of coculture of chondrocytes with mesenchymal stem cells on their cartilaginous phenotype *in vitro*. **Mater. Sci. Engg C** (24) 391-396.

149. Frenkel, S.R., Toolan, B., Menche. D., Pitman, M. and Pachence, J.M. (1997) Chondrocyte transplantation using a collagen bilayer matrix for cartilage repair. **J. Bone Joint Surg.** (79B) 831–836.

150. Nehrer, S., Breinan, H.A., Ramappa, A., Shortkroff. S., Young, G., Minas, T., Sledge, C.B., Yannas, I.V. and Spector, M. (1997) Canine chondrocytes seeded in type I and type II collagen implants investigated *in vitro*. **J Biomed. Mater. Res.** (38) 95–104.

151. Kang, R., Marui, T., Nita, I., Georgescu, H.I., Suh, J.K., Robbins, R.D. and Evans, C.H. (1997) *Ex vivo* gene transfer to chondrocytes in full-thickness articular cartilage defects-a feasibility study. **Osteoarthritis Cartilage** (5) 139–143.

152. Qi, W.N. and Scully, S.P. (1997) Extracellular collagen modulates the regulation of chondrocytes by transforming growth factor-beta 1. **J Orthop Res** (15) 4, 483–490.

153. Freed, L.E., Grande, D.A., Lingbin, Z., Emmanuel, J., Marquis J.C. and Langer, R. (1994) Joint resurfacing using allograft chondrocytes and synthetic biodegradable polymer scaffolds. **J Biomed. Mater. Res. (**28) 891–899.

154. Agrawal, C.M., Niederauer, G., Micallef, D. and Athanasiou, K. (1995) The use of PLA–PGA polymers in orthopaedics. In: *Encyclopedic Handbook of Biomaterials and Bioengineering*, eds. D. Wise, New York, pp 2081–2115.

155. Freed, L.E., Marquis, J.C., Nohria, A., Emmanual, J., Mikos, A.G. and Langer, R. (1993) Neocartilage formation *in vitro* and *in vivo* using cells cultured on synthetic biodegradable polymers. **J. Biomed. Mater. Res.** (27) 11–23.

156. Chu, C.R., Coutts, R.D., Yoshioka, M., Harwood, F.L., Monosov, A.Z. and Amiel, D. (1995) Articular cartilage repair using allogenic perichondrocyte-seeded biodegradable porous polylactic acid (PLA): a tissue-engineering study. **J. Biomed. Mater. Res.** (29) 1147–1154.

157. Athanasiou, K.A., Niederauer, G.G. and Agrawal, C.M. (1996) Sterilization, toxicity, biocompatibility and clinical applications of polylactic acid/polyglycolic acid copolymers. **Biomaterials** (17) 93–102.

158. Sittinger, M., Reitzel, D., Dauner, M., Hierlemann, H., Hammer, C., Kastenbauer, E., Planck, H., Burmester, G.R. and Bujia, J. (1996) Resorbable polyesters in cartilage engineering: affinity and biocompatibility of polymer fiber structures to chondrocytes. **J.Biomed. Mater. Res.** (33) 57–63.

159. Hanff, G., Sollerman, C., Abrahamsson, S.O. and Lundborg, G. (1990) Repair of osteochondral defects in the rabbit knee with Gore-Tex (expanded polytetrafluoreethylene). An experimental study. **Scand J Plast Reconstr Surg** (24) 217–223.

160. Reisses, N., Downes, S. and Bentley, G. (1994) Cartilage repair using a new synthetic polymer. **Transactions of the Orthopaedics Research Society**, New Orleans, LA.

161. Messner, K. and Gillquist, J. (1993) Synthetic implants for the repair of osteochondral defects of the medial femoral condyle: a biomechanical and histological evaluation in the rabbit knee. **Biomaterials (**14) 513–521.

162. Thomson, R.C., Wake, M.C., Yaszemski, M.J. and Mikos, A.G. (1995) Biodegradable polymer scaffolds to regenerate organs. **Adv. Polymer. Sci.** (122) 245–274.

163. Fuss, M., Ehlers, E., Russlies, M., Rohwedel, J. and Behrens, P. (2000) Characteristics of human chondrocytes, osteoblasts and fibroblasts seeded onto a type I/III collagen sponge under different culture conditions. A light, scanning and transmission electron microscopy study. **Anat. Anz.** (182) 303–310.

164. Ronziere, M. C., Roche, S., Gouttenoire, J., Demarteau, O., Herbage, D. and Freyria, A.(2003) Ascorbate modulation of bovine chondrocyte growth, matrix protein gene expression and synthesis in three-dimensional collagen sponges. **Biomaterials** (24) 851–856.

165. Wakitani, S., Goto, T., Pineda, S., Young, R., Mansour, J., Caplan, A. and Goldberg, V. (1994) Mesenchymal cell-based repair of large, full-thickness defects of articular cartilage. **J. Bone Joint Surg.** (76A) 579–592.

166. Samuel, R. E., Lee, C., Ghivizzani, S., Evans, C., Yannas, I., Olsen, B. and Spector, M. (2002) Delivery of plasmid DNA to articular chondrocytes via novel collagen-glycosaminoglycan matrices. **Hum. Gene Ther.** (13) 791–802.

167. Ohno, T, Tanisaka, K, Hiraoka, Y, Ushida, T, Tamaki,T and Tateishi, T. (2004) Effect of type I and type II collagen sponges as 3D scaffolds for hyaline cartilage-like tissue regeneration on phenotypic control of seeded chondrocytes in vitro. **Mat. Sci. Engg.** (24) 3, 407-411.

168. Chen, T., Sato, T., Ushida, R., Hirochika, A. and Tateishi T. (2003) Redifferentiation of dedifferentiated bovine chondrocytes when cultured *in vitro* in a PLGA-type I collagen hybrid mesh. **FEBS Lett.** (542) 95–99.

169. Van Susante, P., Buma, G. and Van Osch. (1995) Culture of chondrocytes in alginate and collagen carrier gels. **Acta Orthop Scand** (66) 6, 549–556.

170. Frondoza, A. Sohrabi, C. and Hungerford, D. (1996) Human chondrocytes proliferate and produce matrix components in microcarrier suspension culture. **Biomaterials** 179, 879–888.

171. Van Damme, M.P., Boyaci, N. and Ilic M. (1998) The size of aggrecan core protein synthesized in monolayer culture and in collagen gels with time in culture. **Int. Cartilage Rep. Soc. Newslett.** (2) 8–9.

172. Lee, C.H., Singla, A.and Lee, Y. (2001) Biomedical application of collagen **Int. J. Pharm.** (221) 1-22

173. Schlag, G. and Redi, H. (1988) Fibrin sealant in orthopaedic surgery. **Clin. Orthop. Rel. Res.** (227) 269-285.

174. Hendrickson, D.A., Nixon, A.J., Grande, D.A., Todhhunter, R.J., Minor, R.M., Erb, H. and Lust, G. (1994) Chondrocytes fibrin matrix transplants for resurfacing extensive articular cartilage defects. **J. Orthop. Res.** (12) 4, 485-497.

175. Ting, V., Sims, C.D., Brecht, L.E., Kasabian, J.G., Connley, A.K., Elisseeff. J, Gittes, G.K. and Longaker, M.T. (1998) *In vitro* prefabrication of human cartilage shapes using fibrin glue and human chondrocytes. **Ann Plast Surg** (40) 4, 413-420.

176. Minas, T. and Peterson, L. (1999) Advanced techniques in autologous chondrocytes transplantation. **Clin. Sports Med.** (18) 1, 13-44

177. Fortier, J.A., Brofman, P.J., Nixon, A.J. and Mohammed, H.O. (1998) Dispartate chondrocyte metabolism in three dimensional fibrin cultures derived from autogenous or commercially manufactured fibrinogens. **Am. J. Vet. Res.** (59) 4 514-520.

178. Kawabe, N. and Yoshinao, M. (1991) The repair of full-thickness articular cartilage defects. Immune responses to reparative tissue formed by allogeneic growth plate chondrocyte implants. **Clin. Orthop**. (268) 279–293.

179. Gelse, K., Von der Mark, K., Aigner, T., Park, J. and Schneider, H. (2003) Articular cartilage repair by gene therapy using growth factor producing mesenchymal cells. **Arthritis Rheum**. (48) 430–441.

180. Piechotta, U. and Flemming, I. (1983) The maximization of wound healing with fibrin glue. **Aesthetic Plast. Surg.** (7) 81–82

181. Rahfoth, B., Weisser, J., Sternkopf, F., Aigner, T., Vonder Mark, K. and Brauer, R. (1998) Transplantation of allograft chondrocytes embedded in agarose gel into cartilage defects of rabbits. **Osteoarthritis Cartilage** (6) 50–65.

182. Diduch, D. R., Jordan, L., Mierisch, C. and Balian, G. (2000) Marrow stromal cells embedded in alginate for repair of osteochondral defects. **Arthroscopy** (16) 571–577

183. Dausse, Y., Grossin, L., Miralles, G., Pelletier, S., Mainard, D., Hubert, P., Baptiste, D., Gillet, P., Dellacherie, E., Netter, P. and Payan. E. (2003) Cartilage repair using new polysaccharidic biomaterials: Macroscopic, histological and biochemical approaches in a rat model of cartilage defect. **Osteoarthritis Cartilage** (11) 16–28.

184. Grigolo, B., Roseti, L. and Fiorini, M. (2001) Transplantation of chondrocyte seeded on a hyaluronan derivative (hyaff-11) into cartilage defects in rabbits. **Biomaterials** (22) 2417-2424.

185. Butnariu- Ephrat, M., Robinson, D., Mendes, D., Halperin, N. and Nevo, Z. (1996) Resurfacing of goat articular cartilage by chondrocytes derived from bone marrow. **Clin. Orthop**. (330) 234-243.

186. Girotto, D., Urbani, S., Brun, P., Renier, D., Barbucci, R. and Abatangelo, G. (2003) Tissue-specific gene expression in chondrocytes grown on three-dimensional hyaluronic acid scaffolds. **Biomaterials** (24) 19, 3265-3275.

187. Chang, C.H., Liu, H.C., Lin, C.C., Chou, C.H. and Lin, F.H. (2003) Gelatin–chondroitin–hyaluronan tri-copolymer scaffold for cartilage tissue engineering. **Biomaterials**, (24) 26, 4853-4858.

188. Sechriest, V.F., Miao, Y.J., Niyibizi, C., Westerhausen-Larson, A., Matthew, H.W., Evans, C.H., Fu, F.H. and Suh, J.K. (2000) GAG-augmented polysaccharide hydrogel: a novel biocompatible and biodegradable material to support chondrogenesis. **J. Biomed. Mater. Res.** (49) 4, 534–541.

189. Solchaga, L.A., Goldberg, V.M. and Caplan, A.I. (2001) Cartilage regeneration using principles of tissue engineering. **Clin. Orthop.** (391S) S161–70.

190. Scully, S.P., Lee, J.W., Ghert, P.M.A. and. Qi, W. (2001) The role of the extracellular matrix in articular chondrocyte regulation. **Clin Orthop** (391S) S72–89.

191. Matthew, H.W., Salley, S.O., Peterson, W.D. and Klein, M.D. (1993) Complex coacervate microcapsules for mammalian cell culture and artificial organ development. **Biotechnol Prog** (9) 510–519

192. Gaserod, O., Smidsrod, O. and Skjak-Braek, G. (1998) Microcapsules of alginate–chitosan-I. A quantitative study of the interaction between alginate and chitosan. **Biomaterials** (19) 1815–1825.

193. Denuziere, A., Ferrier, D., Damour, O. and Domard, A. (1998) Chitosan–chondroitin sulfate and chitosan–hyaluronate polyelectrolyte complexes: biological properties. **Biomaterials** (19) 14, 1275–1285.

194. Lu, F., Prudhommeaux, A., Meunier, L., and Guillemin, G. (1999) Effects of chitosan on rat knee cartilages. **Biomaterials** (20) 20, 1937–1944.

195. Mattioli-Belmonte, M., Gigante, A., Muzzarelli, R.A., Politano, R., De Bendittis, A., Speechia N., Buffa, A., Biagini, G. and Greco, F. (1993) N, N-dicarboxymethyl chitosan as delivery agent for bone morphogenetic protein in the repair of articular cartilage. **Med. Biol. Eng. Comput** (37) 1, 130-134.

196. Chenite, A., Chaput, C., Wang, D., Combes, C., Buschmann, M., Hoemann, C., Leroux, C., Atkinson, B., Binette, F. and Selmani, A. (2000) Novel injectable neutral solutions of chitosan form biodegradable gels *in situ*. **Biomaterials** (21) 2155–2161.

197. Athanasiou, K.A., Schmitz, J.P. and Agrawal, C.M. (1998) The effects of porosity on in vitro degradation of PLA-PGA implants used in repair of articular cartilage. **Tissue Engg.** (4) 53–63.

198. Agrawal, C.M. and Athanasiou, K.A. (1997) A technique to control the pH in the vicinity of biodegrading PLA-PGA implants. **J. Biomed. Mater. Res.** (Appl Biomater) (38) 2, 105–114.

199. Spain, L., Agrawal, C.M. and Athanasiou, K.A. (1998) New technique to extend the useful life of a biodegradable implants. **Tissue Engg.** (4) 343–352.

200. Chu, C. R., Dounchis, J., Yoshioka, M., Sah, R., Coutts, R. D. and Amiel, D. (1997) Osteochondral repair using perichondrial cells. A 1-year study in rabbits. **Clin. Orthop.** (340) 220–229.

201. Dounchis, J. S., Bae, W., Chen, A., Sah, R., Coutts, R. D. and Amiel, D. (2000) Cartilage repair with autogenic perichondrium cell and polylactic acid grafts. **Clin. Orthop.** (377) 248–264.

202. Giurea, A., Klein, T., Chen, A., Goomer, R., Coutts, R. D., Akeson, W., Amiel, D. and Sah, R. (2003) Adhesion of perichondrial cells to a polylactic acid scaffold. **J. Orthop. Res.** (21) 584–589.

203. Caterson, E. J., Li, W., Nesti, L., Albert, T., Danielson, K. and Tuan, R. (2002) Polymer/alginate amalgam for cartilage-tissue engineering. **Ann. N. Y. Acad. Sci.** (961) 134–138.

204. Goomer, R. S., Deftos, L., Terkeltaub, R., Maris, T., Lee, M., Harwood, F. and Amiel, D. (2001) High-efficiency non-viral transfection of primary chondrocytes and perichondrial cells for *ex-vivo* gene therapy to repair articular cartilage defects. **Osteoarthritis Cartilage** (9) 248–256.

205. Frenkel, S. R., Bradica, G., Goldman, S., Kronengold, R., Coutts, R. D. and Brekke, J. H. (2003) Multiphasic device engineered to regenerate articular cartilage. **Trans. Orthop. Res. Soc.** (28) 249.

206. Langer, R. and Vacanti, J. P. (1993) **Tissue Engg. Sci.** (260) 920–926.

207. Cohen, S. B., Meirisch, C., Wilson, H. and Diduch, D. (2003) The use of absorbable co-polymer pads with alginate and cells for articular cartilage repair in rabbits. **Biomaterials** (24) 2653–2660.

208. Sherwood, J. K., Riley, S., Palazzolo, R., Brown, S., Monkhouse, D., Coates, M., Griffith, L., Landeen, L. and Ratcliffe. A. (2002) A three-dimensional osteochondral composite scaffold for articular cartilage repair. **Biomaterials** (23) 4739–4751.

209. Kawanishi, M., Ushida, T., Kaneka, T., Niwa, H., Fukubayashi, T., Nakamura, K., Oda, H., Tanaka, S. and Tateishi, T. (2004) New type of biuodegradable porous scaffolds for tissue engineered articular cartilage. **Mat. Sci. Engg. C** (24) 431-435.

210. Bryant, S.J. and Anseth, K.S. (2002) Hydrogel properties influence ECM production by chondrocytes photo encapsulated in poly (ethylene glycol) hydrogels. **J. Biomed. Mater. Res.** (59) 63-72.

211. Lelah, M.D. and Cooper, S.L. (1986) Polyurethanes in medicine, CRC Press, Boca Raton, FL

212. Gogolewski, S. (1996) Biomedical Polyurethanes. In: *Desk reference of functional polymers. Syntheses and applications,* eds. Arshady, R. (American Chemical Society, Washington) pp. 657–698.

213. Gogolewski, S., Pennings, A.J., Lommen, E., Nieuwenhuis, P. and Wildevuur, Ch.R.H. (1983) Growth of a neo-artery induced by a biodegradable polymeric vascular prosthesis. **Macromol. Chem. Rapid Commun.** (4) pp. 213–219.

214. Gogolewski, S., Walpoth, B. and Rheiner, P. (1987) Polyurethane microporous membranes as pericardial substitutes. **Colloid. Polymer. Sci.** (265) 971–977.

215. Elema, H., de Groot, J.H., Nijenhuis, A.J., Pennings, A.J., Veth, R.P., Klompmaker, J. and. Jansen, H.W. (1990) Use of biodegradable polymer implants in meniscus reconstruction. 2. Biological evaluation of porous biodegradable implants in menisci. **Colloid Polymer Sci.** (268) 1082–1088.

216. Klompmaker, J., Jansen, H.W., Veth, R.P., de Groot, J.H., Nijenhuis, A.J. and Pennings, A.J. (1991) Porous polymer implant for repair of meniscal lesions: a preliminary study in dogs. **Biomaterials** 12 (1991), pp. 810–816.

217. Gogolewski, S., Gorna, K., Rahn, B. and Wieling, R. (2001) Biodegradable polyurethane cancellous bone graft substitute promotes bone regeneration in the iliac crest defects. Transactions 24, 573, **27th Annual Meeting, Society for Biomaterials**, Saint Paul, MN, USA.

218. Gogolewski, S., Gorna, K. and Turner, A.S. (2002) Regeneration of bicortical defects in the iliac crest of estrogen deficient sheep using new biodegradable polyurethane cancellous bone graft substitutes. A pilot study. 48th Annual Meeting, Orthopaedic Research Society, Dallas, TX, USA.

219. Saad, B., Neuenschwander, P., Uhlschmid, G.K. and Suter, U.W. (1999) New versatile, elastomeric, degradable polymeric materials for medicine. **Int. J. Biol. Macromol.** (25) 293–301.

220. Grad, S., Kupcsik, L., Gorna, K., Gogolewski, S. and Alini, M. (2003) The use of biodegradable polyurethane scaffolds for cartilage tissue engineering: potential and limitations. **Biomaterials** (24) 28, 5163-5171.

221. Saeed, K.A., Eribo, B.E., Ayorinde, F.O. and Collier, L. (2002) Characterization of copolymer hydroxybutyrate/hydroxyvalerate from saponified vernonia, soybean, and "spent" frying oils. **J. AOAC. Int.** (85) 4, 917–924.

222. Zinn, M., Witholt, B. and Egli, T. (2001) Occurrence, synthesis and medical application of bacterial polyhydroxyalkanoate. **Adv. Drug Deliver. Rev.** (53), 1, 5–21.

223. Deng, Y, Lin, X.Y., Zheng, Z., Deng, J.G., Chen, J.C., Ma, H. and Chen, G.Q. (2003) Poly (hydroxybutyrate-co-hydroxyhexanoate) promoted production of extracellular matrix of articular cartilage chondrocytes *in vitro*. **Biomaterials** (24) 23, 4273-4281.

224. Lane, J.G., Tontz, W.L. Ball, S.T., Massie, J.B., Chen, A.C., Bae, W.C., Amiel, M.E., Sah, R.L. and Amiel, D. (2001) A Morphologic, biochemical, and biomechanical assessment of short-term effects of osteochondral autograft plug transfer in an animal model. **Arthroscopy** (17) 856–863.

225. Hurtig, M., Evans, P., Clarnette, R., Pearce, S. and Singh, K.M. (2001) Mosaic arthroplasty for resurfacing the equine third carpal bone. **Vet. Surg.** (30) 228–239.

226. Gilbert, J.E. (1998) Current treatment options for the restoration of articular cartilage. **Am J Knee Surg.** (11) 42–46.

227. Pearce, S., Hurtig, M.B., Clarnette, R., Kalra, M. and Miniaci, A. (2001) Investigation of two techniques for optimizing joint surface congruency with mosaic arthroplasty. **J Arthroscopy** (17) 50–55.

228. Pilliar, R.M., Filiaggi, M.J., Wells, J.D., Grynpas, M.D. and Kandel, R.A. (2001) Porous calcium polyphosphate scaffolds for bone substitute applications: *in vitro* characterization. **Biomaterials** (22) 963–972.

229. Waldman, S.D., Grynpas, M.D., Pilliar, R.M. and Kandel, R.A. (2002) Characterization of cartilaginous tissue formed on calcium polyphosphate substrates *in vitro*. **J Biomed. Mater. Res.** (62) 323–330.

230. Bhardwaj, T. Pilliar, R.M., Grynpas, M.D. and Kandel, R.A. (2001) Effect of material geometry on cartilaginous tissue formation *in vitro*. **J Biomed. Mater. Res.** (57) 190–199.

231. Dunn, G., Bellincampi, L.D., Tria, A.J. and Zawadsky, J.P. (1997) Preliminary development of a collagen–PLA composite for ACL reconstruction. **J. Appl. Polym. Sci.** (63) 1423–1428.

232. Chen, G., Ushida, T. and Tatsuya, T. (2000) Hybrid biomaterials for tissue engineering: a preparative method of PLA or PLGA–collagen hybrid sponge. **Adv. Mater**. (12) 455–457.

233. Chen, G., Ushida, T. and Tatsuya, T. (2000) A biodegradable hybrid sponge nested with collagen microsponges. **J. Biomed. Mater. Res.** (51) 273–279.

234. Sato, T., Chen, G., Ushida, T., Ishii, T., Ochiai, N. and Tatsuya, T. (2001) Tissue-engineering cartilage by *in vivo* culturing of chondrocytes in PLGA–collagen hybrid sponge. **Mater. Sci. Engg. C** (17) 83–89.

235. Sellers, R.S., Zhang, R., Glasson, S., Kim, H., Peluso, D., D'Augusta D., Beckwith, K. and Morris, E. (2000) Repair of articular cartilage defects one year after treatment with recombinant human bone morphogenetic protein-2 (rh BMP-2). **J. Bone joint Surg. Am**. (82) 151-160.

236. Marijnissen, W. J., van Osch, G., Aigner, J., Verwoerd-Verhoef, H. and Verhaar, J. (2000) Tissue-engineered cartilage using serially passaged articular chondrocytes. Chondrocytes in alginate, combined *in vivo* with a synthetic (E210) or biologic biodegradable carrier (DBM). **Biomaterials** (21) 571–580.

237. Knudson, W., Casey, B., Nishida, Y., Eger, W., Kuettner, K. and Knudson, C. (2000) Hyaluronan oligosaccharides perturb cartilage matrix homeostasis and induce chondrocytic chondrolysis. **Arthritis Rheumatism** (43) 1165–1174.

238. Grande, D., Pitman, M., Peterson, L., Menche, D. and Klein, M. (1989) The repair of experimentally produced defects in rabbit articular cartilage by autologous chondrocyte transplantation. **J. Orthop. Res.** (7) 208–219.

239. Veldhuijzen, J.P., Huisman, A.H., Vermeiden, J.P. and Prahl-Andersen, B. (1987) The growth of cartilage cells *in vitro* and the effect of intermittent compressive force. A histological evaluation. **Connective Tissue Res.** (16) 187–196.

240. Steinmeyer, S. and Knue, S. (1997) The proteoglycan metabolism of mature bovine articular cartilage explants superimposed to continuously applied cyclic mechanical loading. **Biochem Biophys Res Commun** (240) 216–221.

241. Vunjak-Novakovic, G., Martin, I., Obradovic, B., Treppo, S., Grodzinsky, A.J., Langer R. and Freed L.E. (1999) Bioreactor cultivation conditions modulate the composition and mechanical properties of tissue-engineered cartilage. **J. Orthop. Res.** (17) 130–138.

242. Mauck, R. L., Wang, C., Oswald, E. S. Ateshian, G. A. and Hung, C. T. (2003) The role of cell seeding density and nutrient supply for articular cartilage tissue engineering with deformational loading. **Osteoarthritis and Cartilage** (11) 879–890.

243. LeBaron, R. G. and Athanasiou, K. A. (2000) *Ex vivo* synthesis of articular cartilage. **Biomaterials** (21) 24, 2575-2587.

244. Puelacher, W.C., Kim, S.W., Vacanti, J.P., Schloo, B., Mooney, D. and Vacanti, C.A. (1994) Tissue-engineered growth of cartilage: the effect of varying the concentration of chondrocytes seeded onto synthetic polymer matrices. **Int J. Oral Maxillofacial Surg. (**23) 49–53.

245. Ruoslahti, E. (1984) Fibronectin in cell adhesion and invasion. **Cancer Metastat. Rev.** (3) 43–51.

246. Pierschbacher, M.D. and Ruoslahti, E. (1984) Cell attachment activity of fibronectin can be duplicated by small synthetic fragments of the molecule. **Nature** (309) 30–33.

247. Ruoslahti, E. (1996) RGD and other recognition sequences for integrins. **Ann. Rev. Cell Dev. Bicl.** (12) 697–715.

248. Le Baron, R.G. and Athanasiou, K.A. (2000) Extracellular matrix cell adhesion peptides: functional applications in orthopaedic materials. **Tissue Engg** (6) 2, 85–103.

249. Howlett, C.R., Evans, M.D., Walsh, W.R., Johnson, G. and Steel, J.G. (1994) Mechanism of initial attachment of cells derived from human bone to commonly used prosthetic material during cell culture. **Biomaterials (**12) 697-715.

250. Colombet, P., Allard, M., Bousquet, V., de Lavigne, C., Flurin, P.H. and Lachaud, C. (2002) Anterior cruciate ligament reconstruction using four-strand semitendinosus and gracilis tendon grafts and metal interference screw fixation Arthroscopy. **J. Arthroscopic Relat. Surg** (18) 3, 232-237.

251. Yong, W.J.B. and Kwon O.H. (2004) Cross-pin femoral fixation for hamstring posterior cruciate ligament reconstruction Arthroscopy. **J. Arthroscopic Relat. Surg. (**20) 4, 29-33.

252. Hutmacher, D.W. (2000) Scaffolds in tissue engineering bone and cartilage. **Biomaterials** (21) 2529-2543.

253. Bujja, J. (1994) Humoral immune response against minor collagen types IX and X in patients suffering from cartilage graft resorption after reconstructive surgery. **Ann. Rheum. Dis.** (53) 229-234.

254. Rudderman, R., Guyuron, B. and Mendelsohn, B. (1994) The fate of fresh and preserved crushed and non-crushed autogenous cartilage in the rabbit model. **Ann. Plast. Surg.** (32) 250-254.

255. Langer, F. Gross, A.E. and Greaves, M.F. (1972) The auto-immunogenicity of articular cartilage **Exp. Immunol.**(12) 31-7.

256. Rodrigo, J., Sakovich, L., Travis, C. and Smith, G. (1978) Osteocartilaginous allografts as compared with autografts in the treatment of knee joint defects in dogs. **Clin. Orthop. Relat. Res.** (134) 342-349.

257. Garrett, J.C. (1986) Treatment of osteochondral defects of the distal femur with fresh osteochondral allografts: a preliminary report. **Arthroscopy** (2) 222-226.

258. Czitrom, A., Langer, F., McKee, N. and Gross, A. (1986) Bone and cartilage allotransplantation. **Clin. Orthop. Relat. Res.** (8) 141-145.

Advances in Biochemistry and Biotechnology Volume 1 *Pages 53-80*
Edited by: **Dr. Chiranjib Chakraborty**
Published by: **DAYA PUBLISHING HOUSE**

2

Human Granulocyte-Colony Stimulating Factor: A Therapeutic Protein

Shyam Sundar Nandi, Chiranjib Chakraborty, J.K. Deb & Ragini Gothalwal*
*(*Head of the Dept.), School of Biotechnology,*
Guru Ghasidas University, Bilaspur, Koni, (C.G)
E-mail: ragini_gothalwal@hotmail.com

ABSTRACT

Human Granulocyte Colony Stimulating Factor (G-CSF) is a protein of 175 amino acid residues with a molecular weight of 18.8 kda and is encoded by a 525 bp gene encoding. It is one of a family of hematopoietic growth factors, which play an important role in proliferation, and differentiation of cells of the neutrophill lineage. Neutropenia is the most common dose limiting toxicity of conventional chemotherapy. Clinical trails are going on at many places for various newer indications, such as-bone marrow transplantation, treatment in complicated plasmodium falciparum malaria, malignant hematological disease, peripheral blood stem cell transplantation, mylodyplasia syndrome and aplastic anemia, and primary and idiopathic neutropenia. Because of codon bias between prokaryotes and eukaryotes, there becomes a limitation for the expression of hG-CSF in *E.coli* system.rh-GCSF is new proteins pharmaceutical incorporation's version of

* Correspondent Author

hG-CSF. The protein has an amino acid sequence that is identical to the natural G-CSF sequence predicted from human DNA sequence analysis, except for the addition of an N-terminal methionine necessary for expression in *E.coli*. rhG-CSF is produced in *E.coli* by recombinant DNA technology. An attempt to obtain hG-CSF by good cost effective technology in large quantity, by using high expression level vector (pET32a, pET29a, pMALp2, pTrc99A), the gene encoding it has been cloned, and recombinant hG-CSF were similar to those of the natural glycosylated hG-CSF. Therefore, recombinant hG-CSF is commercially produced by a very innovative method to treat patients with neutropenia in a minimum cost. This paper review contains-importance of G-CSF, cytokines, G-CSF, structure of G-CSF, stability and reengineering of G-CSF, kinetics of G-CSF folding, G-CSF and its receptor, atomic structure of the G-CSF-receptor complex, mode of action of G-CSF, gene structure, gene sequence, amino acid sequence, experimental work of rhG-CSF by recombinant DNA technology, G-CSF and its relationship to other growth factor, clinical application of G-CSF, clinical trails going on in G-CSF therapy and the future of this therapeutic protein.

INTRODUCTION

Before we introduce the readers to the epicenter of the chapter *i.e.*, G-CSF we would like to give a brief introduction of "cytokines", the family to which G-CSF belongs. G-CSF happens to be only one of the many cytokines having a vast array of activities in the body and a unbelievably complex and interwoven network of activity. Though we have tried our best to keep the discussion within the boundaries of G-CSF but at instances when a comparison or analogy had to be established other molecules have also been discussed. As we flow down the chapter that we have tried to discuss all the aspects of G-CSF whether factual or applied in the most lucid way possible.

Almost always we say that the term "cytokines" appears in the context of immune response especially "cell-mediated immunity (CMI)" although "cytokines" are likely to be referred to in many different places within the text. Such a practice gained its ground only because the early soluble mediators–the lymphocyte-derived mediators had close association with CMI responses. At a very later stage of research in this direction was the unraveling of a very important fact that many cells of origin other than that of the immune system were also able to produce such factors. Cytokines are a group of low molecular weight regulatory proteins secreted by white blood cells as well as a variety of other cells in the body in response to inducing stimuli and influences the cellular behavior in a big way. But this statement can have an objection now, as we now know that there are membrane-bound forms of some of the secreted cytokines. Cytokines may sometimes be denoted as a hormone like polypeptide, which activates cellular replication, proliferation, and activation in processes relating to host defense and repair. Its role as a regulatory polypeptide which co-ordinates a vast number of muticellular interaction (1).

To be more precise we should mention here that cytokine are polypeptides or glycoproteins having molecular weight of the order of 30 Kda and which are not produced or only feebly produced constitutively. These are mostly local in their action and hence their action can be better called autocrine or paracrine. Cytokine are generally transiently produced and bind to very high affinity receptors on the cells having Ka ranging 10^9 -10^{12} M. Although the range of the activities of an individual cytokine are very diverge and broad but some of their actions are well targeted especially towards hematopoetic cells.

So we can sum up the actions of cytokines by saying that it is a perfect blend of ambiguity by virtue of which it has multiple target cells and multiple actions; Redundancy, as many different cytokines have similar action. Synergism and antagonism are also very common in their action network. A cytokine may also increase or decrease the production and expression of other cytokines and cytokine receptor respectively.

Importance of G-CSF

G-CSF has been my topic of interest not only because of its application as a very potent weapon against diseases such as neutropenia, but its production in bacterial system has always fascinated us a lot. The degree of biological activity obtained in rhG-CSF is very high and generates interest due to the fact that even though there is always some extent of difference in the pattern of glycosylation as it is a protein of eukaryotic origin but their activity *in vivo* is least effected. Not only this, G-CSF has gained a lot of attention of people round the globe because of its cost effectiveness and an relatively affordable therapeutic protein. G-CSF has been successfully produced in the bacterial system and lot of experiments has been performed in which the newly synthesized protein has been targeted to the periplasmic space of the bacteria for the ease of purification. There are also evidences that the protein can be synthesized in the bacteria as inclusion bodies. Thus it may be already clear to the readers of this chapter that we have a preformed propensity and inclination towards the industrial production of G-CSF but we have tried my level best to give as much insight about the molecule into the as possible.

Human Granulocyte Colony Stimulating Factor (2,3) is a protein of 175 amino acid residues with a molecular weight of 18.8 kda encoded by 525 bp gene. Growth and cytokines are potential therapeutic agents of their key roles of regulating the production maturation, activity of the blood; muscles and bone cells.A number of cytokines are currently employed to treatment of blood deficiencies and cancers. For example, erythropoietin is administered to stimulate red blood cell production in patients suffering from anemia. Human growth hormone is used to stimulate growth in children and adults exhibiting extremely slow or subnormal growth due to hormonal deficiencies and granulocyte colony stimulating factor (G-CSF) is employed to treat cancer patients undergoing chemotherapy in order to alleviate the depression of red blood cell produced by cytotoxic therapeutic agents. It is one of a family of hematopoietic growth factors that play an important role in proliferation and differentiation of cells of the neutrophil lineage(4,5,6). Neutropenia is the most common dose limiting toxicity of conventional chemotherapy. Because of codon bias between prokaryotes and eukaryotes, there becomes a limitation for the expression of hG-CSF in *E.coli* system. In this report, the strategy developed involved the use of entire synthetic gene using oligos to increase the expression level. The biological activity of the non-glycosylated recombinant hG-CSF(7) was found to be similar to those of natural glycosylated hG-CSF(8). Neutropenia is the most common dose, limiting toxicity of conventional chemotherapy. Clinical trials are going on at many places for various newer indication, such as-bone marrow transplantation, treatment in complicated *plasmodium falciparum* malaria, malignant hematological disease, peripheral blood stem cell transplantation, mylodyplasia syndrome and aplastic anaemia, primary and idiopathic neutropenia. Recombinant hematopoietic colony-stimulating factors, which stimulate the production of granulocytes, may be a way of reducing the infectious complications of myelotoxic cancer treatment.

Granulocyte colony-stimulating factor (G-CSF) stimulates proliferation of neutrophils and enhances their phagocytic and microcidal activity. Increasing resistance to existing antibacterials and the dearth of new alternatives have complicated the treatment of Gram-negative infections. The aim of this study was to evaluate the efficacy of G-CSF in the treatment of *Pseudomonas aeruginosa* pneumonia when administered in combination with ceftazidime in both neutropenic and non-neutropenic hosts.

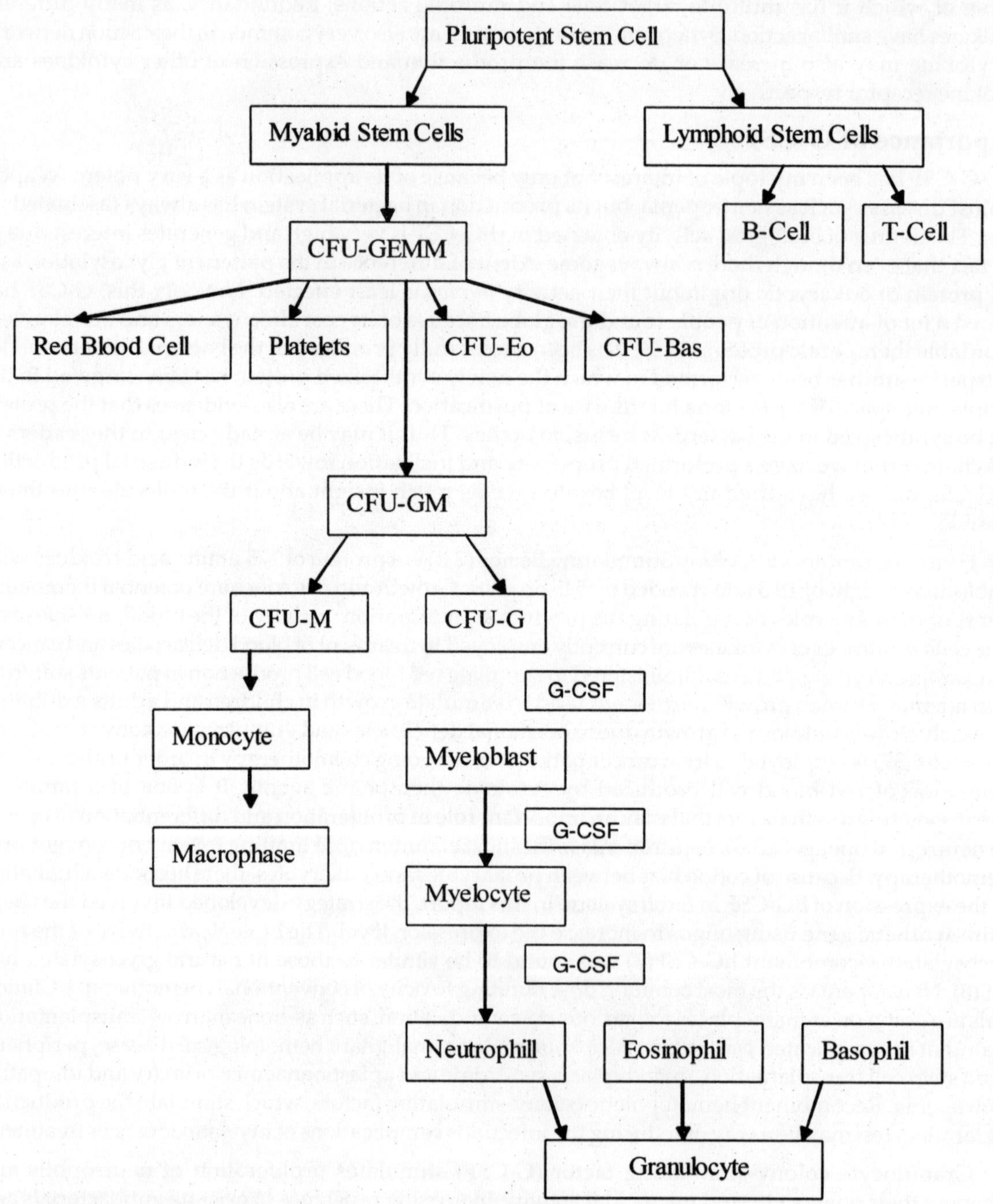

Fig. 2.1: Schematic Illustration of Hematopoiesis. The sites of action of G-CSF in this complex cascade are indicated (11,12)

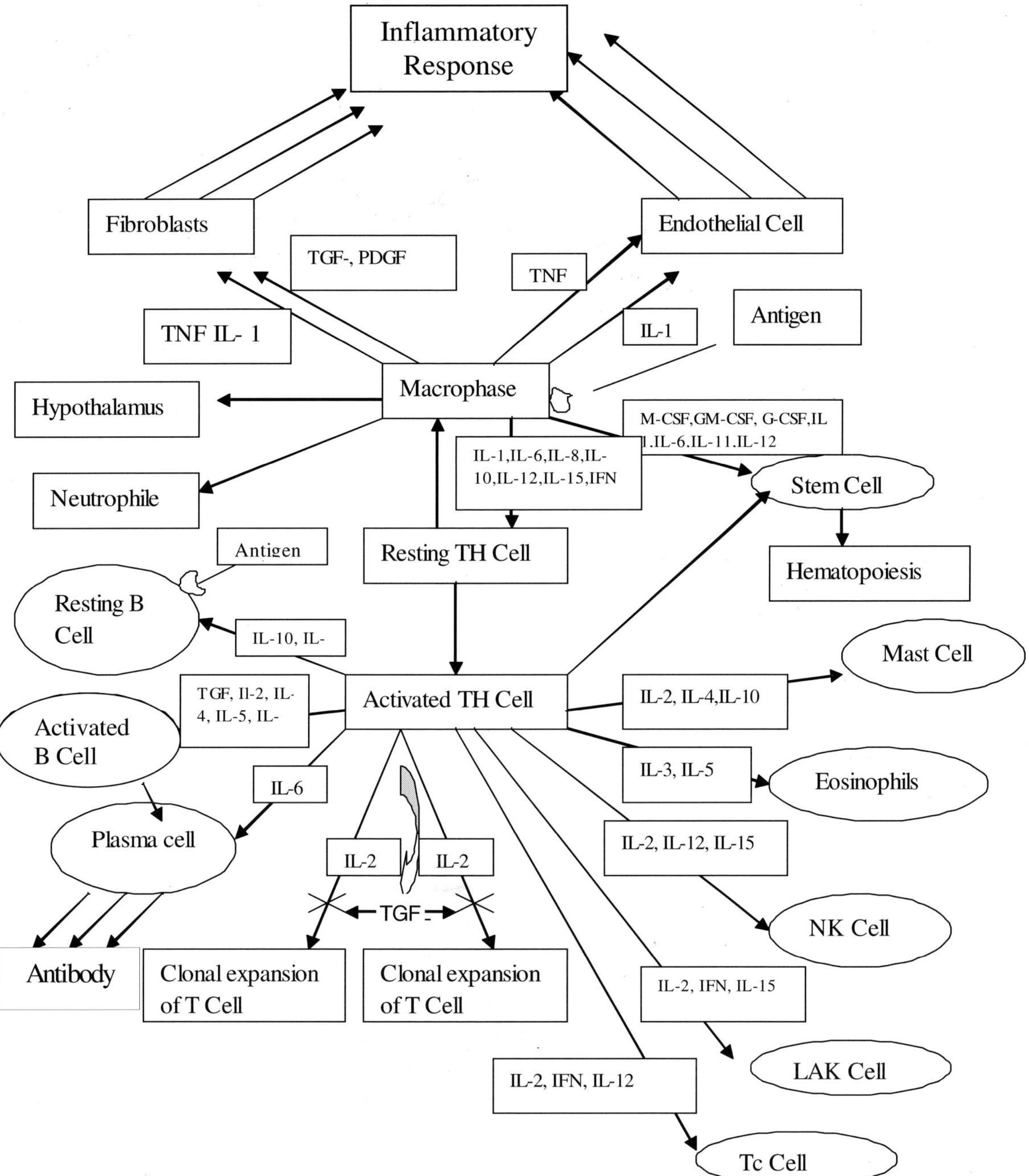

Fig. 2.2: Interaction of Antigen with Nacrophages and the Subsequent Activation of Resting TH Cells Leads to Release of Numerous Cytokines, Generating a Complex Network of Interacting Cells in the Immune Eespons. A small subset of TH cells secretes TGF-β, a cytokine that can inhibit the proliferation of T cells

The efficacy of colony-stimulating factors has been studied in several settings. Initially, attention was focused on preventing infection by starting treatment with the agents immediately after chemotherapy and continuing through the anticipated period of neutropenia. These agents were then used therapeutically in patients with established neutropenia after chemotherapy, although their efficacy in this context had not been established. When given as adjuncts to antibiotics for febrile neutropenia, granulocyte colony-stimulating factor (G-CSF), granulocyte–macrophage colony-stimulating factor (GM-CSF), (9) or both were found to shorten the duration of severe neutropenia, but they had no effect on mortality due to infections or other clinical end points, such as duration of hospital stay and number of days of antibiotic therapy. For this reason, the guidelines of the American Society of Clinical Oncology do not recommend the routine use of colony-stimulating factors in patients with febrile neutropenia. However, a recently published survey revealed that most physicians surveyed, prescribe colony-stimulating factors for patients with febrile neutropenia.The therapeutic use of a colony-stimulating factor in febrile patients with severe chemotherapy-induced neutropenia has not been studied in a large, placebo-controlled trial (10). Theoretically, the administration of colony-stimulating factors before the onset of fever could prevent serious infectious consequences in patients with severe neutropenia.

Structure of Granulocyte Colony Stimulating Factor

The recombinant human G-CSF synthesized in an *E. coli.* expression system is called filgrastim. The structure of filgrastim differs slightly from the natural glycoprotein. Most published studies have used filgrastim and it was the first form of G-CSF to be approved for marketing in Australia.

The three-dimensional structure of recombinant human granulocyte-colony-stimulating factor has been determined by x-ray crystallography (13). Phases were initially obtained at 3.0-Å resolutions by multiple isomorphous replacements and were refined by solvent flattening and by averaging of the electron density of the three molecules in the asymmetric unit. The R factor was 21.5% for all data

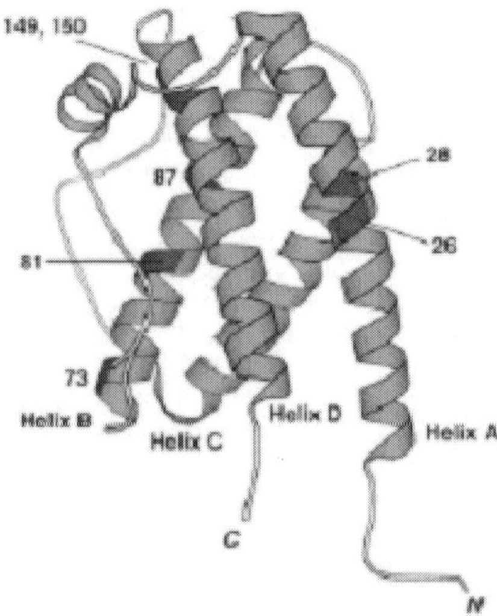

Fig. 2.3: Ribbon Representation of the Structure of G-CSF made in Molscript from NMR Structure (14). A 4-helix bundle motif with long overhand loop connection is shown.The helices are lebeled in order from the N terminus: A,B,C,D. This figure was taken with acknowledgment.

between 6.0- and 2.2-Å resolutions. The structure is predominantly helical, with 104 of the 175 residues forming a four-α-helix bundle. The only other secondary structure is also helical. In the loop between the first two long helices a four-residue 3_{10}-helix is immediately followed by a 6-residue α-helix. Three residues in the short connection between the second and third bundle helices form almost one turn of left-handed helix. The up-up-down-down connectivity with two long crossover connections has been reported previously for five other proteins, which like granulocyte-colony-stimulating factor are all signaling ligands: growth hormone, granulocyte/macrophagecolony-stimulating factor, interferon β, interleukin 2, and interleukin 4. Structural similarity among these growth factors occurs despite the absence of similarity in their amino acid sequences. Conservation of this tertiary structure suggests that these different growth factors might all bind to their respective sequence-related receptors in an equivalent manner.

Stability and Reengineering of Granulocyte Colony-Stimulating Factor

Granulocyte colony-stimulating factor is a long-chain cytokine that has both biological and therapeutic applications. It is involved in the production and maturation of neutrophilic progenitor cells and neutrophils and is administered to stimulate the production of white blood cells to reduce the risk of serious infection in immunocompromised patients. Granulocyte colony-stimulating factor has been reengineered to improve the thermodynamic stability of the protein (13), focusing on enhancing the alpha-helical propensity of residues in the antiparallel 4-helix bundle of the protein.

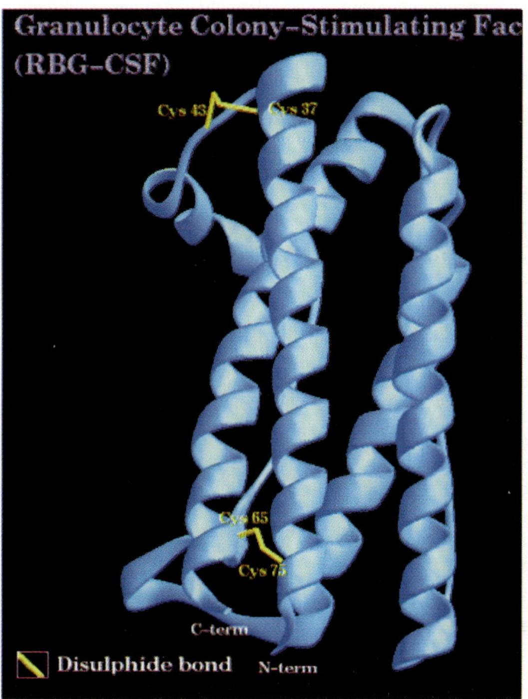

Fig. 2.4
(This figure was taken with acknowledgment)

These redesigns resulted in proteins with substantially enhanced stability while retaining wild-type levels of biological activity, measured as the ability of the reengineered proteins to stimulate the proliferation of murine myeloid cells transfected with the granulocyte colony-stimulating factor receptor

Kinetics of G-CSF Folding

The folding kinetics of G-CSF have been determined by trp-fluorescence and far-UV circular dichroism(15). Folding and unfolding has been achieved by rapid dilution and mixing of the some protein denaturant.G-CSF has a four-helical bundle protein with two long loops between the first and second helices and between the third and fourth helices. The whole conformational change expected by fluorescence was observed by stopped-flow technology, but due to rapid refolding kinetics only a portion was observed by circular dichroism. G-CSF contains two trp residues, and their contribution to the fluorescent-detected kinetics were deciphered through the use of single-site trp mutants. The trp moieties are probes of the local conformation surrounding their environment. One trp at residue 118 is located within the third helix while the other trp at residue 58 is part of the long loop between the first and second helices. The refolding results were most consistent with the following mechanism: $U \leftrightarrow I_1 \leftrightarrow I_2 \leftrightarrow N$; where U represents the unfolded protein, I_1 represents intermediate state 1, I_2 represents intermediate state 2, and N represents the native state. I_1 is characterized as having approximately one-half of the native-like helical structure and none of the native-like fluorescence. I_2 has 100% of the native helical structure and most of the trp-118 and little of the trp-58 native-like fluorescence. Thus refolding occurs in distinct stages with half of the helix forming first followed by the remaining half of the helix including the third helix and finally the loop between the first and second helices folds.

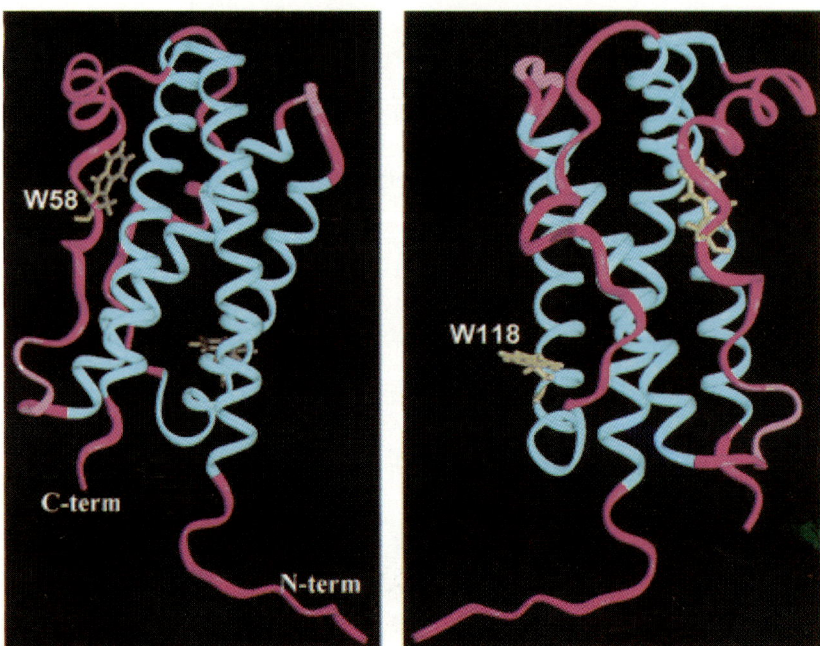

Fig. 2.5: Depiction of the Three-dimensional Structure of the Main Chain of G-CSF as Determined by NMR Spectroscopy (Zink *et al.*, 1994). This figure was taken with acknowledgment.

The four main helices are shown in blue and the loops in purple. The left panel shows the location of W58 within the first long loop, and for the right panel the structure was rotated 180° to illustrate the location of W118 in the third helix.

Granulocyte Colony-stimulating Factor (G-CSF) and its Receptor

Granulocyte colony-stimulating factor (G-CSF) forms a tetrameric complex with its receptor, comprising two G-CSF and two receptor molecules (16). The structure of the complex is unknown, and it is unclear whether there are one or two binding sites on G-CSF and the receptor. The immunoglobulin-like domain and the cytokine receptor homologous module of the receptor are involved in G-CSF binding, and Arg^{288} in the cytokine receptor homologous module is particularly important. To identify residues in G-CSF that interact with Arg^{288}, selected charged residues in G-CSF were mutated to Ala. To clarify whether there are two binding sites, a chimeric receptor was created in which the Ig domain was replaced with that of the related receptor gp130. This chimera bound G-CSF but could not transduce a signal, consistent with failure of dimerization and loss of one binding site. The G-CSF mutants had reduced mitogenic activity on cells expressing wild-type receptor. When tested with the chimeric receptor, all G-CSF mutants except one (E46A) showed reduced binding, suggesting that Glu^{46} is important for interaction with the Ig domain. On cells expressing R288A receptor, all the G-CSF mutants except E19A showed reduced mitogenic activity, indicating that Glu^{19} of G-CSF interacts with Arg^{288} of the receptor.

Identification of Ligand-binding of the Granulocyte Colony-stimulating Factor Receptor

The N-terminal Ig-like domain of the G-CSF-R is required for receptor dimerization, but it is not known whether it binds G-CSF or interacts elsewhere in the complex(17). Alanine scanning mutagenesis was used to show that residues in the Ig-like domain of the G-CSF-R (Phe^{75}, Gln^{87}, and Gln^{91}) interact with G-CSF. This binding site for G-CSF overlapped with the binding site of a neutralizing anti-G-CSF-R antibody. A model of the Ig-like domain showed that the binding site is very similar to the viral interleukin-6 binding site (site III) on the Ig-like domain of gp130, a related receptor. To further characterize the G-CSF-R complex, exposed and inaccessible regions of monomeric and dimeric ligand-receptor complexes were mapped with monoclonal antibodies. The results showed that the E helix of G-CSF was inaccessible in the dimeric but exposed in the monomeric complex, suggesting that this region binds to the Ig-like domain of the G-CSF-R. In addition, the N terminus of G-CSF was exposed to antibody binding in both complexes. These data establish that the dimerization interface of the complete receptor complex is different from that in the x-ray structure of a partial complex. A model of the tetrameric G-CSF·G-CSF-R complex was prepared, based on the viral interleukin-6·gp130 complex, which explains these and previously published data.

Atomic Structure of the GCSF-receptor Complex

GCSF is a member of the long-chain subtype of the class 1 cytokine superfamily, which includes growth hormone, erythropoietin, interleukin 6 and oncostatin M. The crystal structure of GCSF complexed to the BN-BC domains, the principal ligand-binding region of the GCSF receptor (GCSFR) has also been determined (18). The two receptor domains form a complex in a 2:2 ratio with the ligand, with a non-crystallographic pseudo-twofold axis through primarily the interdomain region and secondarily the BC domain. This structural view of a gp130-type receptor-ligand complex presents a new molecular basis for cytokine receptor recognition.

The Carboxyl Terminus of the Granulocyte Colony-stimulating Factor Receptor

The G-CSF receptor transduces signals that regulate the proliferation, differentiation, and survival of myeloid cells. A subgroup of patients with severe congenital neutropenia (SCN) has been shown to harbor mutations in the G-CSF receptor gene that resulted in the truncation of the receptor's carboxyl-terminal region. SCN patients with mutations in the G-CSF receptor gene are predisposed to acute myeloid leukemia. The truncated receptors from SCN/acute myeloid leukemia patients mediate augmented and sustained activation of Stat transcription factors and are accordingly hyperactive in inducing cell proliferation and survival but are defective in inducing differentiation. Little is known about the molecular mechanisms underlying the negative role of the receptor's carboxyl terminus in the regulation of Stat activation and cell proliferation/survival. In a study, evidence that SH2-containing phosphatase-1 (SHP-1) plays a negative regulatory role in G-CSF-induced Stat activation was showed. Also, it was demonstrated that the carboxyl terminus of the G-CSF receptor is required for SHP-1 down-regulation of Stat activation induced by G-CSF (11). This regulation is highly specific because SHP-1 has no effect on the activation of Akt and extracellular signal-related kinase1/2 by G-CSF. The data together strongly suggest that SHP-1 may represent an important mechanism by which the carboxyl terminus of the G-CSF receptor down-regulates G-CSF-induced Stat activation and thereby inhibits cell proliferation and survival in response to G-CSF.

Mode of Action of GCSF

Very much like other CSF's, G-CSF initiates its activity by binding to specific all surface receptors which are present in low numbers (1000) on normal neutrophils and promyelocytes. Inflammatory

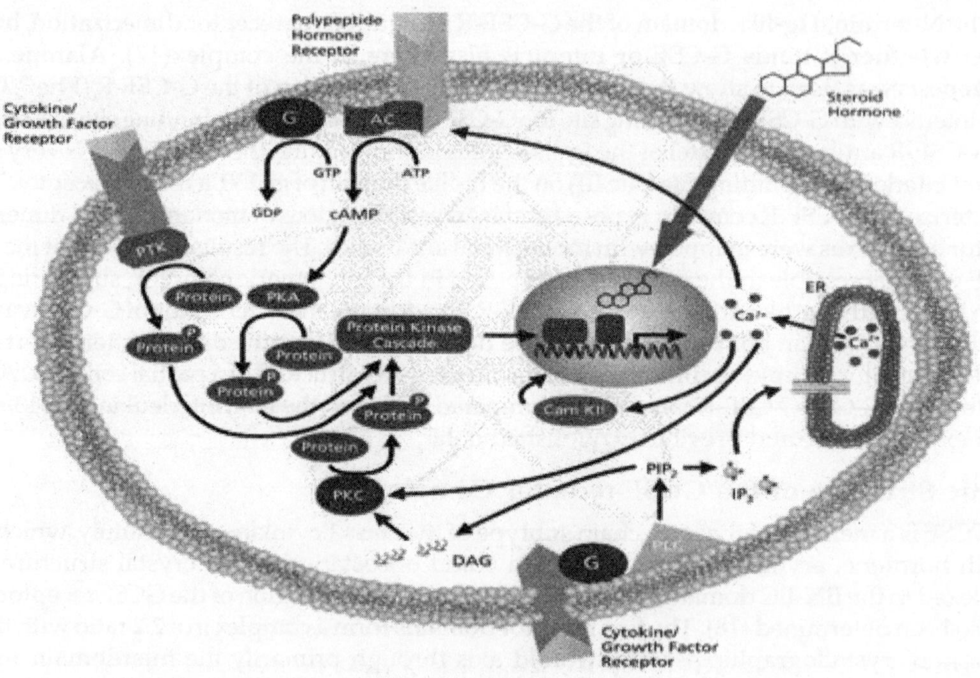

Fig. 2.6: Mode of Action of GCSF
(This figure was taken with acknowledgment)

modulators such as fMet-len-phe,LPS and GM-CSF are the most probable candites which down modulate the G-CSF activity. GCSF receptors of some count and characteristics as that of normal cells are also present on human myeloid leukemia cell line HL60 and have been found consistently on fresh human leukemic cell. most leukemic blasts bearing G-CSF receptors will proliferate in response to the cytokine, but the presence of receptors is not completely predictive of response(19).

G-CSF receptors has only a single high affinity binding site with a dissociation from 60-900 pM is various reports. Cross linking studies suggest that both murine and human G-CSF receptor is a 150 Kda protein and the cytokines, only murine G-CSF can complete to receptor binding of labelled human G-CSF.

G-CSF signal trnsduction mechanism have not yet been fully elucidated. Rapid phophorylation of a 75 Kda protein has been demonatrated after incubation of receptor bearing cells with G-CSF, suggesting tyrosine kinase activity.

Gene Structure

It is reported that the GCSF gene has 4 introns and that 2 different polypeptides are synthesized from the same gene by differential splicing of mRNA.(20) The 2 polypeptides differ by the presence or absence of 3 amino acids. Expression studies indicate that both have authentic GCSF activity.

Some Important Information of GCSF at a Glance

Chromosomal Location

17q11.2-q12.

Micro Array Gene Expression Data

The protein encoded by this gene is a cytokine that controls the production, differentiation, and function of granulocytes. The active protein is found extracellularly. Three transcript variants encoding three different isoforms have been found for this gene. (21)

Swiss Prot Accession No.

P09919 Granulocyte colony-stimulating factor precursor (Homo sapiens); 100% similarity over 207 a.a.

Function

granulocyte/macrophage colony-stimulating factors are cytokines that act in hematopoiesis by controlling the production, differentiation, and function of 2 related white cell populations of the blood, the granulocytes and the monocytes-macrophages. this csf induces granulocytes.

Subcellular Location

Secreted

Subunit

Monomer

Experimental Work of rhGCSF by Recombinant DNA Technology

Granulocyte colony-stimulating factor (G-CSF) is a cytokine that stimulates hematopoiesis and induces proliferation and differentiation of granulocyte progenitor cells as well as production of bone marrow neutrophilic granulocyte colonies. Nowadays, hu-man recombinant G-CSF(hr G-CSF) is used

3' ACC CCC CTG GGC CCT GCC AGC TCC CTG CCC CAG AGC TTC CTG

Thr Pro Leu Gly Pro Ala Ser Ser Leu Pro Gln Ser Phe Leu

CTC AAG TGC TTA GAG CAA GTG AGG AAC ATC CAG GGC GAT GGC

Leu Lys Cys Leu Glu Gln Val Arg Lys Ile Gln Gly Asp Gly

GCA GCG CTC CAG GAG AAG CTG TGT GCC ACC TAC AAG CTG TGC

Ala Ala Leu Gln Glu Lys Leu Cys Ale Thr Tyr Lys Leu Cys

CAC CCC GAG GAG CTG GTG CTG CTC GGA CAC TCT CTG GGC ATC

His Pro Glu Glu Leu Val Leu Leu Gly His Ser Aeu Gly Ile

CCC TGG GCT CCC CTG AGC TCC TGC CCC AGC CAG GCC CTG CAG

Pro Trp Ala Pro Leu Ser Ser Cys Pro Ser Gln Ala Leu Gln

CTG GCA GGC TGC TTG AGC CAA CTC CAT AGC GGC CTT TTC CTC

Leu Ala Gly Cys Leu Ser Gln Leu His Ser Gly Leu Phe Leu

TAC CAG GGG CTC CTG CAG GCC CTG GAA GGG ATA TCC CCC GAG

Tyr Gln Gly Leu Leu Gln Ala Leu Glu Gly Ile Ser Pro Glu

TTG GGT CCC ACC TTG GAC ACA CTG CAG CTG GAC GTC GCC GAC

Leu Gly Pro Thr Leu Asp Thr Leu Gln Leu Asp Val Ala Asp

TTT GCC ACC ACC ATC TGG CAG CAG ATG GAA GAA CTG GGA ATG

Phe Ala Thr Thr Ile Trp Gln Gln Met Glu Glu Leu Gly Met

GCC CCT GCC CTG CAG CCC ACC CAG GGT GCC ATG CCG GCC TTC

Ala Pro Ala Leu Gln Pro Thr Gln Gly Ala Met Pro Ala Phe

GCC TCT GCT TTC CAG CGC CGG GCA GGA GGG GTC CTG GTT TCT

Ala Ser Ala Phe Gln Arg Arg Ala Gly Gly Val Leu Val Ala

AGC CAT CTG CAG AGC TTC CTG GAG GTG TCG TAC CGC GTT CTA

Ser His Leu Gln Ser Phe Leu Glu Val Ser Tyr Arg Val Leu

CGC CAC CTT GCG CAG CCC TGA 5'

Arg His Leu Ala Gln Pro OPA

Fig. 2.7: Natural Human G-CSF, DNA Sequence and Amino Acid Sequence

for the treatment of chemotherapy- and radiotherapy-induced neutropenia, and also in patients with bone marrow transplantation(22).

A cDNA of human G-CSF (hG-CSF) was synthesized by PCR fromrecombinant cloning vector, with two altered nucleotides for increasing mRNAstability and overexpression, then inserted into a pET expression vector under the control of T7 promoter and cloned in *E. coli* strain BL21 (DE3).

After culture and induction of recombinant *E. coli* with IPTG, we achieved a high level expression of the hG-CSF, where it represented approximately 35% of the total protein as determined by SDS-PAGE and confirmed by western blotting with polyclonal and monoclonal hG-CSF antibodies.

rhG-CSF was produced in a significantly high quantity with a yield of 35% of total protein as determined by SDS-PAGE. Since simple purification steps easily obtain it, it may be cost-effective, even on an industrial scale.

The synthesis and secretion of recombinant human granulocyte colony-stimulating factor (rhG-CSF) are investigated in fed-batch cultures at high cell concentration of recombinant Saccharomyces cerevisiae, and some important characteristics of the secreted rhG-CSF are demonstrated. Transcription of the recombinant gene is regulated by a GAL1-10 upstream activating sequence (UASG), and the rhG-CSF is expressed in a hybrid fusion protein consisting of signal sequence of Kluyveromyces lactis killer toxin and N-terminal 24 amino acids of human interleukin 1beta. The intracellular KEX2 cleavage leads to excretion of mature rhG-CSF into extracellular culture broth, and the cleavage process seems to be highly efficient. In spite of relatively low copy number the plasmid propagation is stably maintained even at nonselective culture conditions. The rhG-CSF synthesis does not depend on galactose level, whereas the production of extracellular rhG-CSF was significantly enhanced by increasing the inducer concentration above a certain level and also by supplementing the nonionic surfactant to the culture medium, which is notably due to the enhanced secretion efficiency. Various immunoblotting analyses demonstrate that none of the rhG-CSF is accumulated in the cell wall fraction and that a significant amount of intracellular rhG-CSF antibody-specific immunoreactive proteins is located in the ER. A core N-glycosylation at fused IL-1beta fragment is likely to play a critical role in directing the high-level secretion of rhG-CSF, and the O-glycosylation of secreted rhG-CSF seems nearly negligible. (23) Also the extracellular rhG-CSF is observed to exist as various multimers, and the nature of molecular interaction is evidently not the covalent disulfide bridges. The CD spectra of purified rhG-CSF and Escherichia coli-derived standard show that the conformations of both are similar and are almost identical to that reported for natural hG-CSF

Expression and Simple Purification of Recombinant Human Granulocyte Colony-Stimulating Factor in *E. coli*

A human granulocyte colony stimulating factor(hG-CSF) gene was synthesized and inserted the synthetic gene encoding the 175 amino acids of hG-CSF including the initiation codon into a trp expression vector for over expression in E.Coli.A strong expression vector was constructed by using pUC18,digested with ClaI and SalI restriction enzymes, and a 540 bp fragment containing the encoding region and a simple purification procedure including in vitro refolding was established. It is reported that the final productivity of hG-CSF was 500~600 mg/l culture, and purified hG-CSF by cation exchange and chromatofocusing chromatography (24).

Production of Human Granulocyte Colony Stimulating Factor Fused to the Pelb Signal Peptide in *Escherichia coli*

It is reported that to induce the secretion of human granulocyte colony stimulating factor (hG-CSF) gene was fused to the gene encoding the pectate lysase B signal peptide (PelB) of *Erwinia caratovora* (25), and attempt to induce the secretion of hG-CSF by *E.coli.*In addition, the pH-stat fed batch culture was adopted to over produce hG-CSF by the high cell density-culture of recombinant *E.coli.* and consequently,did not succeed in directing the secretion of hG-CSF into the periplasmic space of *E.coli,* but obtained inclusion bodies of hG-CSF fused to the PelB signal peptide in the intracellular compartment.hG-CSF oligomers were synthesized, annealed and ligated to form a synthetic hg-CSF gene were disgned to contain BamHI and NcoI sites, respectively.The gene encoding the hG-CSF terminator was placed downstream of the hG-CSF gene. Plasmid pET-22b(+) was used to insert the hG-CSF fragment into the NcoI/BamHI. In case Dh5α was used for the cloning of the hG-CSF gene. The plasmid and host strain used for the expression of the hG-CSF gene were pET-22(+) and E.coli BL21(DE3) respectively, obtained from Novagen.The expression experiments were carried out as described by Studier *et al.* The recombinant cells were grown in LB midium containg ampicillin at a concentration of 100μg/ml at 37^0 C. 0.25 mM IPTG was added when the cell concentration reached an optical density(600nm) to induce the T7 promoter, and the cell were harvested after a further 7-9 hours induction. The induced protein was readily detectable on 12% SDS-PAGE of the cell lysate after staing of the gel with Coomassie blue.No secretion of hG-CSF and no processing of the signal peptide from the fusion protein was observed. Instead, the fusion protein, pelB-hG-CSF, at a concentration of 1.7g/l was produced in an inclusion body form. Interestingly, the biological activity of the purified fusion protein was found to be identical to that of mature hG-CSF, indicating that the pelB signal peptide attached to the N-terminal of hG-CSF had no effect on its biological activity.

The human granulocyte colony-stimulating factor (hG-CSF) was efficiently secreted at high levels in fed-batch cultures of recombinant Saccharomyces cerevisiae. However, the secreted recombinant hG-CSF (rhG-CSF) was shown to exist as large multimers in the culture broth due to strong hydrophobic interaction. It was hardly monomerized even by urea at high concentration. This multimer has been reported to diminish specific receptor-binding activity of hG-CSF and causes undesirable problems in the downstream process. When the rhG-CSF was secreted to extracellular broth in the presence of a non-ionic surfactant (Tween 80) in the culture media, the multimerization of the secreted rhG-CSF was efficiently prevented in the fed-batch cultures. Also, the monomer fraction and secreted efficiency of rhG-CSF were significantly increased at the higher culture pH (6.5). Without using any denaturing agents, the secreted rhG-CSF monomer was easily purified with high recovery yield and purity via a simple purification process under acidic conditions, consisting of diafiltration, cation exchange, and gel filtration chromatography. A lyophilization process devoid of intermonomer aggregation was also designed using effective stabilizing agents. (26)

G-CSF and Other Colony Stimulating Factor

Granulocyte-Colony Stimulating Factor (G-CSF), Macrophage-Colony Stimulating Factor(M-CSF), Granulocyte-Macrophage-Colony Stimulating Factor(GM-CSF).

Cytokine protein that stimulate growth and reproduction of certain kinds of blood cells in the bone marrow. Also referred to as growth factors. The production of white blood cells is controlled by colony stimulating factors. (27) Cancer chemotherapy and inherited disorders are among the causes of low white blood counts, which lower resistance to infection. Thus, CSFs are being investigated not only as a way to counteract low white blood cell counts, but also as a way to produce specific types of white blood cells. In addition, there is hope that CSFs can stimulate the body to produce additional bone marrow as well as cause some cancer cells to stop dividing. (28)

GCSF and its Relationship to Other Growth Factor

The three dimensional structure of recombinant granulocyte colony stimulating factor has determined by x-ray crystallography (29). Phases were initially obtained at 3.0-A resolution by multiple isomorphous replacements and were refined by solvent flattening and by averaging of the electron density of the three molecules in the asymmetric unit. The current R factor is 21.5% for all data between

6.0- and 2.2-A resolutions. The structure is predominantly helical, with 104 of the 175 residues forming a four-alpha-helix bundle. The only other secondary structure is also helical. In the loop between the first two long helices a four-residue 3(10)-helix is immediately followed by a 6-residue alpha-helix. Three residues in the short connection between the second and third bundle helices form almost one turn of left-handed helix. The up-up-down-down connectivity with two long crossover connections has been reported previously for five other proteins, which like granulocyte-colony-stimulating factor are all signaling ligands: growth hormone, granulocyte/macrophage-colony-stimulating factor, interferon beta, interleukin 2, and interleukin 4. Structural similarity among these growth factors occurs despite the absence of similarity in their amino acid sequences. Conservation of this tertiary structure suggests that these different growth factors might all bind to their respective sequence-related receptors in an equivalent manner.

Mapping

By somatic cell hybridization and in situ chromosomal hybridization, mapped the GCSF gene to 17q11 in the region of the breakpoint in the 15; 17 translocation characteristic of acute promyelocytic leukemia. Further studies indicated that the gene is proximal to the said breakpoint and that it remains on the rearranged chromosome 17. Southern blot analysis using both conventional and pulsed field gel electrophoresis showed no rearranged restriction fragments. By use of a full-length cDNA clone as a hybridization probe in human-mouse somatic cell hybrids and in flow-sorted human chromosomes, mapped the gene for GCSF to 17q21-q22 by in situ hybridization. Also assigned CSF3 to 17q21 by in situ hybridization. CSF3 and CRYB1 are located on the same large restriction fragment (HGM10) (23).

Another form of recombinant human G-CSF called *lenograstim* is synthesised in Chinese hamster ovary (CHO) cells. As this is a mammalian cell expression system, lenograstim is indistinguishable from the 174 amino acid natural human G-CSF. No clinical or therapeutic consequences of the differences between filgrastim and lenograstim have yet been identified, but there are no formal comparative studies. G-CSF should not be confused with granulocyte macrophage colony stimulating factor (GM-CSF), which is a distinctly different haemopoietic growth factor and is also under clinical development.

Granulocyte colony-stimulating factor (or colony stimulating factor-3) specifically stimulates the proliferation and differentiation of the progenitor cells for granulocytes determined the partial amino acid sequence of purified GCSF protein, and by using oligonucleotides as probes, isolated several G-CSF cDNA clones from a human squamous carcinoma cell line cDNA library. Cloning of human GCSF cDNA shows that a single gene codes for a 177- or 180-amino acid mature protein of molecular weight 19,600.

Indication

Patients with neutropenia have significant morbidity due to bacterial infections. These infections require treatment in hospital with intravenous antibiotics. The vulnerability of neutropenic patients to infections dictates that antibiotics are given at the onset of fever, before microbiological confirmation, and continued until the neutropenia resolves to prevent recurrent infection. G-CSF is the first treatment available to Australian patients, which reliably ameliorates the underlying neutropenia.

Clinical Application of G-CSF

Cancer Chemotherapy

G-CSF (filgrastim) is indicated for the prevention of febrile neutropenia in patients receiving myelosuppressive chemotherapy for non-myeloid malignancies. It reduces the duration and severity of post-chemotherapy neutropenia.(30). In 3 randomised placebo controlled studies of G-CSF (filgrastim) given after chemotherapy to patients with small cell lung cancer or lymphoma, the haematological effects of G-CSF (filgrastim) resulted in a 50% reduction in the number of patients experiencing febrile neutropenia. There were associated reductions in the use of intravenous antibiotics, hospitalisation, and the incidence of microbiologically documented infections. These 3 pivotal studies used significantly myelosuppressive chemotherapy regimens: without G-CSF, the incidence of febrile neutropenia for these regimens was 44-77%.

The recommended dose of G-CSF (filgrastim) for the prevention of febrile neutropenia is 5 micrograms/kg/day, commencing the day after chemotherapy and finishing after 10-14 days, when the post-nadir neutrophil level has reached 10x 10^9/L. G-CSF is repeated with each cycle of chemotherapy. Concurrent administration of G-CSF and chemotherapy is not recommended as some combinations may increase myelo suppression. Subcutaneous and intravenous G-CSF has equivalent potency, but subcutaneous injection is generally recommended, as it is suitable for outpatients.

G-CSF (lenograstim) is also approved for use to reduce the incidence of infection associated with established cytotoxic chemotherapy(31). Similar practical considerations apply to its use as for G-CSF (filgrastim).

Some potentially cost-saving variations to the recommended G-CSF regimen are under investigation. They include lower G-CSF doses (in particular, doses based on vial size rather than body mass) and shorter G-CSF courses (in particular, delayed commencement or earlier cessation of G-CSF). Formal assessments of whether or not these other approaches are as effective as the recommended regimen have not been reported.

At present, G-CSF is not recommended for patients with myeloid malignancies because in vitro data suggest that it may stimulate some myeloid malignancies. The safety and efficacy of G-CSF for infection prevention in patients with myeloid leukaemia is currently under investigation. Other research is investigating if the stimulatory effects of G-CSF on myeloid leukaemias will increase the susceptibility of leukaemic cells to cytotoxic drugs.

The patient should be monitored at least twice per week and the full blood count drawn before that day's G-CSF dose. (In the hour immediately following G-CSF administration, there is a transient reduction in numbers of circulating neutrophils.) All febrile neutropenic post-chemotherapy patients must seek urgent medical attention whether or not they have been receiving G-CSF. General practitioners will probably encounter patients who are self-administering G-CSF during outpatient phases of chemotherapy. They should refer febrile patients urgently to hospital.

The greatest benefit of G-CSF is expected with chemotherapy regimens given with intent to achieve cure or prolonged remission and having an incidence of febrile neutropenia above 40% (*e.g.* treatment regimens for patients with non-Hodgkin's lymphoma, relapsed Hodgkin's disease, germ cell tumours and acute lymphoblastic leukaemia). The validity of correlating the degree and duration of neutropenia with the risk of febrile neutropenia or infection has been demonstrated for G-CSF-treated patients in the pivotal studies with G-CSF (filgrastim). Since G-CSF has similar haematological effects for a variety of chemotherapy regimens including some with lower incidences of febrile neutropenic episodes, patients receiving other chemotherapy regimens for different tumour types are also expected to benefit clinically from G-CSF therapy. G-CSF (filgrastim and lenograstim) are approved for use in all non-myeloid tumour types.

The patient should be monitored at least twice per week and the full blood count drawn before that day's G-CSF dose. (In the hour immediately following G-CSF administration, there is a transient reduction in numbers of circulating neutrophils.) All febrile neutropenic post-chemotherapy patients must seek urgent medical attention whether or not they have been receiving G-CSF. General practitioners will probably encounter patients who are self-administering G-CSF during outpatient phases of chemotherapy. They should refer febrile patients urgently to hospital.

The greatest benefit of G-CSF is expected with chemotherapy regimens given with intent to achieve cure or prolonged remission and having an incidence of febrile neutropenia above 40% (*e.g.* treatment regimens for patients with non-Hodgkin's lymphoma, relapsed Hodgkin's disease, germ cell tumors and acute lymphoblastic leukaemia)(32). The validity of correlating the degree and duration of neutropenia with the risk of febrile neutropenia or infection has been demonstrated for G-CSF-treated patients in the pivotal studies with G-CSF (filgrastim). Since G-CSF has similar haematological effects for a variety of chemotherapy regimens including some with lower incidences of febrile neutropenic episodes, patients receiving other chemotherapy regimens for different tumour types are also expected to benefit clinically from G-CSF therapy. G-CSF (filgrastim and lenograstim) is approved for use in all non-myeloid tumour types.

There are many new applications of G-CSF in autologous hematopoietic support under investigation in Australia. G-CSF can be used to increase the numbers of circulating haemopoietic precursor cells which are then harvested by leucapheresis and used either as a supplement to or a substitute for aspirated bone marrow cells in autologous haemopoietic cell transplants. This approach has resulted in significantly accelerated platelet recovery in addition to neutrophil recovery. The role of post-transplant G-CSF after these peripheral blood progenitor cell autologous transplants is being reassessed. Community practitioners should be aware of these new approaches to haemopoietic support after myeloablative chemotherapy as these patients are now often managed as outpatients for a significant proportion of the period of haemopoietic reconstitution. The transplant physician should be contacted if any management questions arise regarding the care of these patients.

Effects of Recombinant Human Granulocyte Colony-Stimulating Factor on Neutropenia in Patients with Congenital Agranulocytosis

Congenital agranulocytosis is a disorder characterized by severe neutropenia and a profound deficiency of identifiable neutrophil progenitors in bone marrow(33). In an attempt to stimulate neutrophil production and thereby reduce the morbidity and mortality associated with this disease, recombinant human granulocyte colony-stimulating factor (rhG-CSF) in doses of 3 to 60 micrograms per kilogram of body weight per day to five patients with congenital agranulocytosis was administered. In all five patients, an increase in the number of neutrophils was noted eight to nine days after the initiation of the effective dosage (the dose at which the neutrophil count reached 1000 cells per microliter or more and the bone marrow showed granulocyte maturation beyond the myelocyte stage). The absolute neutrophil counts rose from less than 100 to between 1300 and 9500 cells per microliter. Marrow aspirates obtained after 14 days at the effective dosage showed maturation to the mature neutrophil stage. The side effects that were observed were medullary pain, splenomegaly, and an elevation of levels of leukocyte alkaline phosphatase. All five patients have had sustained neutrophil counts of 1000 cells per microliter or more for 9 to 13 months while receiving subcutaneous maintenance therapy. Preexisting chronic infections have resolved clinically, and the number of new infectious episodes and the requirement for intravenous antibiotics have decreased. It was concluded that treatment with rhG-CSF can lead to a large increase in the numbers of functional neutrophils in patients with congenital agranulocytosis.

Severe Chronic Neutropenia

The early observations of the efficacy of G-CSF in patients with severe chronic neutropenia and recurrent infections have been confirmed in a multicentre randomised study of patients with congenital, idiopathic and cyclical neutropenia.G-CSF (filgrastim) reduced the incidence and duration of infection-related events and antibiotic use, and the incidence of hospitalisation. G-CSF (filgrastim) is approved for chronic administration to patients with severe chronic neutropenia.

Treatment of Febrile Neutropenia

Not all patients given chemotherapy for cancer receive G-CSF to prevent febrile neutropenia. A randomised placebo-controlled Australian study evaluated the effects of using G-CSF (filgrastim) as an adjunct to empiric intravenous combination antibiotics for post-chemotherapy febrile neutropenia. G-CSF shortened the duration of neutropenia and febrile neutropenia, but did not significantly shorten the duration of fever or hospitalisation, although there were consistent trends towards these benefits. G-CSF halved the relative risk of prolonged hospitalisation. Subset analyses suggested particular benefits for patients with solid tumours and documented infections.

This study supports the hypothesis that there is a real benefit from adding G-CSF to the standard antibiotic management of febrile neutropenia. It seems improbable that this benefit is as significant to patients as the prevention of febrile neutropenia. However, the relative benefit of G-CSF therapy by these two approaches has not been assessed.(34)

Adverse Effect and Safety

G-CSF has been well tolerated by patients. In the randomized studies of patients having chemotherapy for cancer, the only significant adverse effect of G-CSF has been bone pain, which occurred in 15-20% of patients (approximately twice the incidence of placebo-treated patients). This medullary bone pain is effectively managed with simple oral analgesics such as paracetamol with or without codeine. Stronger analgesics are only very rarely required. Thrombocytopenia may be more pronounced in patients who do not have their chemotherapy dose reduced or delayed because of the effect of G-CSF to ameliorate neutropenia (35), which permits intended doses of chemotherapy drugs to be administered on schedule without dose reductions. (Chemotherapy doses are usually decreased if there is a delayed recovery of the neutrophil count.) The thrombocytopenia is probably a consequence of continued delivery of the intended chemotherapy doses rather than an effect of G-CSF itself. G-CSF has not been associated with the flu-like symptoms or fever, which occur with the administration of other cytokines.

The 'low-grade' adverse events related to G-CSF (filgrastim) administration in the randomized study of patients with severe chronic neutropenia were musculoskeletal and bone pain, headache, rash and asymptomatic increase in spleen size. G-CSF (filgrastim) administration to some of these patients for up to 3 years has remained effective and well tolerated. (36)

Infrequent adverse effects reported include vasculitis, Sweet's syndrome (acute febrile neutrophilic dermatosis) and exacerbation of some pre-existing dermatological conditions. Possible anaphylactoid reactions (incidence <0.001%) have been reported, although these have not been confirmed to be allergic reactions. Antibody formation has not been reported, even during G-CSF (filgrastim) courses of up to 9 months.

Cost and Benefit Consideration

G-CSF is an expensive drug. The cost of G-CSF (filgrastim) for the recommended minimum 10-day schedule for prevention of febrile neutropenia after cancer chemotherapy is A$1504 or $2407 per cycle depending on which of the two vial sizes (300 and 480 micrograms/vial) is used. G-CSF (lenograstim) costs $1283.60 for 1O vials (263 micrograms/vial), which is sufficient to treat a 53 kg person for 10 days at the recommended dose. The cost per microgram is $0.50 for G-CSF (filgrastim) and $0.49 for G-CSF (lenograstim). In Australia, a full assessment of the economic impact of G-CSF is difficult because of the different sources of health care funding. A significant proportion of the cost of G-CSF may be offset by the savings to the total health care budget through the reduction in admissions for management of febrile neutropenia. These admissions have a median duration of 8 days and are estimated to cost in the vicinity of A$500-700 per day. Such cost offsetting has been documented within both the private and public health care systems in the U.S.A. For the few patients with severe chronic neutropenia and clinically significant infections for whom there is no reliably effective alternative therapy, the benefits of daily G-CSF administration such as an improved quality of life, the significant reduction in morbidity and the reduced cost of managing infections must be considered in cost-benefit analyses. Cost of per vial (300mg) is Rs.-3000.00 in India. Dr.Reddy's, Intus, and some other companies are marketing G-CSF in India.

G-CSF (filgrastim) is available under Section 100 of the Pharmaceutical Benefits Scheme for out patient use during treatment of patients with selected tumour types (non-Hodgkin's lymphoma(35), germ cell tumours, relapsed Hodgkin's disease, acute lymphoblastic leukaemia, Ewing's sarcoma, rhabdomyosarcoma and neuroblastoma), and after autologous marrow transplantation, and in selected patients with severe chronic neutropenia. As of August 1994, G-CSF (lenograstim) also became available as a pharmaceutical benefit for similar uses.

Although G-CSF reduces the morbidity, which accompanies neutropenia, it is not known whether G-CSF affects the mortality of neutropenic sepsis. Febrile neutropenia has a mortality of 10%, but occurs only in a proportion of patients and varies with the degree of myelosuppression of the various chemotherapy regimens. Although febrile neutropenia is a serious complication of chemotherapy, large numbers of patients would be required before a statistically significant effect of G-CSF on mortality could be detected.

G-CSF is currently used for the haematological support of cancer patients receiving chemotherapy, but it is not an anticancer treatment (37). The use of G-CSF allows patients to receive the intended dose of chemotherapy without reductions or an increased dose. However, it is unknown if this will result in a better outcome for the patient.

Balancing the Benefits and Costs of Colony-stimulating Factors: A Current Perspective

Neutropenia is the most common dose-limiting toxicity associated with cancer chemotherapy. Neutropenic complications, including febrile neutropenia, occur in a substantial proportion of patients with chemotherapy-induced neutropenia. Although neutropenia and its complications can be reduced with colony-stimulating factors (CSFs), it is not considered cost-effective to administer them prophylactically to all patients. The current American Society of Clinical Oncology guidelines recommend primary prophylaxis with the CSFs in patients receiving chemotherapy regimens that produce febrile neutropenia in 40% or more of those treated (38). This recommendation was supported by an early economic analysis based on a decision model incorporating hospital cost data to determine the risk threshold for the cost-saving use of the CSFs. Updating this model to include current estimates of hospitalization costs, indirect costs such as productivity losses, and out-of-pocket patient expenses reduced the risk threshold for cost savings with CSF to 18%. Recent efforts have focused on identifying individual patient characteristics that might be used to target prophylactic CSF in patients who are at greatest risk. Should such factors be identified and validated, assessing each patient's individual risk for neutropenic complications may prove to be a better strategy for the cost-effective use of the CSFs than the risk threshold approach. Preliminary results suggest that the targeted application of CSF support based on a risk model may improve the associated cost savings.

Clinical Trails Going on in G-CSF Therapy

New Treatment Options for Managing Chemotherapy-induced Neutropenia

There are so many benefits of using colony-stimulating factors in cancer patients including prevention of neutropenia and decreasing its severity and duration. However, several areas of controversy regarding chemotherapy-induced neutropenia and the use of colony-stimulating factors remain *e.g.* dose intensity, which colony-stimulating factor to use and when, and administration. Colony-stimulating factors have not been found to increase overall survival time in patients undergoing chemotherapy, but they are used to prevent febrile neutropenia, reduce hospitalizations, and improve quality of life. The methods of using colony-stimulating factors are also evolving, with altered administration showing potential as a means of reducing treatment cost. If the rate of febrile neutropenia for a given combination chemotherapy regimen is < 40%, routine prophylactic use of colony-stimulating factors is not considered cost-effective and is not recommended by the American Society of Clinical Oncology. Pegfilgrastim is filgrastim to which polyethylene glycol has been added to extend the drug's half-life and permit less frequent administration. The issue of dose intensity will continue to be evaluated as new cytotoxic agents and combinations are introduced (39). Better models for identifying patients most likely to benefit from colony-stimulating factors are needed.

Current Management of Chemotherapy-induced Neutropenia: The Role of Colony-stimulating Factors

Cytotoxic chemotherapy suppresses the hematopoietic system, impairing host protective mechanisms. One important consequence, chemotherapy-induced neutropenia (CIN), places patients at risk of developing fever and life-threatening infections. These complications have substantial economic impact and may severely affect the quality of life of patients undergoing treatment of cancer (40). Currently, CIN is managed by delaying and reducing chemotherapy treatment with hematopoietic growth factors and with intravenous antibiotic therapy. Reducing chemotherapy may compromise treatment outcomes in potentially curable malignancies, such as early stage breast cancer and non-Hodgkin's lymphoma. Randomized clinical trials have clearly shown that granulocyte colony-stimulating factor (filgrastim) and the longer-acting pegylated granulocyte colony-stimulating factor (pegfilgrastim), when administered 24 hours after chemotherapy before the occurrence of CIN, are effective in reducing the incidence and severity of neutropenia and its complications, including administration of full doses of chemotherapy without treatment delay.

Colony-stimulating Factors Especially G-CSF for the Management of Neutropenia in Cancer Patients

Neutropenia and its subsequent infectious complications represent the most common dose-limiting toxicity of cancer chemotherapy. Febrile neutropenia (FN) occurs with common chemotherapy regimens in 25 to 40% of treatment-naive patients, and its severity depends on the dose intensity of the chemotherapy regimen, the patient's prior history of either radiation therapy or use of cytotoxic treatment, and comorbidities(41). The occurrence of FN often causes subsequent chemotherapy delays or dose reductions. It may also lengthen hospital stay, increase monitoring, diagnostic and treatment costs, and reduce patient quality of life. A decade after their introduction, colony-stimulating factors (CSFs) such as granulocyte colony-stimulating factor (G-CSF) and granulocyte-macrophage colony-stimulating factor (GM-CSF) are now an integral part of the prevention of potentially life-threatening FN; however, only G-CSF has US Food and Drug Administration approval for use in chemotherapy-induced neutropenia. These adjunctive agents accelerate formation of neutrophils from committed progenitors, thereby reducing the duration and severity of neutropenia. Important uses of CSFs in oncology are prevention of FN after chemotherapy, treatment of febrile neutropenic episodes and support following bone marrow transplantation, and collection of CSF-mobilised peripheral blood progenitor cells. G-CSF is used more frequently than GM-CSF for all of these indications because of fewer associated adverse effects. However, they have demonstrated clinical utility in allowing the delivery of planned chemotherapy dose on schedule, an important clinical goal especially in curative tumour settings. The high cost of these agents limits their widespread use. Current American Society of Clinical Oncology guidelines recommend primary prophylaxis, or first cycle use, with CSFs being confined to patients with > or = 40% risk of FN, which may include elderly patients and other high-risk patients. In addition to the risk of FN, primary prophylaxis should also be considered if the patient has risk factors that place them in the Special Circumstances category. (42) These risk factors may include decreased immune function in patients who are already at an increased risk of infection and pre-existing neutropenia due to disease, extensive prior chemotherapy, or previous irradiation to the pelvis or other areas containing large amounts of bone marrow. Future studies are needed to better define the patients most likely to benefit from CSF therapy, both for prophylaxis and as an adjunct to antibiotics for treatment of FN. Other potential uses include combination therapy with stem cell factors and other cytokines to boost progenitor cell development, maintaining dose intensity of salvage therapy in metastatic cancer patients, and application in patients with pneumonia, diabetic foot infections etc, and a variety of other infectious conditions.

Once-per-cycle Pegfilgrastim (Neulasta) for the Management of Chemotherapy-induced Neutropenia

Filgrastim (r-metHuG-CSF) was approved in the United States in 1991 for use in decreasing the incidence of infection, as manifested by febrile neutropenia, in patients with nonmyeloid malignancies treated with myelosuppressive chemotherapy (43). Colony-stimulating factors such as filgrastim are a significant advance in the supportive care of patients with cancer. However, because of its short half-life, filgrastim requires daily dosing by injection to maintain its effects on the bone marrow. Pegfilgrastim is a longer-acting, self-regulating form of filgrastim created by the covalent linkage of a 20-kd polyethylene glycol molecule to the N-terminal of the filgrastim molecule. The molecular characteristics of pegfilgrastim result in a longer terminal half-life, making once-per-chemotherapy-cycle administration possible. Pegfilgrastim has been found to be 11 times more efficient in chemotherapy as per the following observations.The results from two randomized double-blind phase III clinical trials in patients with breast cancer treated with myelosuppressive chemotherapy showed that a single dose of pegfilgrastim provides neutrophil support comparable with that provided by an average of 11 daily injections of filgrastim. Pegfilgrastim has also been shown to be comparable to filgrastim in reducing neutropenic complications in patients treated with chemotherapy for lymphoma. The safety profile of pegfilgrastim is comparable to that of filgrastim in the clinical settings studied to date. The once-per-cycle administration of pegfilgrastim may improve patient quality of life because it is less disruptive to patients and caregivers, and increase adherence because no doses are missed, thus further advancing the management of chemotherapy-induced neutropenia and its consequences. (44)

Effect of Recombinant Human Granulocyte Colony-stimulating Factor (rhG-CSF) on Chemotherapy-induced Neutropenia in Patients with Lung Cancer

An attempt to evaluate rhG-CSF for preventing and reducing the period of chemotherapy-induced neutropenia, phase II studies of the agent, KRN 8601, have also been conducted in patients receiving chemotherapy for lung cancer. In a cooperative study, 53 patients with lung cancer were fully evaluated. The chemotherapy regimen for the patients enrolled in the study was not specified, but an identical regimen with an identical dose and schedule was mandatory for the first cycle in which the patients did not receive the rG-CSF, and for the following cycles in which they received it after completion of chemotherapy for 14 days consecutively (45). Patients were allocated to receive the agents either at a dose of 100, 200 or 400 micrograms/m2 via intravenous drip infusion; or at as ub cutaneous dose of 25, 75, or 125 micrograms. In the study, all the 9 patients with non-small cell lung cancer received a 3-drug combination of vindesine, ifosfamide, and cisplatin(VIP) at an identical dose throughout the cycles. The rG-CSF was administered on the second and the following cycles at a dose of 100 micrograms/m2, subcutaneously, in the same manner as the above. Myelogram and neutrophil functions, *i.e.*, superoxide anion production, chemotactic, and phagocytic activity, were serially determined in these patients. With intravenous dose of 100 micrograms/m2, the rG-CSF considerably elevated the nadir count of neutrophils and significantly reduced the duration of neutropenia. Subcutaneously administered rhG-CSF at 75 micrograms doses did as with intravenous infusion. The optimal dose of the agent in conventional chemotherapy was estimated to be 100 micrograms/m2 when infused intravenously, and 75-125 micrograms subcutaneously. Subcutaneously administered rG-CSF at a dose of 100 micrograms/m2 did not contribute to spare the nadir count of neutrophils, but contributed toward reducing the period of neutropenia induced by the 3-drug combination which was much more myelosuppressive than conventional regimens. Thus, the optimal dose of the agent should be determined according to the dose-intensity of chemotherapy. Peripheral neutrophils obtained after recovery from VIP-induced neutropenia showed a normal activity in superoxide anion production and mobility when combined with rG-CSF, although the activity showed a trend to remain subnormal in the recovered neutrophils without rG-CSF(46). In conclusion, rG-CSF considerably reduces the neutropenia and possibly reduces infections caused by intensive chemotherapy. Hereafter, clinical trials must determine whether rG-CSF improve the therapeutic outcomes of patients receiving chemotherapy in terms of response rate and patient survival.(47)

Colony-stimulating Factors as an Adjunct to Chemotherapy in Small Cell Lung Cancer

Myelo suppression is the major dose-limiting toxicity of chemotherapy in small cell lung cancer (SCLC). The capacity of colony stimulating factors (CSFs) to stimulate granular neutrophil recovery may be of great value to prevent or cure febrile neutropenia and to increase dose-intensity. Primary CSF administration has been shown to reduce the incidence of febrile neutropenia, hospital admission rate, and antibiotic use subsequent to cyclophosphamidedoxorubicin-high dose etoposide (CDE) chemotherapy (48), without improvement of survival or disease control. Primary CSF administration may be recommended when the expected incidence of febrile neutropenia is at least 40%. This benefit has not been established with less myelosuppressive regimens, such as cisplatin-etoposide (PE), which remains an alternative combination of SCLC when standard doses are used. A trial comparing high-dose CDE + CSF with PE would be of considerable interest. There is currently little clinical basis for the use of CSFs to increase chemotherapy dose-intensity, outside clinical trials. Peripheral blood progenitor cells mobilized with CSFs offer interesting prospects. Further studies, with later initiation, shorter duration or lower doses of CSFs, are warranted to improve the cost-effectiveness of CSFs. CSF therapy in addition to antibiotics is normally not justified in febrile neutropenia, except perhaps in selected patients with sepsis syndromes, hypotension or pneumonia. (49)

Evidence-based Use of Colony-stimulating Factors in Elderly Cancer Patients

Neutropenia and its complications represent the major dose-limiting toxicity of cancer chemotherapy, especially in the elderly. Hematopoietic growth factors have been shown to reduce the severity and duration of febrile neutropenia (FN) and to sustain chemotherapy dose intensity. A systematic review was undertaken of studies of the relationship between age and the risk of neutropenia and its complications. The risk of FN associated with standard regimens increases with age and appears to be greatest during the first cycle of chemotherapy (50). FN continues to have a considerable clinical, economic, and quality-of-life impact on affected individuals. The risk of mortality associated with hospitalization with FN also increases with age but is largely associated with the higher rate of comorbidities observed in the elderly population. Despite increasing evidence that elderly patients experience similar benefit from cancer chemotherapy, reductions in dose intensity often compromise response rates and long-term survival. The hematopoietic growth factors reduce the risk of neutropenic events and the need for reduced dose intensity in elderly cancer patients. Primary prophylaxis with colony-stimulating factors (CSFs) reduces the risk of FN and its complications in elderly patients receiving moderately intensive systemic chemotherapy for responsive malignancies. CSFs also appear to reduce cost and improve quality of life in selected elderly patients receiving chemotherapy.

Recombinant Human Granulocyte-macrophage Colony-stimulating Factor After Combined Chemotherapy in High-grade Non-Hodgkin's Lymphoma: A Randomized Pilot Study

High-grade non-Hodgkin's lymphomas (NHL) can potentially be cured with combination chemotherapy, although the optimum schedules still have to be defined. Clinical trials with intensive chemotherapy are predominantly limited by myelosuppression(51). Here, haematopoetic growth factors open up the possibility of reducing chemotherapy-associated toxicities. In a randomised pilot study, the effects of a recombinant human granulocyte-macrophage colony-stimulating factor (rhGM-CSF) following combined chemotherapy with vincristine, doxorubicin, cyclophosphamide, prednisone and etoposide (VACPE) was investigated. A total of 35 patients with high-grade NHLs were randomised to receive either rhGM-CSF or placebo during the first two chemotherapy cycles and rhGM-CSF for all following cycles. rhGM-CSF was administered at a dosage of 5 micrograms/kg for 10 days or until neutrophils were > 1/nl following chemotherapy. The analyses revealed a significant reduction of neutropenia and duration of neutropenia in the rhGM-CSF group. Adverse events were rare and generally mild apart from one anaphylactoid reaction. No effects of rhGM-CSF were observed concerning the platelet nadir or duration of thrombocytopenia. The benefit of rhGM-CSF for response induction and survival via rhGM-CSF-supported dose intensification remains to be determined.(52)

Effect of Granulocyte-macrophage Colony-stimulating Factor on Neutropenia and Related Morbidity Induced by Myelotoxic Chemotherapy

A phase Ib/II clinical study was undertaken to assess the efficacy of recombinant human granulocyte-macrophage colony-stimulating factor (GM-CSF) to attenuate neutropenia and associated morbidity caused by high-dose anticancer chemotherapy administered in the presence or absence of autologous bone marrow support (53). Around 22 patients were treated with various solid tumors and lymphoid neoplasias with a single daily subcutaneous dose of GM-CSF (250 micrograms/m2)

48 h after a second cycle of highly myelotoxic chemotherapy for a period of 10 days and compared intraindividually neutropenia-related clinical and laboratory variables with data obtained from the same patients having previously received a first neutropenia-inducing cycle of identical chemotherapy in the absence of GM-CSF. The clinical trial showed that GM-CSF is active in neutropenic patients by significantly increasing the neutrophil nadir, reducing the time of relevant neutropenia, and reducing the duration of the patient's hospital stay and necessity for parenteral antibiotics. No significant toxicity was encountered with subcutaneous GM-CSF treatment. (54)

Evidence-based Use of Colony-stimulating Factors in Elderly Cancer Patients

Neutropenia and its complications represent the major dose-limiting toxicity of cancer chemotherapy, especially in the elderly. Hematopoietic growth factors have been shown to reduce the severity and duration of febrile neutropenia (FN) and to sustain chemotherapy dose intensity. A systematic review was undertaken of studies of the relationship between age and the risk of neutropenia and its complications. The risk of FN associated with standard regimens increases with age and appears to be greatest during the first cycle of chemotherapy (50). FN continues to have a considerable clinical, economic, and quality-of-life impact on affected individuals. The risk of mortality associated with hospitalization with FN also increases with age but is largely associated with the higher rate of comorbidities observed in the elderly population. Despite increasing evidence that elderly patients experience similar benefit from cancer chemotherapy, reductions in dose intensity often compromise response rates and long-term survival. The hematopoietic growth factors reduce the risk of neutropenic events and the need for reduced dose intensity in elderly cancer patients(55). Primary prophylaxis with colony-stimulating factors (CSFs) reduces the risk of FN and its complications in elderly patients receiving moderately intensive systemic chemotherapy for responsive malignancies. CSFs also appear to reduce cost and improve quality of life in selected elderly patients receiving chemotherapy.

Conclusion

There are preliminary data suggesting that G-CSF will be useful in the following conditions:

1. Neutropenia due to infections (such as occurs in viral infections)
2. Neutropenia due to non-chemotherapeutic myelosuppressive treatments (such as gancyclovir)
3. Drug-induced agranulocytosis.
4. Primary marrow failure (such as aplastic anaemia)

It will be interesting to see if a role for G-CSF can be demonstrated in non-neutropenic states, such as burns or diabetes (where there are adequate numbers of neutrophils but they are functionally defective), or in the management of infections in patients with normal neutrophil counts.

Growth factors and cytokines are potential therapeutic agents of their key roles of regulating the production maturation, activity of the blood; muscles and bone cells.A number of Colony Stimulating Factor including GCSF are currently employed to treatment of blood deficiencies and cancers. Now we are trying for high yield and cost effective production of rh-GCSF by modern innovative method, which will be beneficial for Indian population.

Acknowledgements

I am very much thankful to Prasantada, Mr. Akash Verma, Mr. Snehasis Jana, Mr. Rajib. Mr. Swapan Patra, Miss Usha, Miss Rupali, Miss Nidhi, Mr. Santosh, Mrs. Sipra for their noble encouragement and help to write this article.

REFERENCES

1. http://www.umdnj.edu/pathnweb/genpath/lec_1/Features_Philosophy/Features/features.htm
2. Metcalf, D. (1985) The granulocyte-macrophage colony stimulating factors. **Science**, 229, 16-22.
3. Wingfield.P., Benedict.R., Turcatti.G., Allet. B., Mermod. J.J., Delamarter. J., Simona. M.G., and Rose, (1988) K.Characterization of recombinant derived granulocyte colony stimulating factor (G-CSF**). Biochem. J**.(256), 213-218.
4. Soo-Hyung Kang, Kyu-Heum Na, Jang-Hyeon Park, Choong-II Park, Se-Yong Lee and Young-Ik Lee (1995) High level expression and simple purification of recombinant human granulocyte colony stimulating factor in E.coli. **Biotechnology letters**, (17) 687-692.
5. Johnston EM, Crawford J. (1998) Hematopoietic growth factors in the reduction of chemotherapeutic toxicity.**Semin Oncology**.
6. Stanley.E.R., Jubinsky.P.T. (1984) Factors affecting the growth and differentiation of haemopoietic cells in culture. **Clin. Haematol**. (13).
7. Souza, L.M., Boone, T.C., Gabrilove, J., Lai, P.H., Zsebo, K.M., Hurdock, D.C., Chazin, V.R., Bruszewski, J.,Moore, M.A.S., Mertelsmann, R., and Welte, K. (1986)Recombinant human granulocyte colony stimulating factor: effects on normal and leukemic myeloid cells. **Science**, 32, 61-65.
8. Bong Hyun Chung, Mi-Jin Sohn, Su-Wanoh, Ui-Sun Park, Haryoung Poo, Beom Soo Kim, Mi Jin Yu and Young Ik Lee(1998) Overproduction of Human Granulocyte colony stimulating factor fused to the PelB signal peptide in E.coli. **Journal of Fermentation and Bioengineering.** (85) 443-446.
9. Lieschk*e.g*.J, Burgess. A.W. (1992) Granulocyte colony-stimulating factor and granulocyte-macrophage colony-stimulating factor (1 and 2). **N Engl J Med.**(327) 28-35,99-106.
10. Chouaid.C., Bassinet.L, Fuhrman.C., Monnet.I., Housset.B. (1998) Routine use of granulocyte colony-stimulating factor is not cost-effective and does not increase patient comfort in the treatment of small-cell lung cancer: an analysis using a Markov model, **J Clin Oncol.** 16 (8). 2700-7.
11. Aritomi. M., Kunishima. N., Okamoto. T., Kuroki. R., Ota. Y., Morikawa. K. (1999) Atomic structure of the GCSF-receptor complex showing a new cytokine-receptor recognition scheme, **Nature.** (6754).713-7.
12. Babior.B.M, Stossel. T.P. (1994) **Hematology,A pathological Approach** (3rd Ed).279-281.
13. Hill.C.P, Osslund. T.D., Eisenberg.D. (1993) The structure of granulocyte-colony-stimulating factor and its relationship to other growth factors. **Proc Natl Acad Sci U S A**. 90(11). 5167-71
14. Barney.Bishop., Debbie.C.,Koay, Alan. C., Sartorelli., Lynne. Regan. (2001). Reengineering Granulocyte Colony-stimulating Factor for Enhanced Stability, (276) 33465-33470
15. David. N. Brems. (2002) The kinetics of G-CSF folding, **Protein Science.** (11) 2504-2511.
16. Layton.J.E, Shimamoto.G., Osslund.T., Hammacher.A., Smith.D.K., Treutlein.H.R., Boone.T. (1999) Interaction of granulocyte colony-stimulating factor (G-CSF) with its receptor. Evidence that Glu19 of G-CSF interacts with Arg288 of the receptor, **J Biol Chem.** 274(25). 17445-51

17. Layton. J.E., Hall. N.E., Connell.F., Venhorst.J., Treutlein.H.R. (2001)Identification of ligand-binding site III on the immunoglobulin-like domain of the granulocyte colony-stimulating factor receptor, **J Biol Chem**. 276 (39). 36779-87

18. Dong. F., Qiu. Y., Yi. T., Touw. I.P., Larner. A.C. (2001) The carboxyl terminus of the granulocyte colony-stimulating factor receptor, truncated in patients with severe congenital neutropenia/ acute myeloid leukemia, is required for SH2-containing phosphatase-1 suppression of Stat activation, **J Immunol.** (11).6447-52

19. Kenneth. Kaushansky., Christopher. B.Brown., Stephen. Petersdorf., Hematopoetic colony stimulating factor, 381-382.

20. Kanda, N., Fukushige, S., Murotsu, T., Yoshida, M. C., Tsuchiya, M., Asano, S., Kaziro, Y., Nagata, S. (1987) Human gene coding for granulocyte-colony stimulating factor assigned to the q21-q22 region of chromosome. **Somat. Cell Molec. Genet**. (13) 679-684.

21. Nagata, S., Tsuchiya, M., Asano, S., Yamamoto, O., Hirata, Y., Kubota, N., Oheda, M., Nomura, H.,Yamazaki, T. (1986)The chromosomal gene structure and two mRNAs for human granulocyte colony-stimulating factor. **EMBO J**. (5) 575-581.

22. Fallah. M. J,Akbari., B.Saeedinia., A.R. Karimi.M., *et al.* (2003) Overexpression of Recombinant Human Granulocyte Colony-Stimulating Factor in E. coli,Vol 28,No.3

23. Nagata, S., Tsuchiya, M., Asano, S., Kaziro, Y.,Yamazaki, T., Yamamoto, O., Hirata, Y., Kubota, N., Oheda, M., Nomura, H., Ono, M. (1986) Molecular cloning and expression of cDNA for the human granulocyte colony-stimulating factor. **Nature** (319) 415-418.

24. Soo-Hyung Kang., Kyu-Heum Na., Jang-Hyeon Park., Choong-II Park., Se-Yong Lee and Young-Ik Lee. (1995) High Level Expression and Simple Purification of Recombinant Human Granulocyte Colony-Stimulating Factor in *E.coli*.**Biotechnology letters**,(17).687-692.

25. Bong. Hyun. Chung., Mi-Jin. Sohn., Su-Wan Oh., *et al.* (1998) Overproduction of Human Granulocyte Colony Stimulating Factor Fused to the Pelb Signal Peptide in *Escherichia coli*,**Journal of Fermentation and Bioenginnering**, (85).443-446.

26. Tweardy, D. J., Cannizzaro, L. A., Palumbo, A. P., Shane, S.; Huebner, K.,Vantuinen, P., Ledbetter, D. H., Finan, J. B., Nowell, P. C., Rovera, G. (1987) Molecular cloning and characterization of a cDNA for human granulocyte colony-stimulating factor (G-CSF) from a glioblastoma multiforme cell line and localization of the G-CSF gene to chromosome band 17q21. **Oncogene Res.** (1) 209-220.

27. Metcalf, D. (1985) The granulocyte-macrophage colony-stimulating factors. **Science** (229)16-22.

28. Lieschke G.J., Burgess.A.W. (1992) Granulocyte colony-stimulating factor and granulocyte-macrophage colony-stimulating factor (1 and 2). **N Engl J Med** (327) 28-35,99-106.

29. Hill.C.P., Osslund.T.D., Eisenberg.D. (1993) The structure of granulocyte-colony-stimulating factor and its relationship to other growth factors, **Proc Natl Acad Sci U S A.** (11).5167-71.

30. Basser. R.L., Underhill.C., Davis.I., Green. M.D., Cebon.J., Zalcberg. J., MacMillan. J., Cohen.B., Marty.J, Fox. R.M., Begley.C. G. (2000) Enhancement of platelet recovery after myelosuppressive chemotherapy by recombinant human megakaryocyte growth and development factor in patients with advanced cancer. **J Clin Oncol.** (15). 2852-61

31. Frampton. J.E., Yarker. Y.E., Goa.K.L., Lenograstim. (1995) A review of its pharmacological properties and therapeutic efficacy in neutropenia and related clinical settings. **Drugs.** 49(5). 767-93.

32. Gemmati. D., Ongaro. A., Scapoli. G.L., Della. Porta. M., Tognazzo. S., Serino. M.L., Di. Bona.E., Rodeghiero. F., Gilli. G., Reverberi. R., Caruso. A., Pasello.M., Pellati. A., De. Mattei. M. (2004).Common gene polymorphisms in the metabolic folate and methylation pathway and the risk of acute lymphoblastic leukemia and non-Hodgkin's lymphoma in adults. **Cancer Epidemiol Biomarkers Prev**. 13(5).787-94.

33. Cottle. T.E., Fier.C.J., Donadieu. J., Kinsey. S.E. (2002) Risk and benefit of treatment of severe chronic neutropenia with granulocyte colony-stimulating factor. **Semin Hematol**.39(2).134-40

34. Trillet-Lenoir. V., Green. J., Manegold.C., Von Pawel. J., Gatzemeier.U., Lebeau.B., *et al.* (1993) Recombinant granulocyte colony stimulating factor reduces the infectious complications of cytotoxic chemotherapy. **Eur J Cancer** (29A) 319-24.

35. Graham. J. (1994) Lieschke Granulocyte colony stimulating factor (G-CSF) (**Aust Prescr**. (17)96-9.

36. Dale. D.C, Bonilla. M.A., Davis. M.W., Nakanishi. A.M., Hammond. W.P., Kurtzberg. J., *et al.* (1993) A randomized controlled phase III trial of recombinant human granulocyte colony-stimulating factor (filgrastim) for treatment of severe chronic neutropenia. **Blood** (81) 2496-502.

37. Metcalf. D., Morstyn. G. (1991) Colony-stimulating factors. **Biologic therapy of cancer.** 417-44.

38. Lyman.G.H. (2003) Balancing the benefits and costs of colony-stimulating factors: a current perspective, **Semin Oncol**. (30). 10-17.

39. Mucenski.J.W., Shogan. J.E. (2003) Maximizing the outcomes in cancer patients receiving chemotherapy through optimal use of colony-stimulating factor, **J Manag Care Pharm**. (2).10-14.

40. Dale.D. (2003) Current management of chemotherapy-induced neutropenia: the role of colony-stimulating factors, **Semin Oncol**. 3-9.

41. Dale. D.C. (2002) Colony-stimulating factors for the management of neutropenia in cancer patients, **Drugs.** (62) 1-15.

42. Sheridan.W.P., Morstyn. G., Wolf. M., Dodds. A., Lusk. J., Maher. D., *et al.* (1989)Granulocyte colony-stimulating factor and neutrophil recovery after high-dose chemotherapy and autologous bone marrow transplantation. **Lancet** (2)891-95.

43. Crawford.J. (2003) Once-per-cycle pegfilgrastim (Neulasta) for the management of chemotherapy-induced neutropenia. **Semin Oncol**. 24-30.

44. Gisselbrecht. C., Prentice. H.G., Bacigalupo. A., Biron. P., Milpied. N., Rubie. H., *et al.* (1994) Placebo-controlled phase III trial of lenograstim in bone-marrow transplantation. **Lancet.** (343) 696-700

45. Faulds.D., Lewis.N.J., Milne.R.J. (1992)Recombinant granulocyte colony-stimulating factor(rG-CSF): pharmacoeconomic considerations in chemothempy-induced neutropenia. **PharmacoEconomics**(1).231-49.

46. Ohnoshi.T., Ueoka. H., Kodani. T., Kamei. H., Kimura. I. (1990) Effect of recombinant human granulocyte colony-stimulating factor (rG-CSF) on chemotherapy-induced neutropenia in patients with lung cancer, **Gan To Kagaku Ryoho**. (17).911-18.

47. Crawford. J., Ozer. H., Stoller. R., Johnson. D., Lyman. G., Tabbara. I., *et al.* (1991) Reduction by granulocyte colony-stimulating factor of fever and neutropenia induced by chemotherapy in patients with small-cell lung cancer. **N Engl J Med**, (325)164-70.

48. Urban.T., Schuller. M.P., Lebeau. B. (1996) Colony-stimulating factors as an adjunct to chemotherapy in small cell lung cancer, **Eur Respir J**. (9),596-602.

49. Sheridan. W.P., Morstyn.G., Wolf. M., Dodds.A., Lusk.J., Maher.D., *et al.* (1989) Granulocyte colony-stimulating factor and neutrophil recovery after high-dose chemotherapy and autologous bone marrow transplantation. **Lancet** (2) 891-5.

50. Lyman.G.H., Kuderer. N., Agboola. O., Balducci. L. (2003) Evidence-based use of colony-stimulating factors in elderly cancer patients, **Cancer Control**. (10).487-99.

51. Bettelheim. P., Valent. P., Andreeff. M., Tafuri. A., Haimi. J., Gorischek. C., Muhm. M., Sillaber. C., Haas. O., L.Vieder., **Recombinant human granulocyte-macrophage colony-stimulating factor after combined chemotherapy in high-grade non-Hodgkin's lymphoma-a randomized pilot study**. (77).700-711.

52. Pettengell. R., Gurney. H., Radford. J.A., Deakin. D.P., James. R.,Wilkinson. P.M., *et al.* (1992) Granulocyte colony-stimulating factor to prevent dose-limiting neutropenia in non-Hodgkin's lymphoma: a randomized controlled trial. **Blood**.(80)1430-36.

53. Guerci. A., Lederlin. P., Reyes. F., Bordessoule. D., Sebban. C., Tilly. H., Kerneis. Y., Biron. P., Gisselbrecht. C., Herbrecht. R., Coiffier. B. (1996) Effect of granulocyte colony-stimulating factor administration in elderly patients with aggressive non-Hodgkin's lymphoma treated with a pirarubicin-combination chemotherapy regimen. Groupe d'Etudes des Lymphomes de l'Adulte, **Ann Oncol**. (9).966-9.

54. Trillet-Lenoir. V., Green. J., Manegold. C., Von Pawel. J., Gatzemeier. U., Lebeau. B., *et al.* (1993) Recombinant granulocyte colony stimulating factor reduces the infectious complications of cytotoxic chemotherapy. **Eur J Cancer**. (29A)319-24.

55. Edgardo.Rivera.,M. HaimErder.,Moshe Fridman.,Debra Frye.,andGabriel N. Hortobagyi. (2003) First-cycle absolute neutrophil count can be used to improve chemotherapy-dose delivery and reduce the risk of febrile neutropenia in patients receiving adjuvant therapy: a validation study. **Breast Cancer Res.** (5). 114–120

FIGURE REFERENCES

Fig. 2.2) http://academic.evergreen.edu/curricular/viruses/

Fig 2.4) http://www.biochem.szote.u-szeged.hu/astrojan/protein1.htm

Fig. 2.5) www.proteinscience.org/content/vol11/issue10/image/large/

Fig. 2.6) http://www.sigmaaldrich.com/Area_of_Interest/Life_Science/Cell_Signaling Scientific_Resources/Pathway_Slides_Charts/Cytokines_Growth_Factors_and_Hormornes.html

Advances in Biochemistry and Biotechnology Volume 1 *Pages 81-94*
Edited by: **Dr. Chiranjib Chakraborty**
Published by: **DAYA PUBLISHING HOUSE**

3

Chemical Analysis of Leaves of *Cestrum diurnum* (L.) (Solanales: Solanaceae) and Seasonal Variation of Some Primary and Secondary Biochemicals

*A. Ghosh & G. Chandra**
Mosquito Research Unit, Department of Zoology, Burdwan University, West Bengal, India

ABSTRACT

The present investigation comprises biochemical analysis of the common shrub *Cestrum diurnum* (Solanales: Solanaceae) leaves at three different physiological stages (young, mature and senescent respectively). Change in the chemical status of some primary biochemicals (such

* Correspondent Author

as total carbohydrate, starch, lipid and protein) and secondary biochemical (phenolics) have been determined to estimate their levels in relation to seasonal growth of the plant. Presence of some secondary biochemicals such as steroids and alkaloids has been ascertained through TLC analysis. The mature leaves of *Cestrum diurnum* were extracted with Hexane-Ether (1:1 in v/v) solvent in Soxhlet apparatus for the isolation of essential oil. The methyl esters of the oil were identified through TLC and analysed by capillary GLC (HP 6890 PLUS GLC AND HP3398 A GC). The main constituents were Palmitic acid (27.62%), Stearic acid (4.62%) and Oleic acid (3.06%). The relative concentrations of some major unidentified fatty acids were also calculated (16.51, 8.16 and 7.69%) according to the retention period in capillary column.

Introduction

Medicinal plants are significant source of synthetic and herbal drugs. The components of the plant that support the disease-preventive and health regulatory roles are termed as functional food ingredients or phytochemicals. For at least five thousand years, humankind has relied on natural products as the primary source of medicine. The drugs were used in crude forms like expressed juice, powder decoction or infusion without knowledge in detailed scientific parameters. Now, at the start of a new millennium it is estimated by the World Health Organisation that 80% of the world's inhabitants must rely on traditional medicines for health care; these traditional medicines are primarily plant based. It is estimated that 74% of the 119 currently most important drugs contain active ingredients from plants used in traditional medicine. Exploring traditional herbal medicines in the context of modern science is the need for optimum and proper utilization of traditional plant drugs. Due to the remarkable development of analytical methodology; *e.g.*, Gas chromatography (GC), High performance liquid chromatography (HPLC) and Mass spectroscopy (MS), it has been possible to characterize even minor molecular weight based phytochemicals.

The genus *Cestrum* (Solanaceae), with more than 300 species, is globally distributed. *Cestrum diurnum* (Solanaceae: Solanales) is a single or multi-stemmed shrub that is also known as Day Jassmine. The bark of the plant is nearly smooth and grey whereas the inner bark is yellow-brown and slightly bitter. The twigs are slender, greenish gray in colour and sometimes drooping. The leaves are oblong, shiny green to yellow green in colour and membranous (15 × 6.5 cm and have petioles about 2.5 cm long). The flowers are small, whitish and are grouped in panicles attached at the bases of the last leaves in each branch. The unripe berries are greenish in colour and develop into elliptical purplish–black when ripe, measuring about 6.5 mm in length (1).

There are several applications of *Cestrum diurnum* in folk medicine but the toxicity of the species to humans and livestock has been reported(2). The leaves contain a calcinogenic glycoside called 1,25–dihydroxyoholocalciferol that leads to a Vitamin D toxicity that results in elevated serum calcium and deposition of calcium in soft tissues. In human the administration may results nausea, vomiting and hallucination effects although the antimicrobial effects of the essential oil of *C.diurnum* have also been established(2).

Quality control of the plant raw materials is the most important challenge in bringing any of the traditional medicines or modern phytomedicines to the acceptance of concerned people. The major problem in quality control is the variation in the quality and quantity of active compounds due to the existence of ecotype pharmacological variations, mainly because of soil condition, nutritional status climatic condition, and seasonal variation. Therefore, pharmaecological evaluation is necessary before therapeutic use and it may bring to light possible development of new pharmacological properties other than the main therapeutic actions under certain ecological or physiological conditions. The objective of the present study was to evaluate the seasonal changes of some important primary and secondary biochemicals of the medicinal plant *Cestrum diurnum* (L.) in different physiological stages and the preparation of a detailed chemical characterization by the application of modern scientific procedures.

Materials and Methods

Collection of Leaves

Random collections of the young, (<1 week old), mature (1-6 weeks old) and senescent (7-9 weeks old) *Cestrum diurnum* (L.) leaves growing in different fields situated at the outskirt of Burdwan (87 °54′ N, 23°16′ E) were done thrice a month for twelve months from January to December 2003. Leaves were initially cleaned in the laboratory with distilled water and soaked with paper towels before each biochemical assay.

Preparation of Leaf Extracts

One gram each of the young, mature and senescent leaves, cut into pieces (0.5–1 cm), was separately plunged in boiling ethanol and ground for 10 min. Leaf extracts were centrifuged at 2000 rpm for 20 min, and the supernatant and pellet were separated. To remove leaf pigment, each supernatant was passed through charcoal powder and filtered through Whatman No-41 (Maidstone,UK) filter paper.

Estimation of Primary and Secondary Biochemicals

Primary Biochzemicals

Estimation of Total Soluble Carbohydrate

Sum of soluble carbohydrate was measured by modified Anthrone method provided by Dubios *et al.* (3) with some modifications. To estimate the total soluble carbohydrates, 1 ml of the clear supernatant plant extract was collected and treated with 4 ml of Anthrone reagent. The carbohydrate content was measured by the preparation of standard curve against a reagent blank of D–glucose in the UV–visible spectrophotometer read at 630 nm.

Estimation of Total Protein

Total protein was measured following the method of Lowry *et al.*, .(4).To estimate the sum of soluble proteins, each kind of pellets separated earlier was suspended in a suitable volume of 5% TCA solution at 0-5 °C in an ice bath for 10 min. Aliquotes (1 ml each) were taken in a centrifuge tubes and 1 ml of 10% TCA was added to each such tube and centrifuged at 5000 rpm for 45 min. The supernatants were discarded and the pellets were re-extracted once with absolute ethanol and twice with hot ethanol–ether mixture, every time discarding the supernatant. The pellet was treated with alkaline copper reagent (alkaline sodium carbonate: copper sulphate–sodium potassium tartarate solution (50: 1 in v/v) and the level of total proteins was measured against a reagent blank of bovine serum albumins by the addition of Folin Ciocaltu reagent in the UV-Visible spectrophotometer read at 750 nm.

Estimation of Total lipid

The lipid content of the leaves of *C.diurnum* was estimated by the procedure of Folch *et al.*, .(5). To determine lipids, 1 g each of the three kinds of fresh weed leaves was separately homogenized in 20 ml chloroform: methanol (CHCl3: MeOH) (1:2 v/v) mixture for 10 min in a cell homogenizer. The extracts were filtered after vigorous shaking and each residue was mixed with 25 ml CHCl3: MeOH mixture and stirred for 30 min. To remove any chlorophyll, the mixture was filtered through activated charcoal. The combined filtrate was shaken with 0.9% aqueous sodium chloride solution, and non-lipid contaminants were eliminated. The lipids thus extracted were dried in a desiccator and weighed in an electrobalance (± 0.01mg)

Estimation of Starch

The amount of starch was measured by modified Anthrone method with necessary modification. To estimate the starch components, 0.5 g of the sample was homogenized in hot 80% ethanol to remove the soluble sugars. The residue was homogenized and centrifuged after washing with hot 80% ethanol (repeated five times) till the washing did not give colour with Anthrone reagent. The residue was dried in a water bath and 5.0 ml of distilled water and 6.5 ml of 52% perchloric acid was added to it. The extraction procedure was continued using fresh perchloric acid until 100 ml of the supernatant was collected. To 1 ml of the supernatant (0.2 ml supernatant + 0.8 ml distilled water solution) 4 ml of the Anthrone reagent was added and the amount of starch was measured in the UV-Visible spectrophotometer read at 630nm.

Estimation of Secondary Biochemicals

Estimation of Phenolics

To determine the level of phenolics the process of Bray and Thrope (1954)(6) was adopted. The level of phenols (hydrolysable and condensed tannins) in the leaves was determined from each of the leftovers in the carbohydrate estimations. The extract was taken in a test tube, and to that 1 ml Folin-Ciocalteu reagent and 2 ml of 20% sodium-carbonate solution were added. The solution was diluted up to 25 ml with water and phenol levels were estimated following standard method against a known level of Pyrocatechol.

Detection of Alkaloid and Steroid

Five grams of each of these weed leaves were homogenized in 100 ml MeOH: H2O (4:1) and filtered. The filtrate was evaporated (<40°C) to 1/10th of its original volume in a Rotary evaporator and acidified with 2M H2SO4. It was further extracted with chloroform (×3). The chloroform and aqueous extracts were analyzed as follows: the chloroform of the chloroform extract was partly evaporated, and the material was subjected to TLC in Silica gel and developed with CHCl3: MeOH (1:1) mixture. To detect any steroid present in the leaves, the Liebermann–Burchard reagent was sprayed on the plate. Further more, the aqueous acid extract was basified to pH 10 with NH4OH and extracted twice with CHCl3: MeOH (3:1). The extract was reduced in volume by air-drying, subjected to TLC in silicagel and developed with CHCl3: MeOH (1:1). Dragendoff reagent was sprayed on the plate and any alkaloid present was observed.

Estimation of Ash and Moisture Contents

Known weights of young mature and senescent leaves were placed separately in a hot-air oven at 50 ± 0.5 C for 72 h, and materials that showed constant dry weights were removed from the oven and weighed. The difference between fresh and dry weights gave moisture contents of each kind of leaves. A portion (140-150 mg/replicate) of each kind of oven-dried leaves was separately ashed in a muffle furnace at 450± 10°C for 20 min. The ash thus produced in each crucible was soaked with 2 ml distilled water and placed in a watch glass for drying at 110°C for 2 h. The respective ash samples were weighed. The percentages of ash and ash-free weight of each sample were determined.

Isolation of Essential Oil

Preparation of Hexane Extract

20 g of freshly colleted mature leaves of *Cestrum diurnum* were plunged in 1 litre of Hexane in Soxhlet apparatus. The temperature during hot extraction was maintained at 40°– 50°C and continued for 72 hours. The leaf extract thus produced was collected and filtered through Whatman No.41 filter paper. The filtrate was evaporated and the essential oil was kept in a desiccator for further study.

Preparation of Methyl Ester for GLC Study

The dried sample was treated with chlorform: perchloric acid mixture (96: 5 v/v) and the solution was placed on a water bath at 50-60°C for 10 min. After cooling, the solution was extracted with PET ether (50 ml at a time) in a separating funnel (repeated thrice). The PET fractions were collected dried over anhydrous Na2SO4 and kept in a rotary evaporator for drying. The concentrated fraction was developed in TLC and methyl esters were identified in Iodine Chamber (Mobile phase–n hexane: ethyl acetate 1: 1 in v/v). The spots were scrapped, dissolved in dehydrated ethanol and the ethanolic solution was boiled on water bath at 55°–60°C for 25 minutes. The ethanol fraction was kept in a 250 ml round bottom flask discarding the silica gel fraction. Above methods were used for the preparation of methyl esters of standard fatty acids such as–Myriastic acid, Palmitic acid, Stearic acid, Oleic acid and Lenoleic acid. The Methyl esters thus prepared were subjected to GCMS study.

Estimation of Fatty Acids

1 gm of the essential oil was taken in a 250 ml round bottom flask and 25 ml of 0.5 M KOH was added in dehydrated ethanol and the flask was kept on a water bath at 55°– 60° C for 20 min. After completion of the saponification procedure, 12.5 ml of concentrated HCL was added and the whole solution was kept in a separating funnel. 50 ml of petroleum ether was added in the separating funnel and after vigorous shaking for 15 min, PET ether fraction was collected in a separate conical flask discarding the aqueous fraction (repeated thrice). The PET fractions were combined and 50 g of anhydrous Na2SO4 was added to it. The whole solution was filtrate and kept in a desiccator. A fraction of the filtrate was dissolved in hexane and undergone TLC on Silica gel G(Sigma,St.Louis.MO,usa) layers (thickness 0.5 cm) which had been prepared using a Unoplan(Shandon,London) coating apparatus with mobile phase hexane: ethyl acetate 1:1, the plate was kept in Iodine chamber and the spot of fatty acid was confirmed by diphenyl pieryl hydrazyl spray.

Analytical Gas Chromatographic Study

The purified fatty acid fraction was analysed directly by GC on a Hewlett Packard (HP;Palo Alto,(A;USA) MODEL HP-6890 PLUS GLC AND HP 3398 A GC CHEMSTATION instrument filled with a column HP-5(Capillary Column-0.32 mm in diameter, length 30 metre).The oven temperature was initially 150°(4 min) and ended with 250° (5 min)

Identification of Components

The identification of each compound was carried out by comparison of RRT (Relative retention time) and mass spectral data obtained with standard methyl esters preparation of Palmitic acid, Stearic acid, Oleic acid, Linoleic acid, Linolenic acid and Myriastic acid (Obtained from Sigma)

Results

Primary Biochemicals

The soluble sum of carbohydrates in the mature leaves was higher than those in the young and senescent leaves during March–August and November–January (Fig. 3.1). The concentrations of soluble carbohydrates in young and senescent leaf were maximum in the month of February and in mature leaves the concentration was highest in the month of March. The concentration of soluble sugar in young and senescent leaves was minimum in the month of May whereas the lowest concentration was found in the month of August for mature leaves.

The concentration of starch in the mature leaves was always higher than those in the young and senescent leaves (Fig. 3.2). The highest concentration was found during the month of May whereas the lowest concentration was found during the month of August in all kind of leaves.

The seasonal distribution of protein revealed that the concentration was always higher in young and mature leaves than that in the senescent leaves (Fig. 3.3). The highest concentration of protein was found in the month of January, February and September for young, mature and senescent leaves respectively. The lowest concentration was found in the month of May in young leaves, in the month of July in mature leaves and in the month of January in senescent leaves.

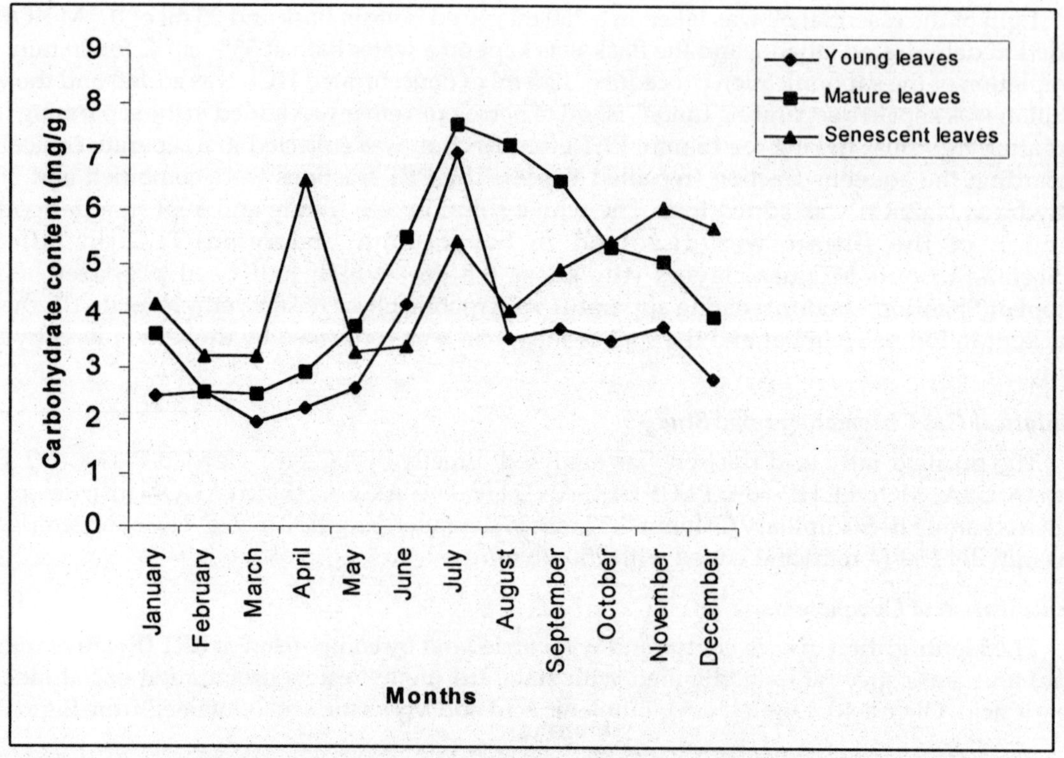

Fig. 3.1: Seasonal Fluctuation in the Level (mg/g) of Carbohydrate Content in the Leaves of *Cestrum diurnum* in Different Physiological Stages

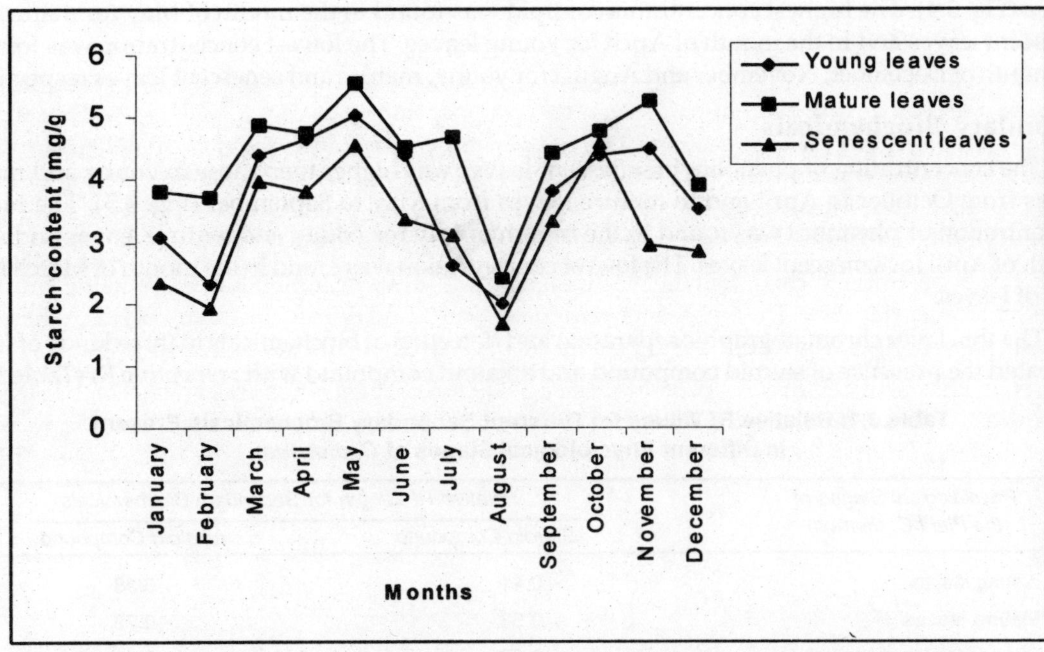

Fig. 3.2: Seasonal Fluctuation in the Level (mg/g) of Starch Content in the Leaves of *Cestrum diurnum* in Different Physiological Stages

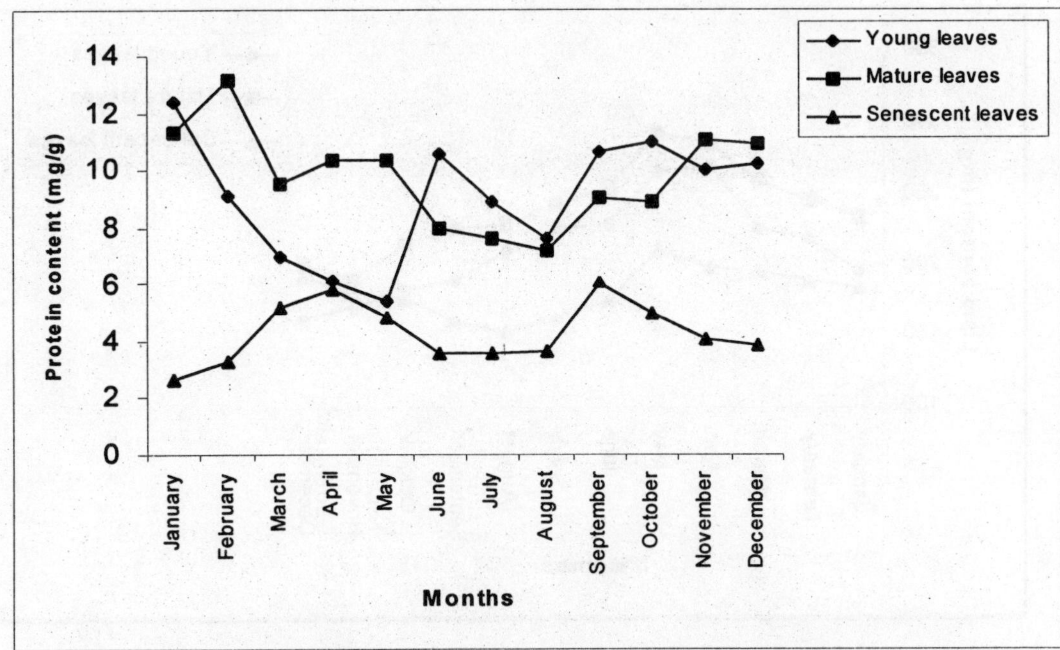

Fig. 3.3: Seasonal Fluctuation in the Level (mg/g) of Protein Content in the Leaves of *Cestrum diurnum* in Different Physiological Stages

The concentration of lipid in mature leaves was always higher than those in young and senescent leaves (Fig. 3.4). The highest concentration of lipid was found in the month of May for mature and senescent leaves and in the month of April for young leaves. The lowest concentration was found in the month of December, November and August for young, mature and senescent leaves respectively.

Secondary Biochemicals

The concentration of phenolics in senescent leaves was higher than those in young and mature leaves from October to April and in mature leaves from May to September (Fig. 3.5). The highest concentration of phenolics was found in the month of July for young and mature leaves and in the month of April for senescent leaves. The lowest concentration was found in the month of March for all kind of leaves.

The thin layer chromatographic separation and detection of biochemicals in three kinds of leaves indicated the presence of steroid compound and alkaloid compound with respective Rf (Table 3.1).

Table 3.1: Relative Rf Values for Different Secondary Biochemicals Present in Different Physiological Stages of *C. diurnum*

Physiological Stages of the Plant C. diurnum	*Relative Rf Length for Secondary Biochemicals*	
	Steroid Compound	*Alkaloid Compound*
Young leaves	0.41	0.58
Mature leaves	0.53	0.73
Senescent leaves	0.67	0.48

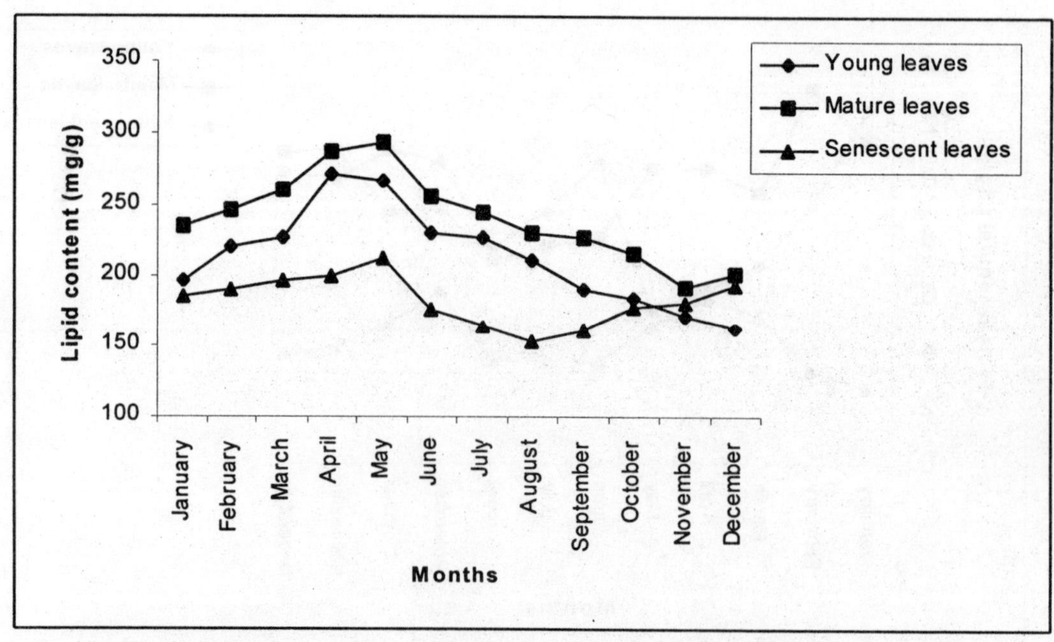

Fig. 3.4: Seasonal Fluctuation in the Level (mg/g) of Lipid Content in the Leaves of *Cestrum diurnum* in Different Physiological Stages

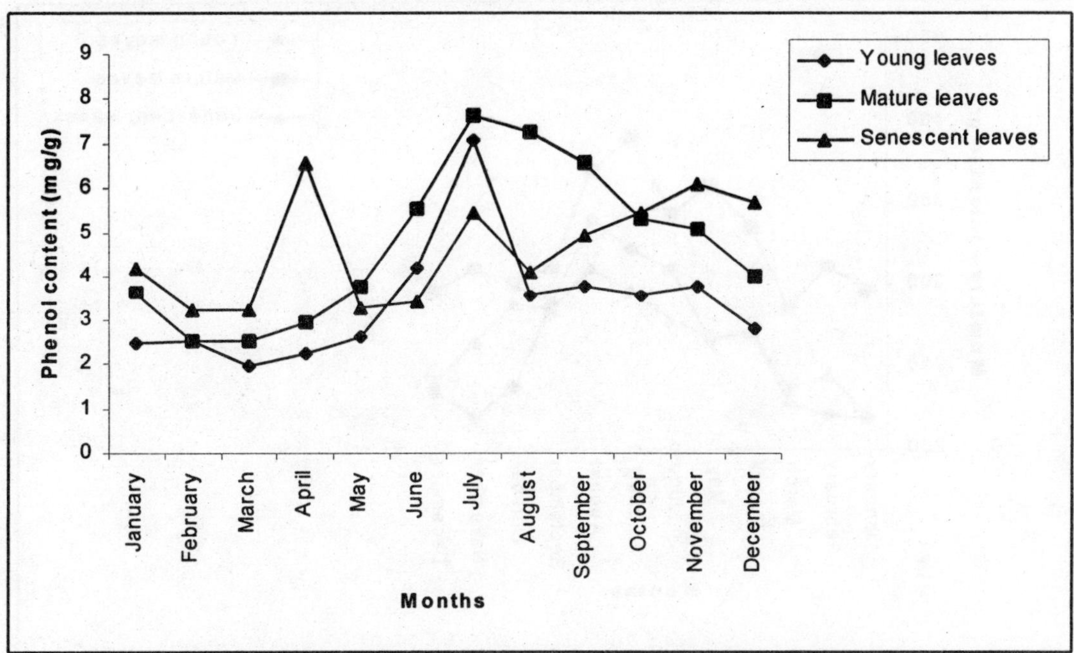

Fig. 3.5: Seasonal Fluctuation in the Level (mg/g) of Phenol Content in the Leaves of *Cestrum diurnum* in Different Physiological Stages

Ash and Moisture Contents

Seasonal changes in the moisture content revealed that the amount of moisture in mature leaves was always higher than young and senescent leaves (Fig. 3.6). The highest concentration of moisture was found in the month of July for mature and senescent leaves and in the month of August for young leaves. The lowest concentration was found in the month of January for all kind of leaves.

The pattern of seasonal changes in the ash content revealed that the higher amount was found in senescent leaves than in young and mature leaves except in the months of September–November when the highest amount of ash content was found in mature leaves (Fig. 3.7). The highest amount of ash content was found in the month of May, February and August in senescent, mature and young leaves respectively. The lowest amount was found in the month of October, May and February for senescent, mature and young leaves respectively.

Essential Oil Analysis

Gas liquid chromatographic analysis of the essential oil of mature leaves of Cestrum, diurnum (L.) revealed the presence of some saturated and unsaturated fatty acids (Fig. 3.8). Analysis of relative amount of major identified fatty acids indicated highest concentration of palmitic acid (27.62%) followed by stearic acid (4.62%) and oleic acid (3.06%).

Discussion

Current study revealed that the levels of primary and secondary biochemicals were extremely variable in the leaves of *Cestrum diurnum* with different physiological stages. Carbohydrate is the most

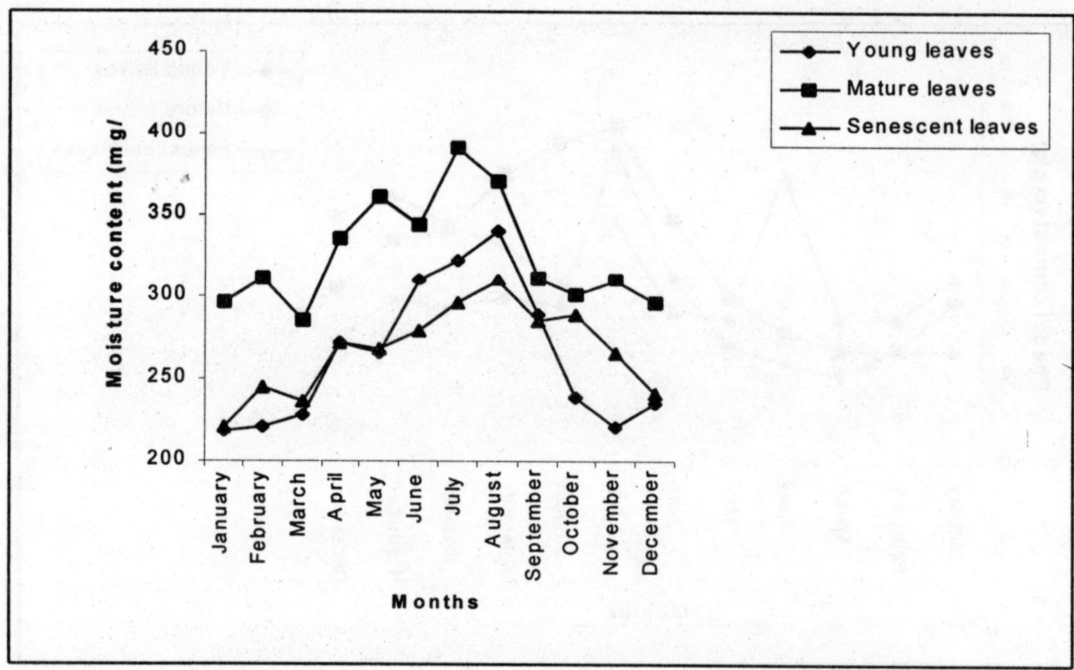

Fig. 3.6: Seasonal Fluctuation in the Level (mg/g) of Moisture Content in the Leaves of *Cestrum diurnum* in Different Physiological Stages

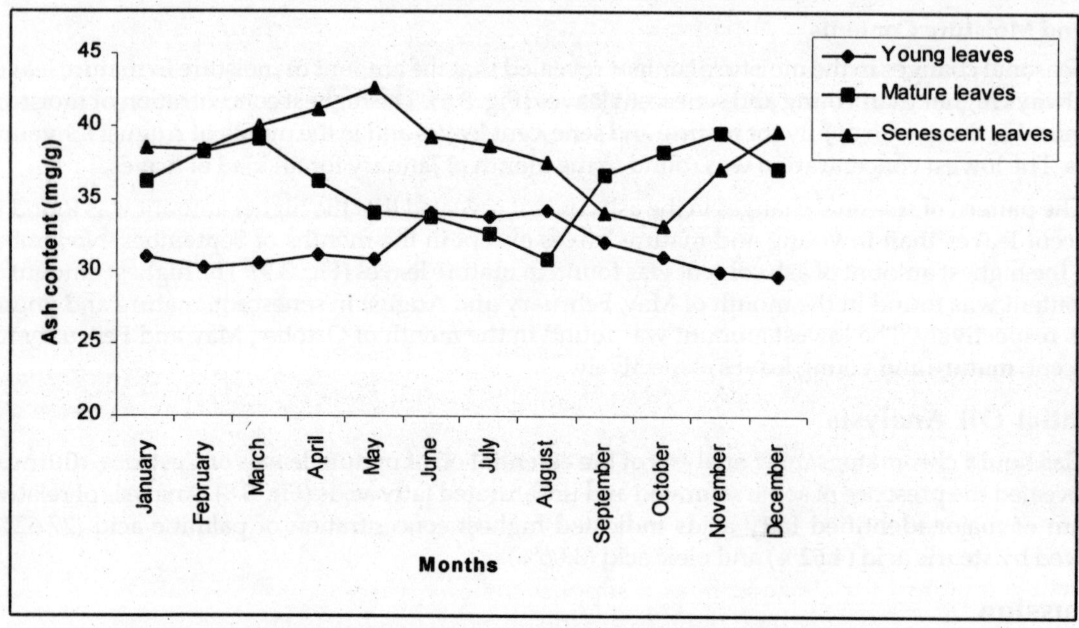

Fig. 3.7: Seasonal Fluctuation in the Level (mg/g) of Ash Content in the Leaves of *Cestrum diurnum* in Different Physiological Stages

HP 6890 Plus GLC and HP3398A GC Chemstation

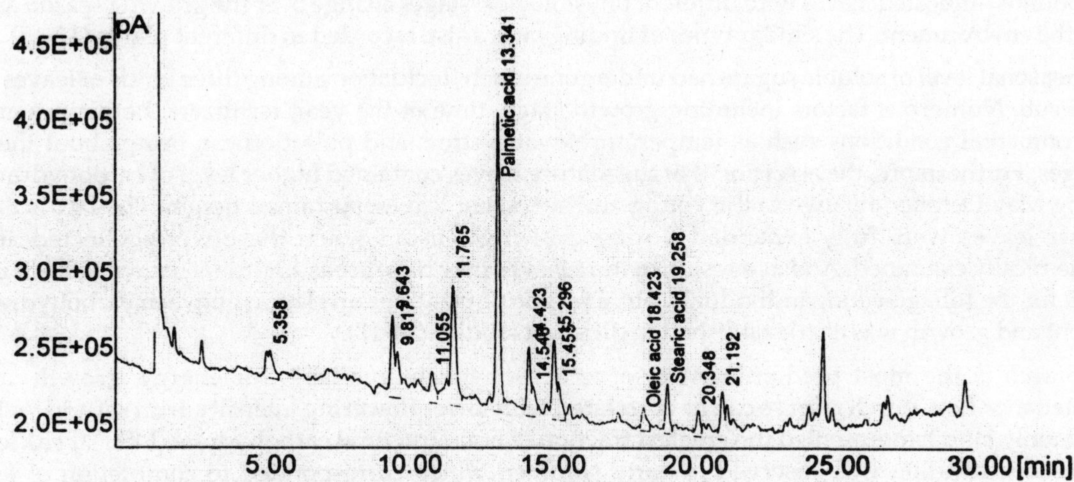

Fig. 3.8: GLC Analysis of the Essential Oil of *Cestrum diurnum*

Fig. Relative Amount of Identified and Unidentified Fatty Acid

Name of Fatty Acids	*Retention Time (Min)*	*Relative Amount*	*Area (pA*s)*
Palmetic Acid	13.341	27.62	1638291
Oleic Acid	18.423	3.06	181198
Stearic Acid	19.258	4.62	274152
Peak1	5.358	3.44	204103
Peak2	9.643	7.69	456133
Peak3	9.817	4.93	292183
Peak4	11.055	4.09	242473
Peak5	11.765	16.51	978866
Peak6	14.423	6.22	369157
Peak7	14.548	2.92	173331
Peak8	15.296	8.16	484091
Peak9	15.455	3.14	185969
Peak10	20.348	2.76	163610
Peak11	21.192	4.84	287249

important component for metabolism as it supplies the energy needed for respiration and other metabolic processes. Carbohydrate reserves are mobilized in trees to fuel maintenance respiration during winter (7), build leaves and support new growth in spring and to provide the energy expended in making adaptive responses to soil water deficits, pathogen attacks and herbivory(8-11).Starch and soluble sugars are the predominant reserve carbohydrate available for energy growth and maintenance. Current study revealed that the the proportion of photosynthecate (including soluble sugar and starch compounds) allocated leaves with different physiological stages change over the growing season and with the environment. The similar types of findings were also recorded in different plants (12-13).

Seasonal level of soluble sugars had undergone much fluctuation among three kinds of leaves of the shrub. Numerous factors including growth stage, time of the year, fertilizers, herbicides and environmental conditions such as temperature, water stress and pollution can bring about these changes. Furthermore, the assertion that the mature leaves contained higher level of carbohydrates during May-October, relative to the young and senescent leaves, sustained herein. The fact is that mature leaves were fully expanded during growing seasons when these were collected and biochemically examined. And as a result, mature leaves may have act as a nutrient source rather than a sink for the foliage-eating individuals. Similar type of positive correlation between carbohydrate content and growth was also established in different studies (14-17).

Starch is the most predominant reserve carbohydrate available for energy, growth and maintenance.Low starch reserve can be correlated with lower flowering intensity, fruit set and yield. Commonly, starch represented the greatest fraction of non-structural carbohydrate (NSC). Peak leaf starch concentration was observed in early summer, which corresponded to completion of leaf expansion and maturation. Active growth in the growing season resulted in a decline in stored starch reserve. During this time, accumulated starch is gradually converted to increased amount of sugars. Starch content decreased following shoot emergence and declined until flower bud formation. Following bud initiation levels of starch increased during flowering and this trend continued until the plant enters into senescent phase.

Protein comprises one of the most important biomacromolecule and a major determinant in controlling tree growth rate and reproduction. The protein content differed substantially amongst young, mature and senescent leaves during growing and reproductive phase of the plant (18-20). Similar to other plant systems (21-25) *C.diurnum* leaves also had higher quantity of soluble protein in mature leaves during growing season and the minimum level was found during the flowering season. This was probably due to the allocation of resource materials in the flower and fruit etc. to act as a sink rather than a source. The results exhibited here suggested that the soluble protein could be used as an indicator of the physiological state of the plant.

The richness in secondary compounds in plants is, at least, partly explicable by the simple fact that plants are rooted in the soil and cannot move; they cannot respond to the environment in ways open to animals. Theories of plant chemical defenses discriminate between quantitative digestibility reducers such as, tannins and phenolics and qualitative toxin such as, steroids and alkaloids.

Phenolics are thought to be the part of the natural defense mechanism of the plant to prevent severe defoliation. The poor livestock performance coincides with a high level of phenolics in the leaf. Present study revealed that the amount of phenolics exhibits an inverse relationship with the total protein contents in early growing season (June to August). Similar relationship has also been observed in several species of vascular plants (26-27) and macro algae (28-29). When nitrogen limits growth, there may be more fixed carbon that can be metabolized into proteins, thus plants accumulate phenolics or other nitrogen free constituents.

Thin layer chromatographic analysis revealed the presence of steroid and alkaloid compounds in the young, mature and senescent leaves (with distinct Rf respectively) may be concluded as the herbivore grazing inhibitor as well as responsible for the calcinogenic activity of the plant.

Estimation of the moisture content and ash content is important to determine the values of other compounds in the leaves. High moisture content can cause over heating of the leaves and deterioration of quality, particularly increase in free fatty acid contents; whereas high ash content is positively related to plant senescence due to an increase in the total crude fibers.

Lipids are unique class of chemical compounds that all organisms need to survive. They are primarily used for energy storage, membrane structure and as hormones. Seasonal variations in lipid content, with the maximum in spring, were observed at all sampling stations. Most of this lipid was tightly bound to the wall fabric. Temperature has a characteristic effect on many plant lipids in which it increases the level of instauration of acyl chains, which slows down both metabolism and transport. Therefore, the highest lipid content of *Cestrum diurnum.* was found at the beginning of the growing season *i.e.* in spring during the study period. Consequently, as the temperature increased, the lipid level decreased and remained almost stable until the end of the growing season. The functional units of lipids are fatty acids. The discovery in the mid-1950's of gas-liquid chromatography (GLC) has revolutionized the analysis of fatty acids and, undoubtedly, this technique is the most frequently used, for the quantification of individual fatty acid in any acylated lipids. The present study indicates a high percentage of saturated fatty acid in comparison to unsaturated fatty acids. This is probably due to the distribution pattern of *Cestrum diurnum*, available in the tropical rather than in temperate zone.

REFERENCES

1. Little, E.L., Jr., R.O. Woodbury, and F.H. Wadsworth. (1974) Trees of Puerto Rico and the Virgin Islands. Vol. 2. Agric. Handb. 449. U.S. Department of Agriculture.
2. Project report-Calcinogenic activity of *Cestrum diurnum*,Animal science division Colonel University,2001.
3. Dubois, M., Gilles, K.A., Hamilton, J.K., Pebers,P.A. and Smith,F.(1956) Colorimetric method for determination of sugar and related substances. Analytical Biochemistry. (28) 350-356.
4. Lowry, O.H., Rosebrough, N. J., Farr,A.L. and Randal,R. J. (1951) Protein measurement with folin phenol reagent. Journal of Biological Chemistry (193). 265-275.
5. Folch,J,Lees, M and Solane-stanley,G.H.(1957). A simple method for the isolation and purification of total lipids from animal tissues.J.Biol.Chem.(226),497-507.
6. Bray,H.G. and Thrope,W.V.(1954). Analysis of phenolic compounds of interest in metabolism.Meth.Biochem.Anal (1) 27-52.
7. Ögren, E. (2000) Maintenance respiration correlates with sugar but not nitrogen concentration in dormant plants. Physiol. Plant (108)295–299.
8. Wargo, P.M. (1981) Defoliation and tree growth. *In* The Gypsy Moth:Research Toward Integrated Pest Management. Eds. C.C. Doane and M.L. McManus. Tech. Bull. (1584) USDA, Washington,pp 225–248.
9. Oren, R.. Schulze, E.D. Werk, K.S. Meyer J. Schneider, B.U. and. Heilmeier, H (1988) Performance of two *Picea abies* (L.) Karst.stands at different stages of decline. I. Carbon relations and standgrowth. Oecologia (75)25–37.
10. Dunn, J.P., D.A. Potter and T.W. Kimmerer.(1990) Carbohydrate re-serves, radial growth, and mechanisms of resistance of oak trees to phloem-boring insects. Oecologia (83)458–468.
11. Canham, C.D., R.K. Kobe., E.F. Latty and R.L. Chazdon.(1999) Inter-specific and intraspecific variation in tree seedling survival: effectsof allocation to roots versus carbohydrate reserves. Oecologia.(121)1–11.

12. Kramer, P.J. and. Kozlowski, T.T. (1979). Physiology of woody plants.Academic Press, London, 811 p.

13. Terziev, N., Boutelje J.and Larsson K. (1997) Seasonal fluctuations of low-molecular-weight sugars, starch and nitrogen in sapwood of *Pinus sylvestris* L. Scand. J. For. Res. (12)216–224.

14. Fischer, C. and Höll. W. (1992) Food reserves of scots pine (*Pinussy lvestris* L.). II. Seasonal changes and radial distribution of carbo-hydrate and fat reserves in pine wood. Trees (6)147–155.

15. Höll, W. (1997) Storage and mobilization of carbohydrates and lipids.*In* Trees-Contributions to Modern Tree Physiology. Eds. H. Ren-nenberg, W. Eschrich and H. Ziegler.Backhuys Publishers,Leiden, The Netherlands, pp 197–211.

16. Priestley, C.A. (1960) Seasonal changes in the carbohydrate resources of some six-year-old apple trees. East Malling Res. Sta. Annu. Rep.(1959) pp 70–77.

17. Saranpää, P. and Höll, W.(1989). Soluble carbohydrates of *Pinus sylvestris* L. sapwood and heartwood. Trees (3)138–143.

18. Rao, C.H.K.; Chauhan,V.D. and Indusekhar, V.K. Elemental (1983) Composition of certain green seaweeds of Porbandar reef. Indian Journal Marine Sciences. (12) 197-199.

19. Fadeel, A.A., Metwalli,A., and Al-houty, F.A. (1987) The effect of seasonal variation in temperature and light intensity on the nitrogenous contents in four species of the chlorophyceae of Kuwait. Dirasat(VIX) 107-118.

20. Ochiai, Y., Katsuragi T. and Hashimoto, K. 1987 Proteins in three seaweeds "aosa" Ulva lactuca, "arame" Eisenia bicyclis, and "makusa"Gelidium amansii. Nippon Suisan Gakkaishi (53) 1051-1055.

21. Munda,I. M., Gubensek F. (1986) The amino acid composition of some benthic marine algae from the Northern Adriatic. Bot. Mar..(29) 367–372.

22. Munda I. M., Gubensek F. (1976) The amino acid composition of some common marine algae from Iceland, Bot. Mar (19) 85–92.

23. Owens, N. J. P., Stewart,W. D. (1983) Enteromorpha and the cycling of nitrogen in a small estuary, Estuar. Coast. Shelf Sci 17 (3), 287–296.

24. Tkachenko F. P., Koval V. T. (1990) Biochemical composition of abundant benthic seaweeds of the Black Sea, Hydrobiol. J., 26 (6), 39–43.

25. Wheeler P. A., Bj¨orns¨ater B. R. (1992) Seasonal fluctuations in tissue nitrogen, phosphorus and N:P for five macroalgal species common in the Pacific northwest coast, J. Phycol. (28) 1–6.

26. Mattson, WJ. (1980)Herbivory in relation to plant nitrogen content.Ann.Rev. Ecol. Syst. (11) 119-161.

27. Buchsbaum, R.F.T., Short and D.P.Cheney.(1990)Phenolic-nitrogen interactions in eelgrass,Zostera marina L.:possible implications for disease resistance.**Aquat.Bot.**(37)291-297.

28. Neish.Shacklock, A.P. Foc, C and.Simpson F (1977) The cultivation of Chondrus crispus.Factors affecting growth under greenhouse conditions.**Can.J.Bot** (55) 2263-2271.

29. MaclerB. (1986) Regulation of carbon flow by nitrogen and light in red algae,*Gelidium coulteri*.Plant Physiol.(82)136-141.

Advances in Biochemistry and Biotechnology Volume 1 *Pages 95-106*
Edited by: **Dr. Chiranjib Chakraborty**
Published by: **DAYA PUBLISHING HOUSE**

4

Protein Folding Pathways: Its Kinetics and Related Functions

*Surajit Sinha, Shyam Sundar Nandi, Akash Verma, Ragini Gothalwal & T.K.Maiti**
**Dept. of Biotechnology, I.I.T.-Kharagpur (W.B)*
E-Mail: tkmaiti@hijli.iitkgp.ernet.in

ABSTRACT

Among the nanobiomolecules, proteins are the most mysterious because of their hugely diversified functions ranging from cellular growth, cell-cell interactions to cellular signaling which are governed by their biologically active 3D structure. Moreover, proteins function in a fluctuating cellular environment, stabilized by disulphide bonds, which influence the thermodynamics of protein folding. Proteins that misfold and exhibit decreased solubility under physiological conditions results in protein disorders such as Parkinson's and Alzheimer's disease. These set of disorders are characterized by either local or systemic extracellular tissue deposition of specific proteins under physiologically attainable conditions. This deposition can lead to a compromise in organ function, which in most cases, is fatal without treatment. Innovative use of

* Correspondent Author

computational methods along with experimental data are thus needed to understand the rate and kinetics of protein folding and protein folding pathways which can provide insights in to the basis of such protein disorders and help develop novel therapeutic approaches.

Keywords: *Nanobiomolecules; kinetics; protein refolding; breaker peptides.*

Introduction

Protein function is a measure of protein stability and protein structure. Proteins are the center of the action in biological processes and their function can only be known based on the structure of their constituent polypeptide chains. Therefore, understanding the protein folding mechanism (1-3) is critical in molecular biology not only because it is one of the fundamental problems remaining in protein science, but also because several fatal diseases are directly related to protein folding/misfolding, such as the Alzheimer's disease, Mad cow disease, Prion disease and cystic fibrosis disease. Studies on protein folding poses two fundamental problems. One is the Levinthal's Paradox which refers to how proteins find their correctly folded state on a non-astronomical time scale given the myriads of possible conformations (4,5). Second, how this efficient process of protein folding is carried out in vivo taking in to account the adverse conditions in the crowded environment of the cell. The failure of molecular chaperones (6-8) in preventing aggregation in these fatal cases is still a question to be answered. Despite the enormous efforts made by various groups, the problem of protein folding still remains largely unsolved. Experiments that probe proteins at different stages of the folding process have helped to elucidate kinetic mechanisms and the thermodynamics stabilities of folding (9-11). For example- the observation that early stages of protein folding involves single-exponential kinetic phases, can be interpreted as evidence that these phases are rate limited by significant energy or entropy barriers. Computer stimulation (12,13) performed at various levels of complexity ranging from simple lattice models to all atom models with explicit solvent are used to supplement experiment and fill in some of the gaps in our knowledge about folding pathways. Because explicit solvent stimulations typically require enormous CPU time, many recent studies have been carried out with implicit solvent models. However, it is still largely an open question as to how well these implicit solvent models can predict the thermodynamics as well as the kinetics of protein folding. We are not very close to realizing this goal, and so the protein-folding problem remains one of the most basic unsolved problems in computational biology. Here, we review how disulphide bonds and contact order have helped us in elucidating the protein folding pathways, its function and kinetics with special reference to protein conformation disorders (PCD). Lastly, we look at protein refolding therapy, which if possible can have enormous applications ranging from therapeutic benefits to artificial modeling of the cell.

Disulphide Bond as Probes for Conformational Transitions

The common belief is that disulphide bonds have been added during evolution to increase the stability of proteins that function in a fluctuating cellular environment (14,15). Disulphides bonds help stabilize the folded states by increasing favourable enthalpy interactions in the folded states and by lowering the entropy of the unfolded states (16,17). But recent evidence indicates that disulphide bonds are not inert structural motifs but play important role in protein conformation and function. Not only because the interaction among thiol groups of cysteine residues are of reductive/oxidative nature and therefore requires the presence of an appropriate electron acceptor or donor, which can be controlled experimentally as can the intrinsic stability of the disulphide bond interaction but because of the fact that the rates of forming or breaking of protein disulphide bonds reflects the protein conformational transitions that occur. Therefore, the equilibrium between folded and unfolded conformations can be altered simply by changing only the thiol/disulphide intrinsic stability (18-20). As a result, all disulphide bonds present in a protein at any time may be trapped in a stable form and be chemically identified. A number of different ways of forming and breaking protein disulfides are known, but the most suitable for studies of protein folding involves the use of thiol-disulphide exchange between the protein and a small molecule reagent in either thiol or disulphide form. The most commonly used reagent is glutathione (GSSG).

The disulfides present at any instant in a protein can be trapped in a stable form by simply blocking, rapidly and irreversibly, all thiol groups in the solution with the help of a quenching reagent (Fig. 4.1). Therefore, in the absence of other nucleophiles, disulfide bonds may no longer be made, broken, or interchanged. The quenching reaction should be very rapid so that it faithfully traps the species present at the time of addition of the quenching reagent. Trapping reactions with reagents like iodoacetamide or iodoacetic acid has the advantage of being irreversible, but the reaction with a free thiol group has a half-time of 0.3 seconds with 0.2M iodoacetate at pH 8.7 and 25°C. Therefore, disulphide rearrangements could occur during the trapping reaction. The other problem being the fact that buried thiol groups do not react at a sufficient rate during the trapping procedure. Nevertheless,

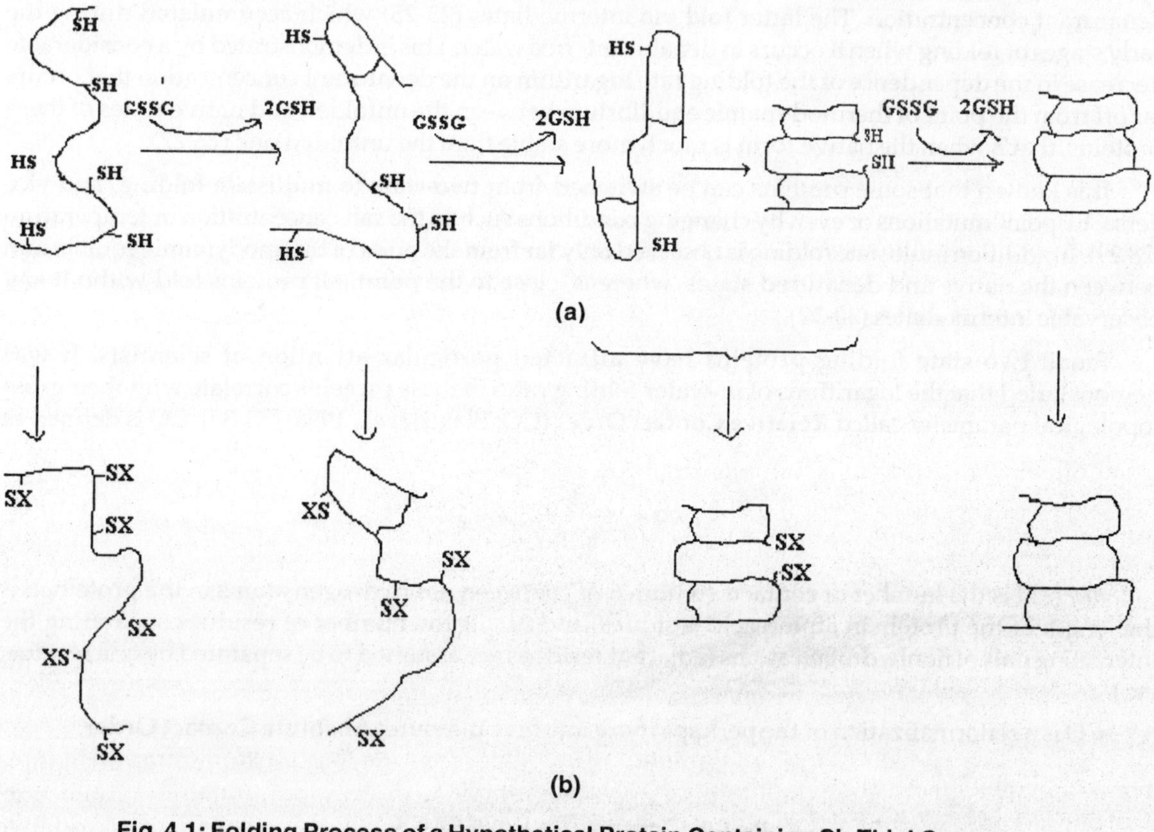

(a)

(b)

Fig. 4.1: Folding Process of a Hypothetical Protein Containing Six Thiol Groups. (a) The folding pathway is illustrated using an intermolecular disulphide reagent-gluthathione(GSSG). (b) The intermediates are trapped in whatever stable conformation they are using a quenching reagent. X denotes the quenching molecule.

once all free thiol groups are blocked irreversibly, the trapped species are stable indefinitely. They can then be separated and characterized in detail. The intermediate species trapped by either acid, iodoacetate or iodoacetamide can be analyzed by nondenaturing polyacrylamide gel electrophoresis, ion-exchange chromatography and by reversed-phase HPLC. Since, the quenching reagents affect the equilibrium between folded and unfolded conformations and the rates of disulfide bond formation, the kinetics of unfolding and refolding can be determined unambiguously under all unfolding and refolding conditions. Some of the proteins whose folding pathways have been elucidated using disulfide bonds includes Bovine Pancreatic Trypsin Inhibitor (BPTI), dendrotoxins I and K from the venom of black mamba snakes, Ribonuclease T_1 and Ribonuclease A.

Kinetics and Contact Order

The ultimate goal of kinetic folding studies is the molecular elucidation of the successive structure formation that takes place during folding. But there is a significant difference in the folding behavior of small proteins with simple two-state kinetics, and of larger proteins having a three-state folding kinetics (21,22). The former have no visible intermediates in the course of folding, which therefore occurs as an " all-or-none " process under all experimental conditions. This peculiarity is demonstrated by linear dependence of logarithms of the folding (and unfolding) rates of these proteins on the denaturant concentration. The latter fold via intermediates (23-25) which accumulated during the early stages of folding when it occurs in denaturant- free water. This is demonstrated by a considerable decrease in the dependence of the folding rate logarithm on the denaturant concentration that occurs far off from the point of thermodynamic equilibrium between the unfolded and native states of these proteins, that's when the native form is much more stable than the unfolded one (26,27).

It is known that some proteins can be switched from two-state to multistate folding, and vice versa, by point mutations or even by changing conditions such as the salt concentration or temperature (28,29). In addition multistate folding is observed only far from the point of thermodynamic equilibrium between the native and denatured states, whereas, close to the point, all proteins fold without any observable intermediates (30-32).

Small two-state folding proteins have attracted particular attention of scientists. It was demonstrated that the logarithms of in-water folding rates of these proteins correlate with their gross topological parameter called **Relative Contact Order** (CO: Plaxco *et al.*, 1998) (33-35). CO is defined as

$$CO = \frac{1}{L.N} \sum^{N} \Delta L_{ij}$$

where N is the number of contacts (within 6 A^0) between nonhydrogen atoms in the protein, L is the length of the protein in amino acid residues, and ΔL_{ij} is the number of residues separating the interacting pair of nonhydrogen atoms (adjacent residues are assumed to be separated by one residue, etc.).

CO is a renormalization of the perhaps more intuitive measure, **Absolute Contact Order,**

$$Abs\_CO = \frac{1}{N} \sum^{N} \Delta L_{ij} = CO \times L$$

Which however, was found to be less correlated than CO with folding rates of the two-state folders (36). The CO was invented to compare differences in topology (rather than in size) between proteins of different length (37). This parameter is small for proteins stabilized mainly by local interactions and is large when residues in a protein interact frequently with partners far away in the protein sequence.

However, CO parameter cannot predict the folding rates of proteins with three state kinetics and for short peptides. Instead, the three state folding proteins demonstrate a strong anticorelation between their size and the folding rate logarithm. It seems likely that CO takes in to account topology only and plays no explicit attention to the protein size. Therefore, a much more specific approach is needed for understanding the kinetics of three state folders.

Energy Profiling

The kinetic reaction involving intermediates in a protein folding process is generally represented as

$$U \xrightarrow{K_{UI}} I \xrightarrow{K_{IN}} N$$

Where the formation of an intermediate (I) causes a lag in the appearance of the product (N-native). The magnitude of the lag period depends upon the ratio of the two rate constants. Spectroscopic techniques are usually not well suited to detect such lag phases since intermediates are often intermediate in their spectral properties between U and N (38). The lag that would be observed in the I→N reaction is compensated by the contribution of the U→I reaction and thus is not visible in the kinetics. The formation of native protein molecules via intermediates during folding or refolding can be measured selectively by a technique that is related to the energetic profile of folding (Fig. 4.2). The completely folded species, N, is separated from all other unfolded or partially folded species by a high activation barrier. Therefore, only N molecules unfold slowly, whereas all intermediates that have not yet crossed this final barrier unfold rapidly. Therefore the formation of native protein can be measured by unfolding assays. Folding is initiated inthese experiments by dilution to the desired refolding conditions at time t = 0 in a test tube. Then, after various time intervals of refolding, samples are withdrawn and are transferred to unfolding conditions. The amplitude of the resulting unfolding reaction (by absorbance or by fluorescence) is a measure of the amount of completely refolded molecules that were present at the time when folding was interrupted. The maximum value of the unfolding amplitude is obtained after folding in the first step of the assay has gone to completion and should correspond to the amplitude that is observed when native protein is unfolded under the same conditions. The folding kinetics of great many proteins have been determined in this way including Rnase and α-lactoglobulin. However, many of the details of protein folding pathways including the structure of the intermediates remain unknown and as long as these are not solved treatment of diseases arising out of protein folding/misfolding disorder will be far from reality.

Protein Aggregation and Transition

The biological function of a protein depends on its three dimensional structure, determined by its amino acid sequence during the process of protein folding. Diverse diseases have been shown to arise from protein misfolding and a number of misfolded proteins and peptides aggregate in to insoluble fibrils (Fig. 4.3) (38,39). The aggregates are toxic and can poison the cells around it. The aggregation of these diseases are related to fatal diseases (40,41). That protein conformation plays an essential role in the development of pathological consequences for the organism became clear in the course of studying the molecular basis of diseases like Prion disease, Alzheimer's disease, Cystic Fibrosis, Mad Cow

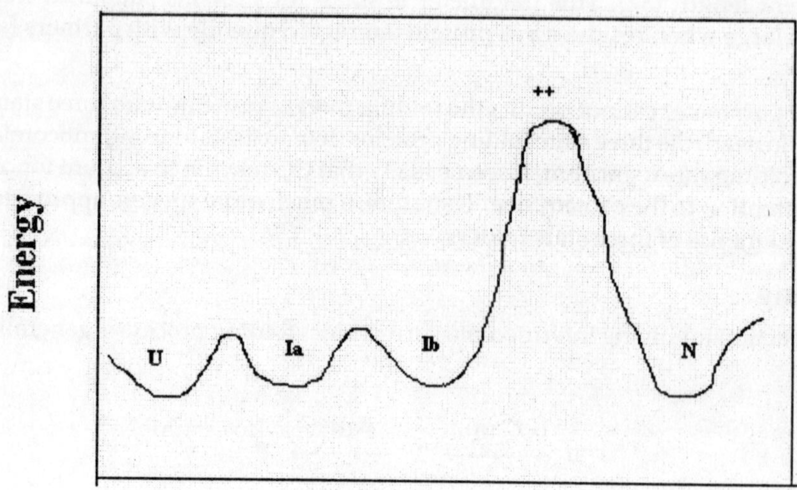

Fig. 4.2: Energy Profile for a Protein Folding Reaction. ++ Denotes activated state close to the native structure N. U is unfolded protein; Ia and Ib are partially folded intermediates.

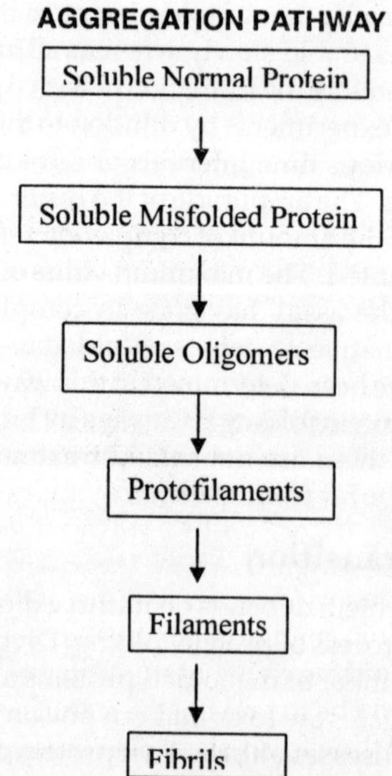

Fig. 4.3: Pathway of Fibril Formation

disease, Parkinson's disease, transmissible spongiform encephalopathies (TSE), Amyotrophic lateral sclerosis (ALS) and others. All these diseases are grouped under the name Protein Conformational Disorders (PCD) (Table 4.1). Diorders of such nature are classified by the accumulation of protein aggregates at specific sites for that particular protein; neurodegenerative diseases, such as Parkinson's or Huntington's disease, form intracellular aggregates, and the systemic amyloidoses form extracellular aggregates. The amyloidoses were the first recognized examples of protein deposition diseases. These deposits bind the dyes thioflavin-S,-T and congo red, the latter with positive birefringence under polarized light (42,43). In Alzheimer's disease, the prion disorders and the islet amyloid polypeptide deposits in pancreatic islets of the elderly, deposition is local, the modes of pathogenesis are less clear, and there is evidence for intracellular aggregation as well as extracellular deposition (44). Electron microscopy of the deposits reveal a discrete fibrillar structure. Recently, proteins not implicated in amyloid diseases were found to form fibril structures in vitro under denaturing conditions (45,46), suggesting that fibril formation is a common feature of destabilized proteins(28). Regardless of sequence and structures of proteins, the fibrils have similar core structures, mainly composed of β-sheets(47). For example, the native Aβ peptide in Alzheimer's disease(48,49) and prion proteins (PrP^c) in prion diseases(50,51) are α–helix rich. The aggregation of α-helix rich proteins or peptides involves a conformational transition from α-helices to β-sheets. A similar transition has also been observed in vitro in some α-helical peptides (52,53) that aggregate in to amyloid fibrils by means of changing the environment, such as varying the organic solvent condition (54), altering the pH, and controlling the redox state. Moreover, similar α-β transitions also occur through the correct folding pathway in proteins with a nonhierarchical folding mechanisms. For example, β-lactoglobulin, a predominant β-sheet protein (55) is observed to form non-native α-helical intermediates on folding. Thus understanding the α-β transition is important for both protein folding and protein aggregation. The secondary structures of proteins, mainly α-helices and β-sheets, are determined by the amino acid sequence and stabilized by hydrogen bonds. However, under denaturing conditions proteins with various organizations of the secondary structure elements can aggregate in to similar β-rich amyloid fibrils. This suggests that sequence non-specific hydrogen bond interaction among the backbone of proteins is an important factor for aggregation (56).

Protein Refolding Therapy

Since the 1950, scientists knew that the information necessary for a protein to fold is encoded in its aminoacid sequence. But do proteins fold by themselves or do they need some machines which assists them in folding?. It became clear only during the late sixties, that cellular protein folding critically depends on the assistance of so called chaperones (57-60). This specialized set of proteins is able to prevent the aggregation of nascent or misfolded substrates by providing microenvironments to protect them from nonproductive interactions (61). These proteins have hydrophobic surfaces that recognizes and bind to exposed hydrophobic surfaces of improperly folded proteins and prevents non-native interactions and thereby aggregation. It is this very concept that has been utilized in targeting refolding therapy. Considering the fact that protein misfolding and aggregation are central in the pathogenesis of PCD, any therapy directed to the cause of the disease, therefore, should aim to inhibit or reverse the conformational changes that result in the formation of the pathological protein conformer. Since, protein misfolding is triggered by conformational changes stabilized by oligomerization, the strategy should be to prevent the stabilization of the misfolding or to destabilize the monomeric intermediates and the early β-sheet oligomers (62,63).

It has been postulated that short synthetic peptides containing self-recognition motif of the protein and engineered to destabilize the abnormal conformation might be useful to correct protein folding.

These peptides are called as synthetic mini-chaperones. These are designed to be similar to the sequence of the protein region responsible for oligomerization and contain residues that specifically favor or disfavor a particular structural motif. Infact such a mini-chaperone known as β-sheet breaker peptide have been designed and has been found to block the conformational changes and aggregation in both Alzheimer's disease (Aβ-peptide) and in Prion disease (PrP^c –prion protein) both *in vitro* and *in vivo.* These results suggests that these synthetic β-sheet breaker peptides or synthetic minichaperones could be used in destabilizing the β-sheet rich abnormal conformation and inducing its conversion in to the normal form (62). These β-sheets breaker peptides are short, synthetic peptides, homologous to the sequence of the protein region thought to play a central role in the conformational change, and containing amino acid residues that disfavor the adoption of a β-sheet conformation. Studies *in vitro,* in cell cultures and *in vivo*, have shown that β-sheet breaker peptides, homologous to the PrP fragment which contain proline residues to destabilize β-sheet structures, were able to prevent and reverse the PrP^C ? PrP^{Sc} (64-66). Also under investigation is whether molecular chaperones–the cell's own chaperones –can be used to fix protein misfolding problems. It had been suggested that drugs that could boost chaperone activity could do the trick. Recent findings that geldanamycin treatment protected flies against dopaminergic neuron loss had laid a general platform technology to design drugs for treatment of PCD.

Table 4.1: List of Some Protein Disorders Associated with Protein Misfolding

Protein Involved	*Associated Diseases*
β-Amyloid	Alzheimer's Disease
α- Synuclein	Parkinson's Disease
Huntingtin	Huntington's Disease
Prion Protein(PrP^c)	Prion Disease
CFTR Protein	Cystic Fibrosis
Hemoglobin	Sickle cell anemia
SOD	Amyotrophic lateral sclerosis
Amyloid- A	Reactive amyloidosis

Conclusion

Solving the folding problem has enormous applications: exact drugs can be designed theoretically on a computer without a great deal of experimentation. Genetic engineering experiments to improve the function of particular proteins will be possible. Moreover, stimulating protein folding can allow us to go forward with the modeling of the cell. Catching hold of intermediates in protein folding pathway can help in designing synthetic mini-chaperones that can prevent proteins proceeding through such intermediates and finally aggregating. Therefore, understanding the protein folding pathways can help us develop principles for generating synthetic mini-chaperones that could be used as novel therapeutic approach for disorders where a function of protein needs to be modified. Also, since the phenotypic characteristics of living organisms depend upon the integrative function of thousands of proteins, engineering mini-chaperones that could alter the structure of specific protein could enable us manipulate phenotypic features of living organisms and could reveal a new therapeutic strategy for a variety of diseases.

REFERENCES

1. Dobson, D.M., Sali,A., and Karplus, M. (1998). Protein Folding: A Perspective from theory and experiment. **Angew. Chem. Int. Ed.** (37),868-893.
2. Ali, A. (1998). 100,000 protein structures for the biologist. ***Nature Struct. Biol.*** (5) 1029-1032.
3. Karplus, M., and Ali, A. (1995). Theoritical studies of protein folding and unfolding. **Curr. Opin. Struct. Biol.** (5) 58-73.
4. Gutsche, I., Essen, O. L., and Baumeister, W. (1999). Group II Chaperonins: New Tric(k)s and Turns of a Protein Folding Machine. **J. Mol. Biol.** (293) 295-312.
5. Callender, R.H., Dyer, R.B., Gilmanshin, R., *et al.* (1998). Fast events in protein folding: the time evolution of primary processes. **Annu. Rev. Phys. Chem.** (49) 173-202.
6. Frydman, J., Nimmesgern, E., Erdjument-Bromage., *et al.* (1996). Principles of chaperone-assisted protein folding: differences between *in vitro* and *in vivo* mechanisms. **Science** (272) 1497-1502.
7. Frydman, J., Nimmesgern, E., Ohtsuka, K., *et al.* (1994). Folding of nascent polypeptide chains in a high molecular mass assembly with molecular chaperones. ***Nature*** (370) 111-117.
8. Netzer, WJ. and Hartl, FU. (1998). Protein Folding in the cytosol: chaperonin-dependent and independent mechanisms. **Trends Biochem. Sci.** (23) 68-73.
9. Ferrara, P., Apostolakis, J. and Caflisch, A. (2000). Thermodynamics and kinetics of folding of two model peptides investigated by molecular dyanamics stimulations. **J Phys Chem** (104) 5000-5010.
10. Daggett, V. and Levitt, M. (1992). Molecular dyanamics stimulations of helix denaturation. ***J. Mol. Biol.*** (228) 1121-1138.
11. Borreguero, J.M., Dokholyan, N.V., Buldyrev, S.V., *et al.* (2002). Thermodynamics and folding kinetic analysis of the SH3 domain from discrete molecular dynamics. **J. Mol. Biol.** (318) 863-876.
12. Takada, S., Luthey-Schulten, Z. and Wolynes, P.G. (1999). Folding dynamics with nonadditive forces: a stimulation study of a designed helical protein and a random heteropolymer. **J. Chem. Phys.** (110) 11616-11629.
13. Dokholyan, N.V., Buldyrev, S.V., Stanley, H.E., *et al.* (2000). Identifying the protein folding nucleus using molecular dynamics. **J. Mol. Biol.** (296) 1183-1188.
14. Buxbaum, N.J. (2003). Diseases of protein conformation: what do in vitro experiments tell us about in vivo diseases?.**Trends in Biochemical Sciences** (28) No. 1, 585-592.
15. Hogg, J.P. (2003). Disulphide bonds as switches for protein function. **Trends in Biochemical Sciences** (28) No. 4, 210-214.
16. Chuang, C.C., Chen, Y.C., Yang, M.J., *et al.* (2003). Relationship between protein structures and disulphide bonding patterns. **Proteins** (52) 1-5.
17. Abkevich, V.I. and Shaknovich, E.I. (2000). What can disulphide bond tell us about protein energetics, function and folding: stimulations and bioinformatics analysis. **J. Mol. Biol.** (*300*)975-985.
18. Clarke, J., Hounslow, A.H., Bond, C.J., *et al.* (2000). The effects of disulphide bonds on the denatured state of barnase. **Protein Sci** (9)2394-2404.

19. Cavagnero, S., Dyson, H.J. and Wright, P.E. (1999). Effect of helix destabilizing mutations on the kinetic equilibrium folding of apomyoglobin. **J. Mol. Biol.** (285) 269-282.

20. Laurents, D.V., Corrales, S., Elias-Arnanz, M., *et al.* (2000). Folding kinetics of phage 434 Cro protein. **Biochemistry** (39), 13963-13973.

21. Oxana, V.G., Garbuzynskiy, O.S. and Ivankov, N.D. (2003). Chain length is the main determinant of the folding rate for proteins with three-state folding kinetics. **Proteins** (51) 162-166.

22. Fersht, A. R. (2000). Transition-state structure as a unifyingbasis in protein-folding mechanisms: contact order, chain topology, stability, and the extended nucleus mechanism. **Proc. Natl. Acad. Sci.** *USA* (97) 1525-1529.

23. Eaton, W.A., Munoz, V., Hagen, S.J., *et al.* (2000). Fast kinetics and mechanisms in protein folding. **Annu. Rev. Biophys. Biomol. Struct**. (29) 327-359.

24. Plaxco, K.W., Simmons, K.T., Baker, D., *et al.* (1998). Contact order, transition state placement and the refolding rates of single domain proteins. **J. Mol. Biol.** (227) 985-994.

25. Thirumalai, D. (1995). From minimal models to real proteins: Time scales for protein folding kinetics. **J. Phys.** (5) 1457-1467.

26. Galzitskaya, O.V., Ivankov, D.N. and Finkelstein, A.V. (2001). Folding nuclei proteins. **FEBS Lett,** (489) 113-118.

27. Finkelstein, A.V. and Badretdinov, A.Y. (1997). Rate of protein folding near the point of thermodynamic equilibrium between the coil and the most stable chain fold. **Des.** (2) 115-121.

28. Ivankov, N.D., Garbuzynskiy, O.S., Alm, E., *et al.* (2003). Contact order revisited:Influence of protein size on the folding rate. **Protein Science** (12) 2057-2062.

29. Burns, L.L., Dalessio, P.M., and Ropson I.J. (1998). Folding mechanisms of three similar b-sheet proteins. **Proteins** (33) 107-118.

30. Cavegnero, S., Dyson H.J., and Wright, P.E. (1999). Effect of helix destabilizing mutations on the kinetic and equilibrium folding of apomyoglobin. **J. Mol. Biol.** (*285*) 269-282.

31. Jackson, S.E. and Fersht, A.R. (1991). Evidence for a two-state transition. **Biochemistry** (30) 10428-10435.

32. Baker, D. (2000). A surprising simplicity to protein folding. **Nature** (*405*) 39-42.

33. Stickle, D.F., Presta, L.G., Dill, K.A., and *et al.* (1992). Hydrogen Bonding in globular proteins. **J. Mol. Biol.** (226) 1143-1159.

34. Chan, H.S. (1998). Matching speed and locality. **Nature** (392) 761-763.

35. Plaxco, K.W., Simons, K.T., Ruczinski, I., *et al.* (2000). Topology, stability, sequence, and length: defining the determinants of two-state protein folding kinetics. **Biochemistry** (39) 11177-11183.

36. Kaya, H. and Chan, H.S. (2003). Simple two-state protein folding kinetics requires near –Levinthal thermodynamic cooperativity. **Proteins** (52) 510-523.

37. Huseyin. K. and Hue, S.C. (2003). Contact order dependent protein folding rates: Kinetic consequences of a cooperativity interplay between favourable nonlocal interactions and local conformational preferences. **Proteins** (52) 524-533

38. Ding, F., Borreguero, M.J., Buldyrey, V.S., *et al.* (2003). Mechanism for the a-Helix to b-Hairpin

Transition. **Proteins** (*53*) 220-228.

39. Kelly, J.W.(1998). The alternative conformations of amyloidogenic proteins and their multi-step assembly pathways. **Curr. Opin. Struct. Biol**. (*8*)101-106.

40. Tan, S.Y. and Pepys, M.B. (1994). Amyloidosis. **Histopathology** (*25*) 403-414.

41. Westermark, P. *et al.* (2002). Amyloid fibril protein nomenclature. ***Amyloid*** (*9*) 197-200.

42. Buxbaum, J.N. and Tagoe, C.E. (2002). The genetics of amyloidoses. *Annu. Rev.* **Med.** (51) 543-569.

43. Buxbaum, J.N. (2003). Diseases of protein conformation: what do in vitro experiments tell us about in vivo diseases?. **Trends in Biochemical Sciences** (*28*) No.11, 585-592.

44. Guijarro, J.I., Sunde, M., Jones, J.A., *et al.* (1998). Amyloid fibril formation by an SH3 domain. **PNAS** (*95*) 4224-4228.

45. Chiti, F., Webster, P., Taddei, N., *et al.* (1999). Designing conditions for in vitro formation of amyloid protofilaments and fibrils. **PNAS** (*96*) 3590-3594.

46. Dobson, C.M. (1999). Protein misfolding, evolution and disease. **Trends in Biochemical Sciences** (*24*) 329-332.

47. Sunde, M., Serpell, L.C., Bartlam, M., *et al.* (1997). The common core structure of amyloid fibrils by synchrotron X-ray diffraction. **J. Mol. Bio.** (273) 729-739.

48. Selkoe, D.J. (1996). Amyloid beta-protein and the genetics of Alzheimer's disease. **J. Biol. Chem.** (271) 18295-18298.

49. Lansbury, P.T. (1996). A reductionist view of Alzheimer's disease. **Acc.Chem. Res.** (*29*) 317-321.

50. Prusiner, S.B. (1997). Prion diseases and the BSE crisis. **Science** (*278*) 245-251.

51. Harrison, P.M., Bammorough, P., Daggett, V. *et al.* (1998). The Prion Folding Problem. **Curr. Opin. Struct. Biol**. (7) 53-59.

52. Takahashi, Y., Ueno, A., and Mihara. H. (1998). Design of a peptide undergoing a- b structural transition and amyloid fibrillogenesis by the introduction of a hydrophobic defect. **Chem. Eur. J.** (*4*) 2476-2484.

53. Fezoui, Y., Hartley, D.M., Walsh, D.M., *et al.* (2000). A de novo designed helix-turn-helix peptide forms nontoxic amyloid fibrils. **Nat. Struc. Biol**. (7) 1095-1099.

54. Ono, S., Kameda, N., Yoshimura, T., *et al.* (1995). Supersecondary structure with amphiphilic β strands probed by phenylanaline. *Chem. Lett.* (10) 965-966.

55. Hamada, D., Segawa, S., and Goto, Y. (1996). Non-native alpha-helical intermediate in the refolding of beta-lactoglobulin, a predominantly beta-sheet protein. **Nat. Struct. Biol.** (3) 868-873.

56. Ding, F., Dokholyan, N.V., Buldyrev, S.V., *et al.* (2002). Molecular dynamics simulation of the SH3 domain aggregation suggests a generic amyloidogenesis mechanism. **J. Mol. Biol**. (324) 851-857.

57. Hamada, D., Segawa, S. and Goto, Y. (1996). Non-native alpha helical intermediate in the refolding of beta-lactoglobulin, a predominantly beta-sheet protein. **Nat. Struct. Biol.** (3) 868-873.

58. Selkoe, D.J. (1996). Amyloid beta-protein and the genetics of Alzheimer's disease. **J. Biol. Chem.** (271) 18295-18298.

59. Bradbury, J. (2003). Chaperones: keeping a close eye on protein folding. **The Lancet** (361) 1194-1195

60. Vainberg, *I.E.*, Lewis, S.A., Kellermann, J., *et al.* (1998). Prefoldin, a chaperone that delivers unfolded proteins to cytosolic chaperonin. **Cell** (93) 863-873.

61. Thulasiraman, V., Yang, C.F. and Frydman, J. (1999). In vivo newly translated polypeptides are sequestered in a protected folding environment. **EMBO J.** (18) 85-95.

62. Cohen, E.F. (1999). Protein misfolding and prion disease. **J. Mol. Biol.** (293) 313-320.

63. Helen, R. S. and Neil, A.R. (2002). The chaperonin folding machine. **Trends in Biochemical Sciences** (27) No. 12, 627-632.

64. Claudio, S. (2001). Protein misfolding and disease; protein refolding and therapy. **FEBS letters** (*498*) 204-207.

65. Claudio, S. and Gabriela, P.S. (2001). Prions: disease propagation and disease therapy by conformational transmission. **Trends in Molecular Medicine** (7) 109-114.

66. Soto, C., *et al.* (2000). Reversion of prion protein conformational changes by synthetic b-sheet breaker peptides. **Lancet** (355) 192-197.

Advances in Biochemistry and Biotechnology Volume 1 *Pages 107-121*

Edited by: **Dr. Chiranjib Chakraborty**

Published by: **DAYA PUBLISHING HOUSE**

5

Recombinant Therapeutic Proteins from Bacteria, Animal and Plant Cells

J.K. Deb

Department of Biochemical Engineering and Biotechnology, I.I.T., Delhi, New Delhi – 110 016

ABSTRACT

In the past two decades, the growth of biotechnology industries has witnessed great progress in the marketing of protein therapeutics which include cytokines, growth factors, hormones, blood clotting factors, enzymes, peptide drugs, antibodies and the list is growing rapidly. Spectacular advances in gene expression technology have made this progress possible. Beginning with the first recombinant human insulin produced by Eli Lilly in 1980s, majority of the recombinant proteins were produced by *Escherichia coli*. The trend has changed during the last decade. Due to the limitations of *E. coli* in folding protein, secretion of proteins in the medium and to carry out glycosylation, alternative gene expression systems have been developed. Gram positive, nonpathogenic bacteria hold promise as alternative bacterial expression systems in situations where glycosylation is not so vital for the biological activities of the therapeutic proteins. At present bacteria, yeast and animal cells constitute the most common expression systems for the production of such therapeutics. Increasing interest is now focussed on transgenic animals and plants because of their ability to produce complex proteins at low cost.

Introduction

Biopharmaceuticals have come to occupy centre stage globally in the area of therapy of both human and animal. The actual boost came towards the last 4 decades of the century just gone by when biomolecules with therapeutic potential such as insulin, antibodies, were used for treatment of diseases and ever since the demands were growing. The basic problem was the availability of these and other biomolecules since these were obtained by direct extraction of from the native sources in exceedingly low quantities. This approach was not practically viable. Besides, presence of contaminants and immunologic reactions could also pose problems in specific cases. Although synthetic and semisynthetic approaches were available at a practical level for small and medium sized molecules, these were not viable for protein therapeutics. Demand for the latter had to wait till important breakthroughs in recombinant DNA technology and the development and production of monoclonal antibodies could be exploited. In 1982, less than ten years after the above breakthrough, the first recombinant protein in the form of human insulin was made available in the market. From 1985 onwards, recombinant proteins have gained enormous importance as therapeutic agents. The total sales of therapeutic proteins was US$9000 in 1994 which is expected to hit 20 billion US dollars in 2000[1] **(TIPS 1996, 17, 450-456)**. During last ten years or so a large number of biopharmaceuticals have received license for medical use. Currently, dozens of biopharmaceuticals are in the market under various trade names[2].(*www.bio.org/aboutbio/drugs.html*)

The area of recombinant protein therapy is getting the limelight in recent times for obvious reasons and there are large number of reviews, series and books available on various therapeutic proteins that description of the entire gamut of such an area of activity in one review is formidable, if not impossible. Therefore, this review will concentrate mainly on various expression systems currently in use for the production of therapeutic proteins like cytokines, hormones, blood clotting factors, thrombolytic agents and vaccines that are either in use or at various stages of development. Recent developments in the alternative expression strategies will also be reviewed.

Most of the therapeutic proteins currently in the market are naturally occurring human or animal proteins such as insulin, growth hormone, tissue plasminogen activator, various growth factors etc. and their modified derivatives. Proteins of viruses mainly constitute recombinant vaccines. Therefore, it is evident that the success of the technology for the production of these molecules depends on the availability of an efficient expression system. As already mentioned, there are several advantages of production of recombinant proteins. It is now possible to produce drugs like G-CSF, erythropoietin which could not so far be produced by conventional methods. The drugs such as insulin or growth hormone which are available in limited amounts can now be produced in high yield. Also, the provision of a safe product free from pathogens can be better ensured by this approach.

Majority of the currently used therapeutic proteins are glycoproteins. However, a good number of such proteins do not require sugar moiety for their biological activity, *e.g.*, IL-2, GM-CSF, G-CSF. The expression system to be used for a given protein, therefore, shall have to be based on this consideration. Other factors that influence the decision are desirability of secretion, the stability and refoldability of the protein and the projected dose of protein per patient. The latter determines whether the cost of the drug becomes critical.

Bacterial Expression Systems

Escherichia coli

Among bacteria, *E. coli* occupies the position of prominence for sound scientific reasons. No other organism is better understood in terms of genetics, physiology or biochemistry than *E. coli*. It is this wealth of information that has led to the present high level of sophistication in *E. coli* recombinant gene expression. Overexpression of a protein is required to be approached systematically taking stock of major issues such as choice of a vector, optimization of expression by making sequence changes and experimenting with genetic background of the host, and finally optimization of growth conditions. It is likely that maximization of all aspects of expression is unattainable because eventually one step will become rate limiting. Thus optimization may be achieved through a combination of choices that can be changed or substituted as the situation demands.

A well designed prokaryotic expression system contains a set of optimally configured genetic elements that both transcriptional and translational aspects of protein production. One of the most important elements is the vector, which is normally a plasmid. High copy number vectors normally ensures high level of production but issues such as toxicity resulting from high gene dosage of certain genes may render such an approach counterproductive. In a situation like this, expression from a strong promoter may be tightly regulated by repression followed by induction. Excellent review has appeared recently on the fundamentals of *E. coli* expression vectors (Beneyx, 1999, Curr. Opin. Biot., 10, 411-421). Many strong promoters are now available in *E. coli*. They belong to different categories. Some are natural, others are hybrid, synthetic or mutagenized. By and large, all the systems are regulated by various inducers. Various forms of induction have been in use with *E. coli*. Common among these are *lacI*, *trp*, *pho* based on induction respectively by IPTG, tryptophan starvation and phosphate starvation. Alternatively, introduction of temperature sensitive mutant of *lacI* gene that encodes a thermosensitive *lac* repressor provides a convenient method to induce *lac* based promoters[3,4]. A disadvantage of this strategy is the upregulation of proteases following heat shock response. This problem is minimized by the use of host strains with a deficient *rpoH* (*htpR*) locus. Another regulatable promoter now being used is *araBAD* inducible by arabinose. Using this promoter, tissue plasminogen activator (tPA) has been produced in *E. coli*[4]. Of all these promoters the one that is inexpensive and subject to automatic induction (*i.e.*, requires no operator intervention) is *phoA* as it desirable in case of large scale production. Since induction requires phosphate starvation, it could limit the duration of protein synthesis. Expression yields can be improved from *phoA* promoter wothout loss of control when PhoS phosphate sensing protein was mutated to allow promoter induction at higher phosphate concentrations. Phosphate can then be fed continually (to avoid phosphate starvation) without repressing the promoter.

Another problem experienced with expression in *E. coli* has to do with its inability to form disulphide bonds in proteins with multidisulphide bonds. The formation of disulphide bonds in *E. coli* is a complex process involving several genes *viz.*, *DsbA*, *B*, *C and D* of which *DsbA* is a oxidoreductase and *DsbC* is a disulphide isomerase. In spite of this it has been observed that that several proteins, especially of eukryotic origin, such as tPA, various growth factors, protease inhibitors which contain substantial number of disulphide bonds can be folded properly in this organism and one has to rely on their expression in eukaryotic system. However, coexpression of *DsbA* and *DsbC* has made it possible to obtain recombinant tPA having as high as 17 disulphide bonds in physiologically active form in *E. coli*. Overproduction of proteins very often leads to the misfolding of proteins in the cytoplasm leading to the formation of insoluble aggregates called inclusion bodies. Although this simplifies purification of proteins from the cytoplasmic milieu, the recovery of native protein requires further manipulation. Proper folding of protein from the inclusion bodies requires control of aggregation. Several approaches have been employed to achieve this. Recent review has dealt with this topic[5]. Tran-Moseman *et al.*, obtained renaturation of human macrophage colony stimulating factor (hM-CSF) from the reduced, denatured monomers using thiol mixture glutathione and oxidized glutathione and a low concentration of denaturing agent (urea or guanidinium chloride). They obtained higher renaturation yield using purified reduced, denatured protein. A maximum of 95% refolding was obtained under this condition[6].

Several chaperones have been implicated in the folding process. In vivo, these chaperones participate by transient interaction with folding intermediates during the folding thereby promoting proper isomerization and cellular targeting. The complex process involves participation of ATP dependent DnaK-DnaJ-GroE and GroEL-GroES systems of chaperones[8]. Finding the optimal refolding conditions is still relatively empirical, with the best results obtained from evaluating a variety of conditions affecting solubility and disulphide bond formation and isomerization. The use of either soluble or immobolized folding aids can sometimes be effective. Altimarano *et al.*, .[9] could improve the folding yield of scorpion toxin Cn5 from less than 5% to more than 85% by immobilization of a minichaperone, prolyl isomerase and the periplasmic oxidoreductase DsbA. Interestingly, Berges *et al.*, have reported export of GM-CSF and Interlukin 13 to periplasm by fusing the coding sequences of these cytokines with two synthetic signal sequences and observed that both the total amount of protein and the fraction of soluble protein in the periplasm depended on the signal peptide-protein combinantion. Overexpression of SecB or DnaK plus DnaJ resulted in marked increase in the amount of protein in the periplasm, but it also depended on the signal peptide-cytokine-chaperone combination[10].Although periplasm is where the proteins can better be folded in contrast to the cytoplasm of *E. coli*, it has not evolved to support folding of large, complex proteins with many disulphide bonds. However, modification of folding environment can sometimes give results as mentioned above for tPA[11]. In the case of IGF-I, with three disulphide bonds, periplasmic expression of DsbA or DsbC actually decreased the yield of folded product, although it significantly increased its accumulation in the inclusion bodies[12].

Inability of *E. coli* to secrete proteins has been a limitation from downstream processing viewpoint. Attempts to selectively release recombinant proteins from this organism into the surrounding medium have not met with much success. Recently, it was shown that overexpression of third topological domain of transmembrane protein TolA (TolAIII) in the periplasm makes the cell leaky by disrupting the integrity of the outer membrane and releasing periplasmic proteins into the medium[13]. Unfortunately, the bacteria became fragile and suffered loss in viability.

For high level production and purification of proteins in *E. coli*, several sophisticated protein-fusion systems have been developed. Fusion partners offer several advantages, such as prevention of inclusion body formation, improved folding characteristics, limited proteolysis and easy protein purification schemes. The most common fusion proteins used are glutathione-S-transferase (GST), staphylococcal protein A, streptococcal protein G, maltose binding protein (MBP). Sometimes, affinity tag, such as His_6, FLAG etc. are also used. All these permit purification of fused proteins using specific ligand.

Gram Positive Bacteria

Bacteria with Low GC Content

Many gram positive bacteria are useful for their medical, food and biotechnological potential. Another important property of these organisms is their secretory power. Industrially important, nonpathogenic bacteria, such as *Bacillus, Streptomyces*, are well studied as far as their genetic manipulation is concerned. Efficient expression vectors are available for these bacteria. Gram positive bacteria such as these belong to two distinct groups based on the GC content of their DNA. While *Bacillus* and related genera have low GC content, *Streptomyces* has high GC content. Among the low

GC content gram positive bacteria, those belonging to the genera *Lactococcus, Lactobacillus,* some nonpathogenic species of *Streptococcus* and *Staphylococcus* are useful for the production of therapeutic proteins and vaccines. Among high GC content gram positive bacteria, the nonpathogenic *Corynebacteria* are also well researched as far as the genetic manipulation is concerned. Bacteria such as *Bacillus subtilis, Bacillus brevis, Bacillus licheniformis, Bacillus megaterium, Corynebacterium glutamicum, Streptomyces, Lactobacillus and Lactococcus* are safe orgainisms, and hence, suitable for the production of therapeutic proteins. Although *B. subtilis* is the most important member of this group, its main drawback is high level secretion of various proteases. Several mutants lacking in two of the major proteases and some lacking other minor proteases have been isolated to get around the above limitation. Using a *B. subtilis* mutant, WB600, deficient in 6 extracellular proteases, Wong *et al.,*[14] secreted streptokinase, the popular thrombolytic drug synthesized by *Streptococcus equisimilis,* with the help of inducible SacB promoter and its signal sequence fused to the streptokinase gene and also with its own promoter and signal sequence. The mature protein was found to have the same N-terminal sequence as that of the native streptokinase from *S. equisimilis. B. subtilis* was also used to produce staphylokinase, a 136-amino acid long thrombolytic protein secreted by *Staphylococcus aureus.* Dramatic improvement in the production of this protein was observed when the mutant in which the gene for subtilisin was disrupted was used. This strategy also improved the stability of the extracellular staphylokinase[15]. An alternative bacillus species having negligible protease activity is *Bacillus brevis.* Significant importance has been attached to this strain which also produces large amount of S layer proteins. Based on these proteins, expression-secretion vectors have been constructed that allowed for the extracellular production of large amount (more than 1g/l) of functional Epidermal Growth Factor (EGF)[16]. In another study, erythropoietin (EPO) and its extracellular receptor were also produced, albeit less efficiently[17]. Bacteria belonging to the genuses *Streptococcus, Staphylococcus, Lactococcus, Lactobacillus* have mostly been used in vaccine development. Among the organisms belonging to the first two classes are those which are food grade bacteria, such as *Staphylococcus carnosus, Staphylococcus xylosus, Streptococcus thermophilus* or *Streptococcus gordonii,* the latter being a commensal bacterium of human oral cavity. All these strains were used to display recombinant protein on the surface using the anchor domain of staphylococcal protein A. Samuelson *et al.,* reported the display of 80 amino acid peptide from a malaria blood stage antigen on the surface of *S. carnosus* and verified the presence on the surface by immunogold staining and immunofluorescence[18]. The aim was to use nonpathogenic staphylococci with a long tradition as novel vaccine carriers. Similarly, strains of both *S. carnosus* and. *xylosus* appeared to be excellent hosts for surface production of functional anti-human IgE single-chain Fv antibodies and mouse IgG1(κ)[19]. Pozzi *et al.,* used *S. gordonii* to display immunodominant epitope of HIV-1 gp120 and it was found to be recognized by human T helper cells[20]. Lactic acid bacteria such as *Lactobacillus* and *Lactococcus* assume great importance in oral vaccine development due to their presence as normal microflora of gut. Among lactobacilli *Lactobacillus plantarum* has received considerable attention due to its capacity to secrete protein to a reasonable degree. With this bacterium, α-amylase and levanase have been secreted using homologous expression-secretion signals both located on the chromosome and on plasmids[21]. In recent times, several constitutive and regulated expression systems have been developed in *Lactococcus lactis,* widely used as a starter in cheese production. A T7 promoter-T7 RNA polymerase based expression system of this bacterium has been used to express murine IL-2 extracellularly[22]. Both lactobacilli and lactococci secretion systems based on S layer protein of *lactobacillus brevis* have been developed[23]. Great deal of studies on the adjuvanticity and immunogenicity of these two classes of bacteria have been carried out. *Lactobacillus casei* var. Shirota and *L. plantarum* ATCC 8014 showed strong adjuvant effects, equivalent to that of common water-in-oil emulsion adjuvant, Specol[24]. Administartion of lactobacilli and lactococci by parenteral route induce the formation of antibodies of

IgG class in experimental animals. Both types of bacteria have been used to express antigens intracellularly, extracellularly or on the surface. Mice immunized with live recombinant bacteria by parenteral, oral or nasal routes and significant immune response was observed following mucosal administartion when the priming reaction was followed by a booster immunization. Parenteral immunization resulted in antibody production following priming immunizations only[25]. High level of expression of tetanus toxin fragment C and protection against lethal challenge was observed with recombinant *L. lactis*[26,27].

Gram Positive Bacteria with High GC Content

Major bacteria of this class are members of actinomycete group. The importance of these bacteria are either due to their pathogenicity, as in the case of *Mycobacteria, Nocardia* or due to their therapeutic or industrial uses, as in the case of *Streptomyces,* certain species of *Nocardia, Coryenbacteria,* except several species of the latter, *e.g., Corynebacterium diphtheriae,* which are pathogenic. Of these organisms, *Streptomyces* has been used for the production of several therapeutic proteins in recent times. The advantage of using this organism is its well understood molecular genetic systems. Efficient cloning and expression vectors, including inducible vectors, are available for these bacteria. The best studied of the *Streptomyces* species is *Streptomyces coelicolor* A(3). The other common species is *Streptomyces lividans. Streptomyces* species have been useful for expressing other actinomycete genes. For example, inducible expression of histidine-tagged *Nocardia* proteins has been made possible in *S. lividans* using *tipA* promoter, which can be induced by thiostreptone in *streptomyces*[28]. Although *S. coelicolor* A3(2) is the most extensively characterized species, *S. lividans* 66 is almost exclusively the species of choice as a host for heterologous protein production. The reason for this is that *S. coelicolor* A3(2) possesses a very powerful restriction system which *S. lividans* lacks. In addition, *S. lividans* is also unlike most other streptomyces in that it generally displays very low level of endogenous extracellular proteolytic activity and at the same time retains the ability to produce a large quantities of extracellular protein.

Of all the actinomycetes, perhaps *Streptomyces* is the organism in which maximum number of therapeutic proteins have been expressed. Thus, Koller *et al.*,[29] have produced proinsulin with an yield of.300 mg/L. Besides, TNF α[30], GMCSF[32] and IL 1β[33] were produced in quantities ranging from 10–100 mg/L, except in case of IL1β, where the yield was 1-5 mg/L.

Another member of actinomycetes which has the potential for such exploitation is food grade corynebacteria such as *Corynebacterium glutamicum.* This strain and other related species have no detectable extracellualr proteases or toxins[34]. Another advantage of this system is similarities of various regulatory processes between corynebacteria and *E. coli* and efficient functioning of common *E. coli* plasmids like *tac, lacUV, trp* in corynebacteria[35]. Various expression and secretion vectors are now available in this system[36-38]. For easy purification of recombinant proteins, gene fusion vectors with streptavidin and Glutathione-S-transferase (GST) have also been constructed[39,40]. Using GST-fusion vector, pBKGEXm2, streptokinase has been produced in *Corynebacterium acetoacidophilum*[40]. One important finding of this work is the presence of bulk of the streptokinase in the medium, indicating secretory nature of this system. Recently, these authors have developed a corynebacterial secretion vector which efficiently secretes heterologous proteins into the culture medium (unpublished observations).

Yeast Expression Systems

Saccharomyces, Hansenula, Kluiveromyces

Yeasts are simplest of eukaryotic organisms capable of performing eukaryotic processing steps of proteins and thus constitute attractive systems for heterologous gene expression. Of these *Saccharomyces cerevisae* is a GRAS level organism and most widely studied yeast, as regards its basic genetics and biotechnology. Since excellent reviews are available on the expression of heterologous genes by these orgainism, the present discussion will be limited to the brief mention of the promoters and vectors used for the purpose. Most of the vectors used for cloning and expression are *E. coli-S. cerevisae* shuttle vectors. Various versions of vectors are available. Depending upon their mode of replication they are classified into three major types, *viz.*, YIp, YEp and YRp. The former does not replicate independently and gets integrated into the chromosome. Consequently, these plasmids have low copy number. These gene dosage limitation has been resolved either by employing yeast transposon to create multiple insertions or by targeting the cloned gene to ribosomal gene cluster[41]. The vector of Yep type are usually present in high copy number (30 or so) and replicate independently of the yeast chromosome due to the presence of *ARS* sequence and the *STB* locus derived from yeast relicating 2μ plasmid. One disadvantage associated with instability of these vectors have been overcome by introducing *srb1-1* mutation that makes the host strain osmotically sensitive, the wild type gene *SRB1* incorporated in the vector complements the deficiency of the host strain[42]. The expression of heterologous genes have been achieved using either a constitutive or an inducible promoter. Some of these promoters belong to genes of highly expressed glycolytic enzymes, such as *GAPDH, PGK*. Other important *S. cerevisiae* promoters are *ADH1, GAL1/GAL10, PHO5, MFα1*. Several important therapeutic proteins such as 1 antitrypsin, hirudin, HbsAg, HuIFN α, HuEGF, HuSOD, human serum albumin (HSA) have been produced in *S. cerevisiae*. Although initially *E. coli* was tried for production of HbsAg, it failed to make complex particle of antigen necessary for its immunogenicity. Until recently, *S.cerevisiae* occupied the exalted position of being the sole organism for the industrial production of HbsAg and insulin globally. Recently, *Pichia pastoris* is being used for the production of HbsAg.

Pitchia pastoris

P. pastoris has become another popular yeast system for the production of recombinant proteins and nearly 300 foreign proteins have been produced since 1984 using this system. This yeast, besides *Hansuela, Candida and Torulopsis*, share a general methanol utilization pathway. In methanol grown cells, the key enzymes of this pathway constitute 30% of total cellular proteins. Two compelling factors for the popularity of *P. pastoris* are: i) the availability of a strong alcohol oxidase I (*AOX1*) promoter, the strongest and most regulated promoter known, and ii) the ability to integrate expression plasmids at specific sites in its genome. The ability to grow on methanol has given an added advantage to this organism, because mineral medium consisting of methanol, biotin, salts and trace elements are economical and well defined, ideal for large scale production in fermenters. It can produce protein at very high level and carry out higher eukaryotic modification like glycosylation, disulphide formation and proteolytic processing. For this reason, many proteins which could not be produced in active forms in bacteria, *S. cerevisiae* and baculovirus, have been produced in this organism.

The *AOX1* promoter is tightly repressed by glucose and most other carbon sources. Therefore, with this promoter the expression remains repressed by growing cells with any of the above carbon source, especially glucose or glycerol, to a high cell density followed by derepression by adding methanol as sole carbon source whereupon the induction level exceeds 1000 fold. Several therapeutic human proteins, such as insulin, HSA, chitinase, antithrombin II, type I collagen, malaria vaccine antigen, bile salt stimulated lipase, antibody cahin fragment C of botulinum antitoxin etc. have been produced in fermenters by *P. pastoris*. All the above proteins, except the last mentioned one, were secretory products and were produced at levels of several milligrams to gram per litre. The yield of insulin was 1.5 g/L[43].

There is a limitation with *AOX1* promoter because of its induction with methanol, a toxic and hazardous substance. Several other promoters, such as constitutive glyceraldehydes-3- phosphate dehydrogenase (*GAP*) and inducible glutathione dependent formaldehyde dehydrogenase, *FLD1*, promoters have been used. Constitutive and high level expression with these promoters are also toxic to the cells. So, a moderately expressed promoter such as *PEX8* promoter ensures modest induction when cells are shifted from glucose to methanol medium. Even in this case also induction by methanol becomes a necessity.

Transgenic Plants as Expression System

The use of plants as production system for foreign proteins has gained wide spread support due to several reasons. Producing therapeutic proteins in plants has many economic and qualitative benefits, including reduced health risks due to pathogen contamination, comparatively high yield, production in seeds or other storage organs. Overall, transgenic plants are potentially cheap sources of recombinant proteins and it is estimated that the cost of production in plants could be 10- to 50- fold cheaper that in recombinant *E. coli*[44].

Besides economic advantages, there are qualitative benefits in producing recombinant proteins in transgeneic plants. Animal expression systems are expensive and sensitive to environmental changes when cultured on an industrial scale. Animal cells are very sheer sensitive and so require special design of reactors. Careful control of culture conditions is required to ensure product purity[45]. Although microbial systems are robust, they are not ideal for the synthesis of many mammalian proteins because of differences in processing and posttranslational modifications. Such differences are not altogether non-existent between plants and animals. Nevertheless, these issues deserved to be sorted out to make such an enterprise a success. For example, high mannose type N-glycans found in plant derived antibody have β (1,2)-xylose and α (1,3)-fucose residues linked to the core Man3GlcNAc2. Being absent in animals, such linkages are potentially immunogenic and may have limited scope as parenterals or for topical or oral use, specially in patient with severe food allergies. One major development in this direction has been the stable cloning of human β 1,4-galactosyl transferase in tobacco plants[46]. This enzyme, missing in plant, is responsible for conversion of typical plant glycans into human glycans. A cross of a tobacco plant expressing the above enzyme with a plant expressing heavy- and light- chains of mouse antibody resulted in the expression of a an antibody in which galactosyl glycan is 30%, a level that is about as abundant as the same antibody produced by hybridoma cells.

Another major problem in the expression of heterologous protein in plants has to do with the silencing of transgene. Post-transcriptional gene silencing (PTGS). It is often associated with methylation of transcribed region of the silenced gene and with the accumulation of small RNA molecules homologous to the silenced gene. The most efficient way of avoiding PTGS was to design carefully transgene avoiding repeat sequences and analyzing the transformants at the molecular level. Another strategy is to introduce a helper-component proteinase (HC-Pro) of plant poty virus into plant, which results in the loss PTGS. HC-Pro brings this about by targeting a PTGS maintenance component at a point that affects accumulation of small RNAs and methylation of genomic DNA[47]. Notable progress has been achieved in recent years in reducing gene silencing problem and position effects of nuclear transformation by introducing a foreign gene between the functional genes of chloroplast using homologous recombination. High level expression of oligomeric cholera toxin B (4-5% of total soluble protein) and human somatotropin (7% of total soluble protein) has been achieved by such a strategy[48].

Although plant is projected as a low cost production system for protein expression, downstream processing of plant-derived product is potentially costly process. In a recent cost analysis of β-glucuronidase production from transgenic corn seed, milling, protein extraction and purification accounted for 94% of the production cost. One way to reduce this to direct protein synthesis to seed endosperm from where proteins could be extracted. Thus, avidin was extracted from transgenic maize seed endosperm at a cost 10 fold cheaper than that conventionally purified from eggs[49]. Different strategies are used to direct protein accumulation in different parts of the plant. For example, hirudin, an anticagulant used for treating thrombosis, has been produced as oleosin-hirudin fusion protein which allowed for extraction of the complex from oil bodies of recombinant oilseed rape[50]. Using various promoters transgenic proteins have been deposited in specific parts of the plant. Thus, using promoters of genes with aleurone specific expression during germination such as α-amylase and signal export peptide recombinant β-glucanase could be exported to the endosperm. On the other hand, using promoter of structural genes for storage proteins such as D hordein (*Hor3-1*) promoter, the same protein was produced in maturing grain in barley[51].

In recent times, *in vitro* plant cell culture technology has become an important tool for the expression of foreign protein. This may be the most suitable system for medium level production of high price, high purity speciality molecules, such as therapeutic proteins. However, plant cell culture is unlikely to be competitive for the majority of proteins required in large quantities that is otherwise deliverable by agricultural means. It may be useful if the product is accumulated at levels at least one order of magnitude higher than those achieved by agricultural systems. This would reduce the cost associated with product recovery to offset the substantially higher cost of biomass production. Another advantage with plant cell culture is that the time involved for production is much shorter than the growth cycle of whole plants. Therefore, compared to months needed for agriculture, proteins would be manufacture within days or weeks. From the safety and containment points of view, plant cell culture is advantageous as it is amenable to growth in bioreactors under controlled conditions.In addition, when the protein of interest is secreted into the medium of cell cultures, product recovery and purification could be greatly simplified, while eliminating the need to destroy the cells.

Mammalian Expression Systems

During past nearly a decade there has been considerable progress in the development of mammalian expression systems for high level recombinant protein production. The expression can be stable or transient. In stable expression technology, which is based on chromosome-integrated plasmid sequences, good quantities of complex proteins from mammalian cells can be obtained[52]. On the other hand, milligram to gram quantities of protein can be obtained by transient expression strategy. The transient transfer of DNA to mammalian cells serves an important role in the manufacture of products involved in somatic gene therapy. Such a system has several key features like i) simplicity in the construction of expression vectors, ii) short duration for the generation of product (days), iii) intrinsic genetic stability and consistencey due to short period elapsed between generation of vector and product recovery, iv) applicability to a wide range of host cell lines.

Cell types that dominate the field of recombinant mammalian system are chinese hamster ovarian (CHO), COS, baby hamster kidney (BHK) and NSO murine myeloma cell lines. Refinements of vector construction, use of strong and regualatable promoters, choice of selectable markers have made the establishment of recombinant cell lines with high specific productivities. Normally, the vectors are either plasmid based or virus based. Strong promoters commonly used are derived from SV40 or CMV virus[53,54]. One nonviral promoter, the elongation factor (EF-1) promoter, has attracted attention because it appears to be as strong or stronger than some viral promoters[55]. Use of elements that provide extended maintenance of plasmids in cells have now been possible in transient expression of genes. These vectors are used in combination with cell hosts that contain elements for the promotion of episomal plasmid replication, such as large T antigen from SV40. The plasmid would then carry the origin of replication from SV40. Also, identification of 'hot spot of integration' within the chromosome resulting in stable transformation and the enhancement of biomass accumulation has increased productivity of recombinant mammalian cells, *e.g.*, an yield of 1-2 g/L of unpurified antibody has been with NS0 myeloma cells[56]

Alphavirus, adenovirus and vaccinia virus are widely used for construction of viral vectors. Sindbis virus and Semliki Forest virus (SFV) are the most studied alphaviruses used for heterologous expression[57]. The broad host range of SFV has particularly made it attractive. In order to use as a safe vector, two plasmids are made from SFV, one for cloning, pSFV1, which carries the replicase genes and the other, the helper plasmid, that codes for the structural proteins of the virus. To improve the safety condition, the system was modified to allow the production of non-infectious SFV particles, which can be subsequently activated by enzymatic cleavage of the spike protein. The extremely efficient SFV 26S promoter used and high copies of RNA (200,000) per infected cell lead to high expression levels. Recombinant proteins are produced with post-translational modifications conferred by the specific processing capacities of the host cell selected[58, 59]. This system has also been scaled up[60]. The extensive knowledge accumulated over the years on adenovirus transcription regulation allowed the engineering of adenovirus vectors modified for heterologous expression. For this purpose, the early regions of E1 and E3 were deleted to render the virus replication incompetent and this function must be provided in trans by the host cell. An expression cassette set in the deleted E1 region and containing additional (or ectopic) major late promoter, such as CMV was used to clone the foreign gene. A series of vectors, pAdBM5, was constructed with enhancers added to the ectopic major late promoter region[61]. The ability of the virus to be propagated in suspension culture is of considerable advantage for large scale work[62]. One of the problems with adebnoviral vectors is the elicitation of humoral and cellular immune response directed against the viral gene products, albeit at a low level, even in the absence of E1A gene product. Newer vectors which are less immunogenic gave been developed, which are defiant or defective in the E2, 3 and E4 gene regions. These vectors can be propagated in trans complementing cell lines. Some vectors, called "gutless vectors" have been developed in which only the vector inverted terminal repeats (ITRs) and viral packaging signal are present. Theoretically it can accommodate upto 35 kb of insert but requires all the viral gene products in trans that are needed for virus assembly. Vaccinia virus genome is double stranded DNA virus of nearly 200 kb and replicates in the cytoplasm of the host cell. Infection with this virus often leads to 5000viral particles per cell, which leads to high level recombinant protein expression. The recombinant gene is placed under the control of viral promoter. Normally, late promoters are preferred or compound early/late promoter may be chosen to minimize the cytopathic effect. Advantages with vaccinia virus are that integration into the chromosome can be done by homologous recombination, direct cloning and wide host range. A disadvantage with this system is that high safety level is required. Vaccinia system has been successfully used at very large (1000 liter) to produce many different proteins[63].

One of the notable areas of advances in the application of genetic engineering approaches to rationally modify specific features of mammalian host to improve their utility in recombainant protein expression. Foe example, work during the past decade has shown how various reactions in the glycosylation pathway can be influenced by cell culture factors, host cell selection and protein specific features, leading to the production of molecules with variable or suboptimal clearance or bioactivity properties[64,65]. Studies have shown that specific manipulation of oligosaccharide structures on a recombinant protein by overexpression of appropriate glycosyl transferases can enhance glycan quality. For example, overexpression of a galactosyltransferaseand a silayltransferase in CHO cells led to corresponding increases in the galactose and sialic acid content of expressed protein[66].

Another important factor has been to improve the efficiency of central carbon metabolism and to reduce lactate accumulation. In one approach, partial disruption of lactate dehydrgenase A gene led to reduced lactate accumulation and improved cell culture performance in hybridoma production[67]. Perhaps the most interesting of the targets for metabolic control have been that of cell growth and the limit of apoptotic cell death. The ability of chemical inhibitors of apoptosis to extend culture viability has opened up the growing area work directed towards genetic engineering of cell death pathways. Overexpression of Bcl-2 and Bcl_{xL} has been shown to inhibit apoptotic death in several mammalian culture systems[68]. A recent report also showed increased volumetric antibody expression in butyrate treated CHO cultures overexpressing Bcl-2[69].

Conclusion

In recent years, high demand for therapeutically active protein molecules has spawned intense research in the development diverse expression systems for the production of these potent biomolecules. Several organisms spanning bacterial, yeast, fungal, plant and animal cells have been used as systems for the purpose depending upon the regulatory need and bioactivity of the product. Although *E. coli* commanded highest applicability and capacity utilization is currently 90%, not all therapeutic proteins can be produced in *E. coli* because the bacteria are unable to modify proteins after they are produced, which can cause misfolded and inactive proteins. For these reasons, monoclonal antibodies, fusion proteins and other proteins that must be modified must be produced ideally in mammalian cells, such as CHO, NS0, BHK etc. Being more cost intensive, other eukaryotic systems, such as yeast, fungi are sometimes exploited as substitute for mammalian system. Even such a system may not be ideal where glycosylation akin to mammalian system becomes a necessity.

The role of plant tissue cultureas a means for commercial protein production is not clear at present, but is likely to depend upon the ability to express high levels of proteins in vitro by manipulating culture conditions. Extracellular localization and the stability of secreted protein in the medium of plant cultures may be important keys to economic feasibility, allowing for easier product recovery without destroying biomass. Because the controlled environment in plant cultures offers advantages for good manufacturing practice (GMP) and regulatory compliance, a potential niche for reactor-scale plant tissue culture is the rapid production of low-to-medium volume therapeutic proteins. Technology for modifying the glycans of plant glycoproteins may be required before systems can be fully exploited for the manufacture of injectable drugs and vaccines.

Mammalian derived cell production of biopharmaceuticals has a good regulatory track record, and allows for the modification of proteins with reduced immunogenicity. Compared to production by *E. coli* or other bacterial system, however, the construction of genetically engineered cells is more time consuming, production is less efficient and more expensive, and purification is more complicated.

REFERENCES

1. Peter Buckel (1996) Recombinant proteins for therapy Tr. in Pharmacology, 17, 450-456.
2. www.bio.org/aboutbio/drugs(html <http://www.bio.org/aboutbio/drugs.html>)
3. Gerhard Hannig and Savvas C. Makrides (1998) Strategies for optimizing heterologous protein expression in Escherichia coli Tr. in Biotechnol. 54, 16

4. Hasan, H. and Szybalski. W. (1995) Construction of *lacIts* and *lacI^q^ts* expression plasmids and evaluation of the thermosensitive *lac* repressor Gene 163, 35-40

5. Poole, E.S., Brown, R. E. and Buckel, P. (1995) The identity of the base following the stop codon determines the efficiency of in vivo translational termination in Escherichia coli EMBO J. 14, 151-158

6. Qui J, Swartz, J.S. and Georgiou, G. (1998) Expression of Active Human Tissue-Type Plasminogen Activator in *Escherichia coli* Appl. Env. Microbiol. 64, 4891-4896.

7. Lilie H, Schwarz, E and Rudolph. R. (1998) Advances in refolding of proteins produced in *E. coli* Curr. Opin. Biotech., 9, 497

8. AnhDao Tran-Moseman, Neil Schauer and Eliana De Bernardez Clark (1999)

9. Renaturation of*Escherichia coli*-Derived Recombinant Human Macrophage Colony-Stimulating Factor Protein Express Purif, 16, 181-189

10. Richardson. A, Landry. S.J. and Georgopoulos. C (1998)The ins and outs of a molecular chaperone machine Tr. in Biochem. Sc., 23, 138-143.

11. Myriam M. Altamirano, Consuelo García, Lourival D. Possani, Alan R. Fersht (1999) Nat. Biotechnol. 17, 187-191

12. H Berges, E Joseph-Liauzun, and O Fayet (1996) Combined effects of the signal sequence and the major chaperone proteins on the export of human cytokines in Escherichia coli Appl. Env. Microbiol., 62, 55-60

13. John C. Joly, W. S. Leung, and James R. Swartz (1998) Overexpression of *Escherichia coli* oxidoreductases increases recombinant insulin-like growth factor-I accumulation Proc Nat. Acad. Sci., USA 95, 2773-2777

14. Wan E.w. and Baneyx. F. (1998) TolAIII Co-overexpression Facilitates the Recovery of Periplasmic Recombinant Proteins into the Growth Medium of *Escherichia coli* 14, 13-22

15. SL Wong, R Ye, and S Nathoo (1994) Engineering and production of streptokinase in a Bacillus subtilis expression-secretion system Appl. Env. Microbiol. 60, 517-523

16. Hee Bae, Si Myung Byun, and Jae Hoon Chung (2000) Enhancement of Secretion and Extracellular Stability of Staphylokinase in *Bacillus subtilis* by *wprA* Gene Disruption Appl. Env. Microbiol. 66, 476-480

17. Ebisu. S., Takagi. H., Kadowaki. K., Yamagata. H. and Udaka. S. (1996) The efficient production of epidermal growth factor by *Bacillus* brevis ANN NY Acad Sci, 782, 115-122.

18. Nagao. M., Inoue. K., Moon. S.K., Masuda. S., Takagi. K. and Udaka. S. (1997) Secretory production of erythreopoeietin receptor by *Bacillu brevis* Biosci Biotechnol Biochem, 61, 670-674.

19. P Samuelson, M Hansson, N Ahlborg, C Andreoni, F Gotz, T Bachi, TN Nguyen, H Binz, M Uhlen, and S Stahl (1995) Cell surface display of recombinant proteins on Staphylococcus carnosus J. Bacteriol. 177, 1470-1476

20. E Gunneriusson, P Samuelson, M Uhlen, PA Nygren, and S Stahl (1996) Surface display of a functional single-chain Fv antibody on staphylococci J. Bacteriol. 178, 1341

21. Gianni Pozzi, Marco R. Oggioni, Riccardo Manganelli, Donata Medaglini, Vincent A. Fischetti, Daniela Fenoglio, Maria T. Valle, Annalisa Kunkl and Fabrizio Manca (1994) Human T-helper

cell recognition of an immunodominant epitope of HIV-1 gp120 expressed on the surface of *Streptococcus gordonii* Vaccine 1994, 12, 1071-76

22. P Hols, T Ferain, D Garmyn, N Bernard, and J Delcour (1994) Use of homologous expression-secretion signals and vector-free stable chromosomal integration in engineering of Lactobacillus plantarum for alpha-amylase and levanase expression Appl. Env. Micro. 60, 1401
23. L Steidler, JM Wells, A Raeymaekers, J Vandekerckhove, W Fiers, and E Remaut (1995) Secretion of biologically active murine interleukin-2 by *Lactococcus lactis* subsp. Lactis Appl. Env. Micro, 1995, 61, 1627
24. Kirsi Savijoki, Minna Kahala and Airi Palva (1997) High level heterologous protein production in *Lactococcus* and *Lactobacillus* using a new secretion system based on the *Lactobacillus brevis* S-layer signals Gene 186, 255-262
25. Boersma.W.J.a., Bogaerts. W.J.A., Bianchi. A.T.J., and Classen. E. Adjuvant properties of stable water-in-oil emulsions:Evaluation of the experience with Specol. (1992) Res Immunol. 143, 503
26. Peter H. Pouwels, Rob J. Leer, Michael Shaw, Marie-Joan Heijne den Bak-Glashouwer, Frans D. Tielen, Egbert Smit, Beatriz Martinez, Jan Jore and Patricia L. Conway (1998) Lactic acid bacteria as antigen delivery vehicles for oral immunization purposes Int J Food Microb. 41, 155
27. Wells, J.M., Wilson, P.W., Norton. P.M., Gasson. M.J., and LePage. R.W.F.(1993) *Lactococcus* lactis: High-level expression of tetanus toxin fragment C and protection against lethal challenge Mol. Microbiol 8, 1155-1162.
28. Karen Robinson, Lisa M. Chamberlain, Karin M. Schofield, Jeremy M. Wells, and Richard W.F. Le Page (1997) Oral vaccination of mice against tetanus with recombinant Lactococcus lactis Nat. Biot. 15, 653-658.
29. Enguita. F.J., de la Fuente., J. L., Martín J. F and Liras P. (1996) An inducible expression system of histidine-tagged proteins in *Streptomyces lividans* for one-step purification by Ni^{2+} affinity chromatography FEMS Micro. Lett, 137, 135-140
30. JE Strickler, TR Berka, J Gorniak, J Fornwald, R Keys, JJ Rowland, M Rosenberg, and DP Taylor (1992) Two novel Streptomyces protein protease inhibitors. Purification, activity, cloning, and expression. J. Biol. Chem. 267, 3236-324.
31. Koller. K-P, Reiss. G., Klaus. S., Uhlmann. E., and Wallmeier (1989) Bio/technology 7, 1055-1059
32. Chang S-Y and Chang S 1988, Biology of Actinomycetes '88- Proc. of the 7th Intl. Symopsium on the biology of Actinomycetes (Okami, S-Y., Beppu, T and Ogawara, H.eds) pp 103-107, Japan Sc. Soc. Press
33. Malek L.T. In Pharm. Biotech. Intl., 1995 (Barnacal, P.A. ed), Sterling Publ., London, pp 55-57
34. Malek L.T., *et al.* (1990) Secretion of granulocyte macrophage colony stimulating factor (GM-CSF)in Streptomyces lividans. J. Cell. Biochem. Suppl. 14A, 127.
35. Lichenstein, H *et al.* (1988) J Bacteriol., 170, 3924-29
36. Billman-Jacobe (1996), Expression in bacteria other than *Escherichia coli* Curr. Opin. Biotechnol. 7, 500-504.
37. Eikmanns, B.J. *et al.* (1991) A family of *Corynebacterium glutamicum*/*Escherichia coli* shuttle vectors for cloning, controlled gene expression, and promoter probing. *Gene* 102, 93-98.

38. Patek, M. *et al.* (1996) Promoters from *Corynebacterium glutamicum*: cloning, molecular analysis and search for a consensus motif. *Microbiology* 142, 1297-1309.

39. Cadenas *et al.* (1996) Construction of new cloning vectors for *Brevibacterium lactofermentum* FEMS Microbiol. Lett. 137, 63-68.

40. Jakoby, M. *et al.* (1999) Construction and application of new *Corynebacterium glutamicum* vectors. *Biotechnol. Techniques* 13, 437-441.

41. Srivastava, P. and Deb, J.K. (2002) Construction of fusion vectors of corynebacteria: expression of glutathione-S-transferase fusion protein in *Corynebacterium acetoacidophilum* ATCC 21476. *FEMS Microbiol. Lett.* 212, 209-216.

42. Lopes T. S. *et al.*, 1989, High copy number integration into the ribosomal DNA of *Saccharomyces cerev.*, Gene 79, 199-206

43. Rech, S.B., Stateva, L.I.,and Oliver. S.G. (1992) Complementation of *Saccharomyces cerevisiae srb1* mutation Curr. Genet., 21, 339-344.

44. Cereghino, GPL *et al.*, , Production of recombinant proteins in fermenter cultures of the yeast *Pichia pastoris* (2002) Current Opinion in Biotechnology 13, 329-332

45. Kusnadi, A., Nikolov, Z.L. and Howard, J.A. (1997) Production of Recombinant Proteins in Transgenic Plants: Practical Considerations. Biotech. Bioengg., 56, 473-484

46. Giddings, G., Allison, G., Brooks, D and Carter, A. (2000) Transgenic plants as factories for biopharmaceuticals Nature, Biotechnology, 18, 1151-1155.

47. Bakker, H., Bardor, M., Molthoff, J.W., Gomord, V.V., Elbers II, Stevens, L.H., Jordi, W, Lommen, A., Faye, L, Lerouge, P., Bosch, D. (2001) Galactose-extended glycans of antibodies produced by transgenic plants Proc. Natl. Acad. Sci., USA. 98, 2899-2904.

48. Llave, C., Kasschau, K.D., Carrington, J.C. (2000) Virus-encoded suppressor of posttranscriptional gene silencing targets a maintenance step in the silencing pathway. Proc. Nat. Acad. Sci. USA 97, 13401-13406.

49. Staub, J.M., Garcia, B., Graves, J., *et al.*, High yield production of a human therapeutic protein in tobacco chloroplasts. (2000) Nat Biotechnol., 18, 330-338

50. Hood, E.E., Kusnadi, A., Nikolov, Z.L., Howard, J.A. (1999) Molecular farming of industrial proteins from transgenic maize. Adv. Exp. Med. Biol., 464, 127-147.

51. Parmenter, D.L., Boothe, J.G., van Rooijen. G.J., Yeung, E.C., Moloney, M.M. (1995) Production of biologically active hirudin in plant seeds using oleosin portioning Plant Mol. Biol., 29, 1167-1180.

52. Horvath, H., Huang, J., Wong, O., Kohl, E., Okita, T., Kannagara, C. G., von Wettstein, D. (2000) The production of recombinant proteins in transgenic barley grains Proc. Nat. Acad. Sc., USA 97, 1914-1918.

53. Wurm, F and Bernard, A. 1999 Large scale transient expression in mammalian cells for recombinant protein production. Curr. Top. Biotechnol., 10, 156-159

54. Cachianes, G., Ho, C., Weber, R.F., Williams, S.R. and Goeddel D.V. (1993) Epstein-Barr virus derived vectors for transient and stable expression of recombinant proteins. Biotechniques 15, 255-259

55. Gorman, C.M., Giles, D.R. and McCray, G. (1990) Transient production of proteins using adenovirus transformed cell line. DNA Protein Eng. Techniques 2, 1-28

56. Kim, D.W., Uetsuki, T., Kaziro, Y, Yamaguchi, N and Sugano, S. (1990) Gene 91, 217-223

57. Zhou, W, Chen, C., Buckland, B. and Aunins. J. (1997) Fed Batch Culture of recombinant NS0 myeloma cells with high monoclonal antibody production. Biotechnol. Bioengg., 55, 783-792

58. Lundstorm, P. (1997) Alphavirueses as expression vectors Curr. Opin Biotechnol., 5, 495-500

59. Blasey, H.D., Lundstrom. K., Tate. S., and Bernard. A.R. (1997) Recombinant proteins produced by Semliki Forest Virus expression system Cytotechnology, 24, 65-72

60. Paul. N.L., Marsh. M., McKeating. J.A., Schultz, T.F., Lilejstrom, P., Garoff, H. and Weiss, R.A. (1993) Expression of HIV-1 envelope glycoproteins by Semliki Forest virus vectors AIDS Res Hum Retroviruses 9, 963-970

61. Blasey. H.D., Brethon. B., Hovius. R., Lundstorm. K., Rey. L., Tairi. A.P. and Bernarad. A.R. In New Developments and New Applications in Animal Cell Technology. Merten. O.W., Perrin. P., Griffiths. P, Dordrecht, 1998, Kluwer Academic Publishers, The Netherlands, pp 449-455

62. Massie. B, Dionne. J., lamarche. N., Fleurent. J., and Langelier. Y. (1995) Improved adenovirus provides herpes simplex virus ribonucleotide reductase R1 and R2 subunits very efficiently Bio/ Technology 13, 602-608

63. Garnier. A., Cote. J., Nadeau. I., Massie. B., Chavarie. C., and Kamen. A. (1994) Scale up of the adenovirus expression system for the production of recombinant protein in human 293 cells Cytotechnology 15, 145-155

64. Pialoux. G., *et al.* (1995) A prime-boost approach to HIV preventive vaccine using a recombinant canarypox virus expressing glycoprotein 160 (MN) followed by a recombinant glycoprotein 160 (MN/LAI) AIDS Res Hum Retroviruses 11, 373-381

65. Jenkins. N, Parekh, R.B. and James D.C. (1996) Getting the glysosylation right: implications for the biotechnology industry Nature Biotechnol. 14, 975-981

66. Grabenhorst. E., Schlenke. P, pohl., S., Nimtz. M. and Conradt. H.S. (1999) Genetic engineering of recombinant glycoproteins and the glycosylation pathway in mammalian host cells Glycoconjugate 16, 81-97

67. Weikert.S. *et al.* (1999) Engineering chinese hamster ovary cells to maximize sialic acid content of recombinant glycoprteios Nature Biotechnol. 17, 1116-1121

68. Chen. K, Liu. Q, xie, L., Sharp. P.A. and Wang. D.I.C. (2001) Engineering of a mammalian cell line for reduction of lactate formation and high monoclonal antibody production Biotechnol. Bioengg. 72, 55-61

69. Laken. H.A., Leonard. M.W. (2001) Understanding and modulating apoptosis in industrial cell culture Curr. Opin. Biotechnol. 12, 175-179

70. Kim. N.S. and Lin, G.M. (2001) Overexpression of Bcl-2 inhibits sodium butyrate induced apoptosis in chinese hamster ovary cells resulting in enhanced humanized antibody production Biotechnol. Bioengg. 73, 211-222

Advances in Biochemistry and Biotechnology Volume 1 *Pages 122-157*
Edited by: **Dr. Chiranjib Chakraborty**
Published by: **DAYA PUBLISHING HOUSE**

6

Programmed Cell Death: An Overview

Satyendra Gautam & Arun Sharma*
**Head, Food Technology Division, B.A.R.C., Mumbai- 400 085,*
Email: ksarun@apsara.barc.ernet.in

ABSTRACT

Programmed cell death (PCD) also termed, apoptosis is a genetically encoded cell death program. It ensures equilibrium between cell proliferation and cell death by eliminating damaged or unwanted cells. PCD has been observed in a widely diversified group of organisms. These include prokaryotes such as bacteria, lower eukaryotes like yeasts, *Caenorhabiditis elegans*, *Drosophila melanogaster*, higher organisms, and mammals, including humans. In a bacterial population quorum sensing, or loss of plasmid may lead to PCD. In eukaryotes PCD has been observed to take place during embryonic development, metamorphosis and homeostasis. Uniqueness of the PCD is demonstrated in the conserved nature of molecular and biochemical machinery involved. A caspase-3 like protein from a bacterium (*Xanthomonas*) has been found to have homology with human caspase. Ced-3, ced-4, and ced-9 genes of *C. elegans* have been reported in mammals as homologues of caspase, Apafs (apoptotic protease activating factors), and bcl-2 genes, respectively.

* Correspondent Author

Caspases are cysteinyl aspartate specific proteinases, which are activated by internal cell death signals via mitochondrial pathway by the release of cytochrome-C and formation of apoptosome or by external cell death signals involving cell membrane located death receptors and corresponding ligands association. Activation of caspases lead to degradation of the systems crucial for the survival of the cell such as structural proteins like lamins, or DNA repair enzymes such as Poly ADP Ribose Polymerase (PARP), and activation of endonucleases leading to DNA damage and cell death. Apafs take part in formation of apoptosome with cytochrome-C, and procaspase-9 leading to the cascade of caspase activation. Bcl-2 is a human oncogene. Bcl-2 family member proteins were first observed in B-cell follicular lymphoma. This family contains both pro (bax, bad, bak, bcl-Xs)–and–antiapoptotic (bcl- Xl, bcl-w, and mcl-1) members.

Deregulation of PCD machinery has been reported to trigger the diseased state either due to induced or repressed apoptosis. Induction of apoptosis was found to be the mechanism of various human diseases such as, neurodegenerative disorders (Alzeihmer's and Parkinson's), ischemic injury and so on. Similarly, repression of apoptosis could lead to diseases like cancer, and autoimmune disorders. The tools of apoptosis have the possibilities to be exploited for therapeutic gains. Recently, bcl-2 antisense oligonucleotide (G3139) has shown a promise towards the treatment of cancer. The present article provides an insight into the machinery and mechanisms of apoptosis reported in diversified organisms with special reference to prokaryotes.

Apoptosis: An Introduction

Programmed cell death (PCD) or apoptosis, is currently one of the hottest areas of modern biology. It describes the orchestrated collapse of cell staging membrane blebbing, cell shrinkage, chromatin condensation, protein fragmentation, and DNA degradation followed by rapid engulfment of corpses by neighboring cells. Apoptosis is a genetically regulated process of cell suicide that is central to the development and integrity of multicellular organisms (Rossi and Gaidano, 2003; Wyllie *et al.*, 1980; 1993; Wyllie, 1980). It is a selective process for deletion of damaged or unwanted cells in various biological systems. The apoptotic event similar to proliferation is tightly regulated and plays essential roles in the homeostasis of renewable tissues. It ensures equilibrium between cell proliferation and cell death (Rossi and Gaidano, 2003; Wyllie *et al.*, 1980; 1993).

Apotosis was first discovered by Carl Vogt in 1842 (Peter *et al.*, 1997). However, apoptosis research remained dormant for more than a century. In the year 1972 the process was rediscovered by Kerr, Wyllie, and Currie and termed as "apoptosis" to refer to the peculiar morphology of physiologically occurring cell death (Kerr *et al.*, 1972). It is a Greek word meaning "falling off" of petals from flowers or leaves from trees. There is a distinct and precisely localized control over the fate of specific cells in a mixed cell population that undergo apoptosis. PCD is a basic need for proper development of a living organism and widely distributed throughout the living systems. (Cory and Adams, 1998; Meier *et al.*, 2000; Opferman and Korsmeyer, 2003; Yuan and Yankner, 2000). It is an essential part of life for many organisms and the way in which most cells die is conserved from prokaryotes to mammals. Evolutionarily conserved nature of cell death machinery is beneficial in the sense that analysis of cell death in simple genetic models would provide information necessary for understanding the control of apoptosis in mammals. During development apoptosis helps to sculpture the body, and shape the organs. Prominent examples of apoptosis include the hormonally regulated involution of the tadpole tail during development, negative selection of lymphocytes to delete autoreactive or non-reactive cells, widespread cell death of neuronal cells during the self-assembly of the central nervous system, the formation of digits by involution of interdigital cells in the primitive limb paddle, and sloughing of the inner lining of the uterus (endometrium) at the start of mensuration. Both the nervous and the immune systems arise through overproduction of cells followed by death of those that fail to establish synaptic connections or productive antigen specificities, respectively. PCD is also needed to destroy cells that represent a threat to the integrity of the organism such as cells infected with pathogens, defective cells of the immune system and cells with damaged DNA (Nicholson, 2000).

Various factors regulate the switching on/off in a PCD process (Dragovich *et al.*, 1998). Some of these factors include, withdrawal of positive signals *i.e.* signals needed for continued survival such as growth factors for neurons, and interleukin-2, an essential factor for the mitosis of lymphocytes. Similarly, receipt of negative signals also affect the decision for PCD, for example, increased levels of oxidants within the cell, damage of DNA due to agents like- oxidants, UV rays, X-rays, chemotherapeutic drugs, and death activators (TNFa, TNFb, FasL). There are various human diseases, which occur due to irregular apoptosis. Some of these are because of inhibition of apoptosis. These include cancer, follicular lymphomas, carcinomas with p53 mutations, autoimmune disorders (*e.g.* systemic lupus erythematosus), and viral infections (*e.g.* herpes viruses, poxviruses, adenoviruses) (Katoch *et al.*, 2002; Nicholson, 2000; Schmitt, 2003; Tsuruo *et al.*, 2003). Some diseases are due to induction of apoptosis (Hussein *et al.*, 2003). These linclude AIDS, neurodegenerative disorders (Alzheimer's and Parkinson's diseases), myelodysplastic syndrome, ischemic injury (*e.g.* stroke, myocardial infarction), and toxin-induced liver diseases (Engidawork *et al.*, 2001; Riess *et al.*, 2003; Stadelmann *et al.*, 1999). In many cases, it is unclear whether perturbations in cell death mechanisms are casual or merely a consequence of the disease process. Nonetheless, efforts aimed at treating these diseases by manipulating cell suicide would seem to have great potential, although they are thus far at a relatively early stage.

Apoptosis Vs. Necrosis

Cells die due to physiological and pathological consequences. Many organisms selectively eliminate cells, which show deleterious activities to the host. Cell death occurs by two different pathways: apoptosis and necrosis. These two pathways are very much different to each other

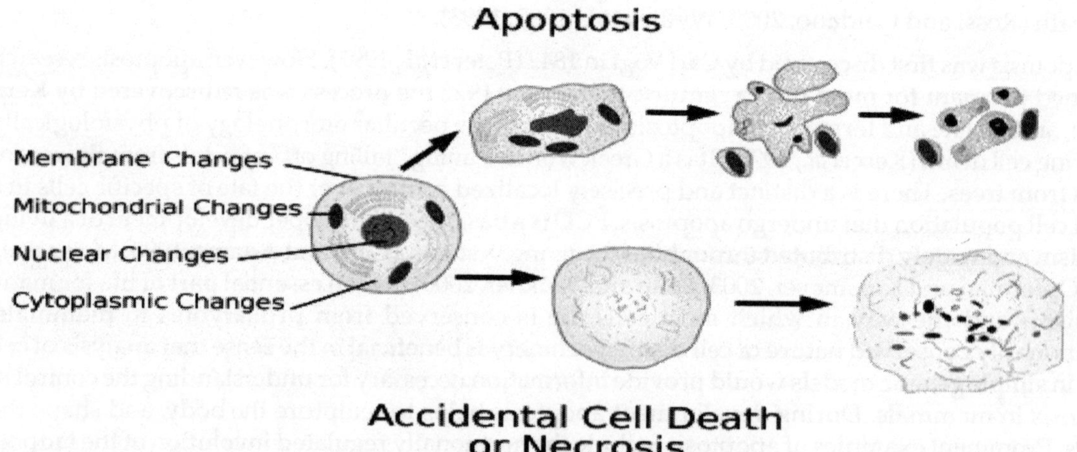

Fig. 6.1: A Eukaryotic Cell Showing Differences in Apoptotic and Necrotic Death Pattern [Reproduced from Molecular probes. Inc. (www.molecularprobes.com <http://www.molecularprobes.com>]

(Table 6.1 and Fig. 6.1) (Proskuryakov *et al.*, 2003; Ueda and Shah, 1994). During necrotic death cell swells and chromatin appears as a flocculent mass that eventually disappears to leave a nuclear ghost. DNA is non-specifically degraded and characteristically appears as a smear on an agarose gel. In apoptotic cell death, chromatin condenses and forms aggregates near the nuclear membrane, which, in turn, becomes convoluted. The nucleolous becomes enlarged and appears abnormally granular. Chromatin is degraded by different activated nucleases that cleave the DNA into 300-50 kb fragments (Brown and Rose, 1992; Oberhammer *et al.*, 1993; Wyllie, 1980). Also during apoptosis cell visibly shrinks, adherent cells round up, and distinct protuberances or membrane blebs become apparent.

Table 6.1: Differences Between Apoptosis and Necrosis

Parameters	*Apoptosis*	*Necrosis*
Stimuli	Physiological	Pathological
Occurrence	Single cell	Group of cells
Reversibility	No	Yes
Cytoplasmic organelles	Last stage swelling	Very early swelling
Lysosomal enzyme release	Absent	Present
Nucleus	Convolution of nuclear outline and breakdown	Disappearance (karyolysis)
Nuclear chromatin	Compaction in uniformly dense masses	Clumping not sharply defined
DNA breakdown	Internucleosomal	Random
Cell	Formation of apoptotic body	Swelling and later disintegration
Phagocytosis by other cells	Present	Absent
Exudative inflammation	Absent	Present

At the cell membrane level, disruption of internal and external membranes is a normal consequence of necrosis. Cells undergoing apoptosis exclude vital dyes such as trypan blue, propidium iodide (PI), rhodamine, acridine orange at early stages indicating membrane integrity. During programmed cell death, loss of cell membrane integrity is a very late event usually preceded by the destructive action of endogenous cellular enzymes (Kroemer *et al.*, 1995). Cell organelles within the shrunken apoptotic cytoplasm retain a largely normal appearance except for some dilation of the endoplasmic reticulum (ER) and swelling of the mitochondria. Recognition and phagocytosis of apoptotic cells is mediated by a variety of independent receptor-ligand interactions. Activation of flippase in apoptotic cells leads to externalization of phosphatidylserine, usually present only in the internal leaflet of the plasma membrane (Fadok *et al.*, 1992).

Fundamental differences between the two death-generating mechanisms are evident in the triggers necessary to initiate the two events. Necrosis is usually the result of an accumulation of toxic agents within cells while apoptosis can be triggered by various environmental stimuli, which lead to the activation of a specific endogenous endonuclease activity (Darzynkiewicz *et al.*, 1992).

Morphological and Biochemical Characteristics of Apoptosis

Apoptotic cell death is marked with some characteristic features (Hengartner, 2000; Raff, 1992; Rich *et al.*, 2000). During apoptosis cell membranes remain intact. However, the shrinkage of cell and blebbing of the cell membrane takes place, which results in the fragmentation of the cell in to membrane bound apoptotic bodies (Fig. 6.2). In the cells undergoing apoptotic death, cytoplasm becomes electron

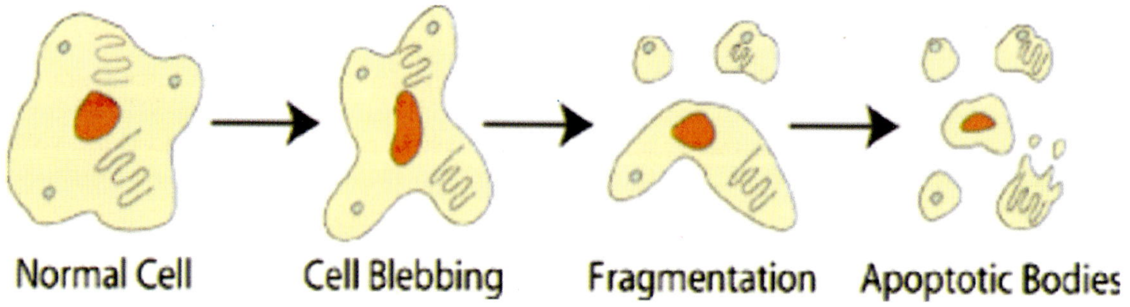

Fig. 6.2: Formation of Apoptotic Bodies in Cells Undergoing Apoptosis [Reproduced from Christine Suh, Missouri Arthritis Rehabilitation and Training center, Columbia (www.muhealth.org)]

dense and highly retractile. The cell rounds up and tends to retract from its neighbors. Cytoplasmic organelles become closely packed and nearly retain their normal appearance. Mitochondrial function however ceases early in the apoptotic process. This was indicated by loss of its ability to oxidize a marker tetrazolium dye. Disruption of Mitochondria results in release of cytochrome-C. Nuclear chromatin condenses in large rounded masses adjoining the nuclear membrane. The plasma membrane remains physically intact while parts of the cell bud off, forming membrane bound apoptotic bodies. These bodies are engulfed by phagocytes. Externalization of the phosphatidylserine (PS) on the surface of the cell membrane has also been reported as an important marker of apoptotic cell death (Koopman *et al.*, 1994; Vermes *et al.*, 1995).

Degradation or damage of DNA appears to be a universal and probably fundamental aspect of apoptosis (Steller, 1995). Endolytic cleavage of the DNA leads to small oligonucleosomal fragments (Steller, 1995). In most cases, the degradation proceeds to the oligonucleosome level where linker regions between nucleosomes are selectively cleaved. Apoptosis never seems to induce DNA degradation to the mononucleotides level. All forms of natural apoptosis generally inactivate the DNA within the cell without releasing it to surrounding environment. DNA degradation together with preservation of plasma membrane integrity probably serves to prevent the release of DNA into tissue spaces. DNA release may elicit an inflammatory reaction. Degradation of nuclear protein is also observed in apoptosis. Increased intracellular free Ca^{2+} was also found to trigger apoptosis. PCD occurs by two different mechanisms, one generated by signal arising within the cell, and the other triggered by death activators binding to receptors at the cell surface (Li *et al.*, 1997b). Fig. 6.3 shows the schematic diagram of the mechanisms of programmed cell death.

PCD: An Evolutionarily Conserved Process

Programmed cell death has been described in diversified organisms including bacteria, slime molds, yeast, plants, nematodes, insects, and vertebrates. However, detailed molecular and genetical findings related to PCD are available for *Caenorhabditis elegans* (a nematode), *Drosophila melanogaster* (an insect) and mammals.

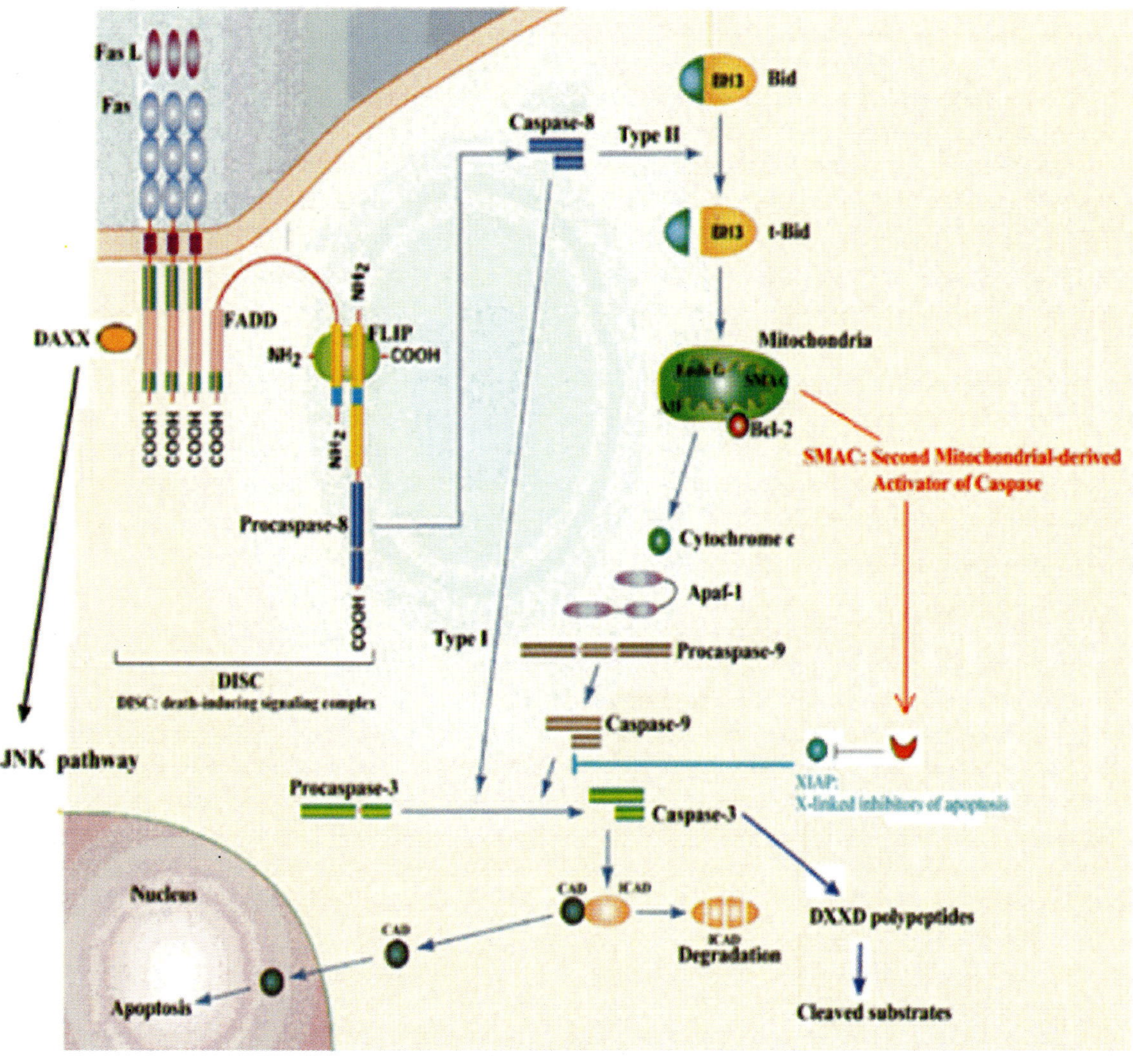

Fig. 6.3: Schematic Diagram Showing Mechanism of Programmed Cell Death [Reproduced from Bobe P, Atlas Genet Cytogenet Oncol Haematol, 2002 (www.infobiogen.fr)]

Programmed Cell Death in Prokaryotes

Although the mechanism of programmed cell death (PCD) or apoptosis is well illustrated in eukaryotic systems, there are only a few reports of its existence in a bacterial system. PCD like processes have been proposed for the microbial world by several investigators (Gerdes *et al.*, 1990; Jaffe, *et al.*, 1985; Lehnherr *et al.*, 1993; Naito *et al.*, 1995; 1998; Roberts *et al.*, 1994; Ruiz-Echevarria *et al.*, 1991; Thisted *et al.*, 1994; Tsuchimoto *et al.*, 1992; Yarmolinsky, 1995). There seems three possible functions for a PCD system in bacteria. First, response to nutritional stress; the death of a sub-population would provide food for the surviving cells. Second, a defense mechanism that would prevent spreading of a phage infection. Third, PCD could act as a guardian of the bacterial chromosome; when other systems fail, PCD systems might lead to the maintenance of genomic stability by causing the elimination of cells carrying genomic defects and mutations in the culture. Sometimes starvation was also found to induce the hyper mutable state in a moribund sub-population of *E. coli* (Nystrom, 2003a; 2003b). In moribund sub-population, members are not recoverable as colony-formers. So, they readily escape notice. They are associated not only with mutability, but also with the carriage of another major source of bacterial adaptability. Some of the possible instances of prokaryotic PCD are described here.

Bacterial Altruism

Altruism is a behaviour that reduces the reproductive success of the altruist while increase the reproductive success of the others. What role could altruism have in a bacterial community? As proposed for eukaryotes, evolution of bacteria might well have selected a programmed cell death mechanism to eliminate crippled cells that have survived damage induced by various physical and chemical factor/s. These cells could consume nutrients but hardly divide. Bacteria appear to be able to exhibit very social behavior. They live in large, complex and organized communities within a biofilm, and even communicate with each other using quorum-sensing pheromones (Bassler, 2002).

Death of Plasmid-free Segregants

The condition prevents the survival of bacterial mutants disabled for plasmid retention that would otherwise overgrow the plasmid carriers (Gerdes *et al.*, 1990; Jaffe, *et al.*, 1985; Lehnherr *et al.*, 1993; Roberts *et al.*, 1994; Ruiz-Echevarria *et al.*, 1991; Thisted *et al.*, 1994; Tsuchimoto *et al.*, 1992). The hypothesis was postulated to explain the death of the plasmid free segregants (Yarmolinsky, 1995). Set of genes carried by a plasmid that is responsible for the lethal consequences of plasmid withdrawl were termed as an "addiction module", which renders the bacterial host addicted to the continued presence of the "dispensable" genetic element it harbors. A plasmid addiction module makes a stable toxin and a labile antidote. Loss of plasmid and hence, death of the bacterial cell occurs when the ratio of antidote to toxin becomes too low. In the plasmid–free cell neither antidote nor toxin is replenished and the antidote is eliminated more rapidly than the toxin, leaving the later to exert its lethal potential (Yarmolinsky, 1995).

Several type II restriction enzymes and their cognate methylases are plasmid encoded toxin-antidote pairs. The DNA methylase offers protection from endonucleolytic attack by the separate restriction enzyme. Plasmids carrying gene pairs encoding type II DNA restriction endonucleases and their cognate modification enzymes have shown increased stability in *E. coli* (Naito *et al.*, 1995). The descendants of cells that had lost these genes appeared unable to modify a sufficient number of recognition sites in their chromosomes to protect them from lethal attack by the remaining restriction enzymes, and consequently died. Thus, these genes showed the behaviour of a selfish symbiont. This could have led to the evolution of restriction-modification gene pairs. The concept was called as selfish behaviour of restriction-modification systems.

Another example of toxin-antidote pairs is related to toxins Gef (alias Hok). In these cases, binding of antisense RNA to a long-lived precursor of the mRNA that encodes the toxin, prevents toxin synthesis. The intracellular toxins of the Gef family kill cells by membrane depolarization. In some other plasmid addiction systems, complex formation between an antidote protein (CcdA protein of plasmid F) and its coordinately synthesized toxin (CcdB) can prevent or even reverse toxin acxtivity. CcdB converts DNA gyrase into a DNA-damaging agent (Yorgey *et al.*, 1994; Bernard and Couturier, 1992). Thus, programmed cell death appears to have a larger role in the life of bacteria than its role in plasmid maintenance.

Many bacteria with a feast-and-famine way of life respond to repeated stresses with offensive and defensive measures that involve self-sacrifice. It is evidenced that various colicin producing plasmids bearing bacteria respond to DNA damage by the lethal synthesis of a colicin, which kill the cells by membrane depolarization.

MazEF Antitoxin/Toxin System and Bacterial Cell Death

It is an example of chromosomally encoded addiction module homologous to that of plasmid, and elicits PCD in response to stress conditions including nutrient deprivation, heat, and exposure to certain antibiotics. It was originally discovered in *Escherichia coli* (Rice and Bayles, 2003). MazF is a stable toxin, and MazE is a labile antitoxin interacting with MazF to form a complex. MazE and MazE-MazF complex can bind to the mazEF promoter region to regulate the mazEF expression. MazF is a sequence specific endoribonuclease that inhibits translation in a manner that is independent of ribosomes. MazF preferentially cleaves mRNA between A and C at the ACA sequence (Zhang *et al.*, 2003). MazF protein was reported to inhibit both initiation of DNA as well as translation. The mazEF genes form an operon with the upstream relA gene, encoding an ATP:GTP 3′-pyrophosphotransferase. RelA is responsible for activating ppGpp synthesis in response to amino acid starvation, and the subsequent increase in intracellular ppGpp concentration acts as a signal for induction of the stringent response. Intriguingly, expression of mazEF is inhibited by high concentrations of ppGpp. Thus, under stressful conditions such as amino acid starvation, thymine starvation, or antibiotics application the labile antitoxin MazE gets degraded rapidly, leading to cell killing by the toxin MazF.

Holin and Bacterial Cell Death

Energized state of the bacterial membrane was found to be a crucial determinant in the regulation of autolysis and murein hydrolase activity in *B. subtilis*. An energized membrane maintains a low localized pH that is inhibitory to the cell wall associated murein hydrolases. Agents that dissipate the proton motive force (PMF) caused the localized pH to increase and enable the murein hydrolases to become active. A similar mechanism of murein hydrolase activation might also be used by bacteriophage to control the timing of host cell lysis. In these systems, the holin family of proteins regulates murein hydrolase activity (*e.g.* S protein holin by bacteriophage Lambda). Many of these holins are presumed to control the access of murein hydrolases to the peptidoglycan by controlling their transport across the membrane. Two biological forms of holins, differing by two or three N-terminal amino acids, were observed (Young, 1992). The shorter form (S105) oligomerizes in the cytoplasmic membrane, whereas the longer form (S107)-the antiholin- inhibits oligomerization, preventing hole formation and murein hydrolase release. The relative proportion of these two forms plays a key role in the regulation and timing of lysis during bacteriophage lambda development. There seems to be a striking functional similarity between the holin family of proteins and the eukaryotic Bcl-2 family of apoptosis regulatory proteins (Bayles, 2003).

PCD in *Bacillus*, Streptomyces, and Myxobacteria

Besides *E. coli*, apoptotic cell death like process has also been observed in *Bacillus, Streptomyces*, and *Myxobacteria*. In these cases terminal differentiation of one portion like mother cell, aerial hyphae, or fruiting bodies, respectively, helped to generate and disseminate members of a complementary portion (spores) that have a greater potential for survival (Losick and Stragier, 1992). In Bacillus, the spore developed entirely within a larger outer mother cell. The mother cell gets lysed after completing its maternal tasks. In Streptomyces, long chains of spores developed from specialized aerial hyphae, which appeared to subsist mainly upon their own vegetative mycelium. In myxobacteria, sometimes-extensive cell lysis occurs during the cellular aggregation that precedes fruiting body formation. The myxospores are borne aloft on a structure consisting largely of dead cells.

PCD in *Xanthomonas*

The molecular mechanism of PCD, similar to that of eukaryotes, has been reported for the very first time in a prokaryote (Bayles, 2003; Rice and Bayles, 2003; Gautam, 2003; Gautam and Sharma, 2002a, 2002b). *Xanthomonas campestris* pv. *glycines* (Xcg), the causal agent of bacterial pustule disease of soybean (*Glycine max*) (Sharma *et al.*, 1999, 1994, 1993), was found to undergo PCD in certain nutritional conditions (Gautam, 2003; Gautam and Sharma, 2002a, 2002b). The PCD process was also found to be present in other *Xanthomonas* strains.

The mechanism of PCD in *Xanthomonas* was found to be similar to that of eukaryotes. During this process the morphological transformations of *Xanthomonas* cells to structures similar to eukaryotic apoptotic body was observed by scanning electron microscopic studies. DNA from apoptotic body was found to be nicked and also devoid of its indigenous plasmids (Gautam *et al.*, 2002). Apoptotic body also displayed the externalization of membrane phosphatidyl-serine, one of the important markers of eukaryotic apoptosis. Interestingly caspase-3 (cysteinyl-aspartate-specific-proteinases) like enzyme activity was also observed in *Xanthomonas* cells undergoing PCD. In eukaryotes, caspase-3 enzyme has been reported as the main executioner of apoptosis. *Xanthomonas* caspase was found to react with human caspase-3 antibody during western blot hybridization. As in eukaryotes, caspase-3 protein biosynthesis in *Xanthomonas* cells was also found to be regulated by nutritional factor/s. Chemical mutagenesis of Xcg resulted in mutants deficient in PCD. These mutants were also observed to be devoid of caspase-3 activity. It indicated a positive co-relation between programmed cell death and caspase-3 activity in Xcg cells.

PCD in Slime Mold

Slime moulds are primitive eukaryotes. The species *Dictyostelium discoideum* grows as free-living amoebae in the litter stratum, feeding on bacteria. Once the food supply is exhausted they undergo a complex developmental cycle in which thousands of individual aggregate to form a multicellular fruiting body. The aggregation takes place by chemotaxis to periodic cAMP signals. Upon starvation a few amoebae acting as an aggregation center start to produce periodic cAMP pulses, which are detected, amplified and relayed by the surrounding amoebae. Individual cells move chemotactically towards increasing cAMP concentrations. The hemispherical aggregate of the amoebae is called as mound, in which the amoebae differentiate into prestalk and prespore cells, which later form the stalk and spore head. The prestalk cells form the slug, which enables cellular slime moulds to migrate towards the surface of the litter stratum providing a better chance for spore dispersal. This organism has been reported to have caspase independent programmed cell death mechanism and have shown high resistance to oxidative stress (Katoch and Begum, 2003).

PCD in Yeast and Other Unicellular Eukaryotes

Recently, a caspase-related protease has been reported to regulate apoptosis in yeast, a unicellular eukaryote (Madeo *et al.*, 2002; Skulachev, 2002). The yeast caspase was reported to mediate the death process within chrolonogically-aged cultures. The yeast caspase was found to have structural homology with mammalian caspases. Yeast cells were also found to undergo PCD following oxidative or chemical (acetic acid) stress (Rice and Wayles, 2003). PCD was also observed in other unicellular eukaryotes, including several species of *Trypanosoma*, *Leishmania*, ciliate like *Tetrahymena thermophila* and Dinoflagellate *Peridinium gatunense* (Rice and Wayles, 2003). These organisms show many features of apoptosis of higher organisms such as DNA fragmentation, and cytoplasmic blebbing in response to environmental stress.

PCD in *Caenorhabditis elegans*

Much of our current knowledge about the molecular genetics of programmed cell death has been achieved by the work carried out on this worm (Hofmann *et al.*, 2000). For the work related to genetic regulation of organ development and programmed cell death in *C.elegans*, Sydney Brenner, John Sulston and Robert Horvitz were awarded Nobel Prize in physiology or medicine for the year 2002 (Knapp, 2003). This organism is well suited for the study of cell death at the cellular, genetic, and molecular levels because it is transparent and developmentally invariant. This has made possible the complete lineage description of every one of the 1090 cells born during development of the hermaphrodite form. The adult hermaphrodite *C.elegans* has 1090 somatic cells, of which 131 die by apoptosis during normal development (Metzstein *et al.*, 1998).

Apoptosis in worms occurs in four phases (Steller, 1995). Decision is the first phase, which indicates whether a cell should die or assume another fate. Decision is followed by death phase, phagocytic phase and degradation phase. During phagocytic phase engulfment of the dead cells takes place. Lastly, engulfed cells are degraded. The analysis of mutants with aberrant patterns of cell death has identified four aspects of cell death process. These aspects are: (1) a determination step, (2) the execution of cell death, (3) engulfment of the dying cell, and (4) degradation of the engulfed cell DNA. A number of genes have been identified that regulate these processes in worms (Steller, 1995). Death execution is mediated by three genes namely, ced-3, ced-4 and ced-9 (ced–cell death defective) (Ellis and Horvitz, 1986; Ottilie *et al.*, 1997a; 1997b). Two of these ced-3 and ced-4 are required for each cell death, and when they are inactivated by mutation, in none of the 131 normal cells death occurs (Ellis and Horvitz, 1986). The other gene, ced-9, acts to antagonize the killing activity of ced-3 and ced-4. Mutations that inactivate ced-9 lead to widespread and lethal embryonic death (Hengartner *et al.*, 1992). The killer gene ced-3 encodes the Ced-3 protein, a cysteine protease that cleaves after aspartate residues, hence called as cysteine aspartyl protease (caspase) (Alnemri *et al.*, 1996). Ced-4 protein physically interacts with both Ced-3 and Ced-9 proteins and appears to act as an adapter protein that facilitates Ced-3 autoactivation (Chinnaiyan *et al.*, 1997; Irmler *et al.*, 1997; James *et al.*, 1997; Spector *et al.*, 1997; Wu *et al.*, 1997). Ced-4 has a primitive ATPase domain that is required for its ability to activate Ced-3. The interaction of Ced-4 with the death inhibiting protein Ced-9 inhibits its ability to activate Ced-3.

A mammalian homologue of Ced-4 was identified as Apaf-1 (apoptotic protease activating factor). It is implicated in mediating caspase activity through mediating cytochrome c dependent activation of caspase-9. Cloning of the apoptotic genes of *C. elegans* and their characterization have led to considerable understanding of the molecular events of apoptosis and to the identification of the mammalian homologues of the apoptotic effectors. The genes ced-9 and ced-3 were found to be homologous to the mammalian proto-oncogene bcl-2 and ice (interleukin-1ß-converting enzyme), respectively (Hengartner and Horvitz, 1994a; 1994b; 1994c; Vaux *et al.*, 1992). Ced-9 prevents apoptosis in *C. elegans* by two distinct mechanisms (Xue and Horvitz, 1997). It may inhibit the Ced-3 protease by an interaction involving Ced-3 cleavage sites within Ced-9 or by protective mechanism similar to that used by mammalian Bcl-2. Cleavage of Ced-9 by Ced-3 generates a carboxy- terminal product that resembles Bcl-2 in sequence and function. *C.elegans* genome was found to encode for 3 long prodomain caspases, namely, ced-3, csp-1, and csp-2. Csp-1, and Csp-2 have active sites that differ in the first two positions.

C. elegans gene egl-1 encodes a small protein containing a nine amino acid domain similar to BH3, which is found in various pro-apoptotic Bcl-2 like mammalian cell death regulators. It has been observed that egl-1 probably acts upstream of ced-3 and ced-4. There is also the report of direct interaction between the Egl-1 and Ced-9 proteins (Conradt and Horvitz, 1998). It seems that Egl-1 may act by interfering with Ced-9 so that it can no longer inhibit Ced-4, which is then free to activate Ced-3, leading to cell death (Liu and Hengartner, 1999; Smits *et al.*, 1999).

PCD in *Drosophila melanogaster*

In this organism PCD was discovered by Herman Stellar, an investigator at MIT, USA. In *Drosophila* cell death is important during development and metamorphosis. During embryonic development, apoptosis occurs in the head, epidermis, and central nervous system. *Drosophila* genome contains a number of genes involved in the cell death such as, reaper, hid (head involution defective), and grim (Lohmann, 2003; Olson *et al.*, 2003).

Reaper is specially expressed in cells destined to die (Schreader *et al.*, 2003). Its expression can be induced by X-ray treatment in embryonic cells that normally does not undergo cell death. The reaper encodes a protein that contains a putative sequence called death domain, which appears to be essential for conferring the cell death promoting activity to the reaper encoded protein. Reaper transcripts are found in doomed cells and their presence precedes morphological manifestations of apoptosis by 1 to 2 h. Deletion of reaper leads to suppression of apoptosis in response to apoptotic stimuli. Thus, reaper mutant embryos contain many extra cells. Reaper is thought to integrate the information from diverse signals and put a cell on the apoptotic pathway. At present, the presence of reaper homologues in other organisms is unknown.

Hid acts in synergy with reaper to regulate cell death in *Drosophila*. Hid mutants have low levels of cell death, particularly in the head region. Hid is predicted to encode a novel protein with limited homology to the putative amino acid sequence of the reaper protein. Hid is also expressed in cells, viable as well as dying, but co-expression of hid and reaper in the midline glia of the nervous system of *Drosophila*, results in a dramatic increase in cell death when compared to that seen due to reaper alone.

Grim seems to have a role similar to hid in modulating reaper activity. It also does not contain any death domain. Ectopic expression of reaper, hid, and grim in the *Drosophila* eye under the control of the eye specific promoter resulted in complete eye ablation at high doses of gene expression while at lower doses various intermediate eye phenotypes were observed (Brachmann and Cagan, 2003).

Besides these three some other apoptosis related genes were also reported in *Drosophila* such as thread gene, which encodes a homologue of the baculovirus IAP (inhibitor of apoptosis) protein. *Drosophila* was also found to encode eight caspases including three long prodomain caspases (Dorstyn, 1999a; 1999b) namely, dcp-2/dredd, dronc, dream and four short domain caspases, (Fraser, 1997) namely, dcp-1, drICE, decay (Dorstyn, 1999a; 1999b), and day dream. An eighth *Drosophila* caspase, a head-to-head partial duplication of daydream, is likely to be nonfunctional because of numerous mutations. Dronc has active site sequence- PFCRG. It cleaves itself after glutamate rather than aspartate, and cleaves tetrapeptide substrates after glutamate as well as aspartate (Hawkins, 2000). Both *Drosophila* and *C. elegans* encode caspases that have long prodomains with unique sequences, as well as a single caspase with a CARD. In *Drosophila*, p53 has also been reported to preserve genomic stability by regulating cell death (Sogame *et al.*, 2003).

PCD in Plants

An emerging query in plant biology is whether plants display analogous elements of mammalian-programmed cell death during development and defense against pathogen attack. In many plant-pathogen interactions, plant cell death occurs in both susceptible and resistant host responses. Certain specific recognition responses in plants trigger hypersensitive response (HR) and activation of host defense mechanisms, resulting in restriction of pathogen growth and disease development. Cell death during HR involves activation of plant-encoded pathway for cell death. Recently, the transgenic tobacco plants were reported to express antiapoptotic animal genes such as bcl-2 and bcl-xl (human), ced-9 (*C.elegans*), and IAP (inhibitor of apoptosis, baculovirus) (Dickman *et al.*, 2001). These transgenes conferred heritable resistance to several necrotrophic fungal pathogens, suggesting that disease development required host- cell death pathways. The findings thus indicated that in incompatible plant-pathogen interactions apoptosis-like programmed cell death could occur. A homologue of human bax inhibitor-1 has been reported from *Arabidopsis thaliana* (AtBI-1). Its expression was found to be unregulated in plants in response to biotic and abiotic stress (Sanchez *et al.*, 2000). Plants also exhibit PCD during the development processes. Programmed cell death occurs during differentiation of procambium in to tracheary elements, which are a major component of vessels or tracheids (Hiroo, 2000). Oxidative stress caused by hydrogen peroxide was also reported to trigger PCD in the form of HR in the lentil (*Lens culinaris*) plants (Maccarrone *et al.*, 2000).

Very recently, a human caspase-3 like protease activity was reported from tobacco plants during the hypersensitive response upon pathogenic challenge by tobacco mosaic virus (TMV) (Chichkova *et al.*, 2003). A specific tetrapeptide aldehyde inhibitor inhibited the substrate cleavage by plant and human caspases in vitro and counteracted TMV-triggered HR in vivo.

Apoptosis in Mammals

Molecular Machineries

In mammals several constituents regulating apoptotic cell death have been discovered. The major one includes, death receptors, caspases, Bcl-2 family proteins, Apafs (apoptotic protease activating factors), tumor suppressing transcription factors like p53, and oncoproteins. Interplay between these components lead to switch ON/OFF of apoptotic cell death program. Various environmental and physiological factors trigger the coupling and uncoupling of these molecular components.

Death Receptors

Apoptosis can be induced either from the cell surface, by ligand-dependent triggering of death receptors, or by the stimulation of intracellular receptor proteins (Armstrong *et al.*, 1996; Guohua *et al.*, 1997; Jones *et al.*, 2000; Krammer, 2000; MacFarlane, 2003; Scaffidi *et al.*, 1998; Smith *et al.*, 1994). Activation of a specific group of transmembrane receptors of the tumor necrosis factor (TNF) receptor superfamily can lead to the direct activation of caspases. Some of the members of TNF receptor superfamily includes, TNF-R1 (P55), TNF-R2 (P75), TNF-R3 (TNF-RP), CD-27, -28, -30, -40, -95, DR6, and TRAIL receptors [TRAIL-R1 (DR4), -R2 (DR5), -R3 (DcR-1 and DcR-2)]. A subgroup of these receptors shares a common intracellular protein-protein interaction domain called as death domain. These receptors are referred as death receptors that include, TNF-R1, CD95, DR-3, -4, -5, and –6. Activation of members of this receptor family triggers a variety of cellular responses, including apoptotic cell death (Baker and Reddy, 1996; Wallach, 1997; Ware *et al.*, 1996).

Death systems transmit apoptotic signals through protein-protein interactions that are mediated by motifs called the death domain (DD), the death effecter domain (DED) and the caspase recruitment domain (CARD) (Tibbetts *et al.*, 2003). Death domains (DD) are 80-100 residue long motifs found both in cytoplasmic proteins such as Fas-associated protein with death domain (FADD); TNF receptor-associated protein with death domain (TRADD); and receptor interacting protein (RIP) and in transmembrane proteins including members of the TNF-receptor superfamily, like Fas, TNF-R1, TRAIL-R1 and R2 (Guohua *et al.*, 1997; Smith *et al.*, 1994).

Mammalian TNF-R family members are type I transmembrane protein. Members of the TNF-R1 family contain one to five conserved extracellular cysteine-rich domains, and a death domain (DD) in their cytoplasmic tail, which is essential for transduction of the apoptotic signal. Death domain serves as recruiting modules through their ability to heterodimerize with the death domain of distinct proteins, including adaptor proteins such as FADD, TRADD and RIP. A functional TNF superfamily receptor is typically a trimeric or multimeric complex stabilized by disulfide bonds. Some members such as CD95, TNF-R1, TNF-R2 also exists in a soluble form *i.e.* generated by proteolytic cleavage (Hughes and Crispe, 1995).

The receptor's ligands comprise another related family, which includes, TNF, CD95L/FasL, TRAIL etc. The ligands are type II transmembrane protein. They share a characteristic 150 amino acid region towards the C-terminus by which each ligand interacts with its cognate receptor. Recently, a domain N-terminal to the ligand-binding domain in the extracellular region of TNF-R1, -R2, and CD 95 was identified, and called as pre-ligand assembely domain (PLAD). It mediates receptor self-association before ligand binding (Chan *et al.*, 2000).

CD95

It is also called as Fas or Apo-1. The Fas protein is a type 1 transmembrane glycoprotein with three extra-cellular cysteine-rich domains (CRD). Its molecular mass is ~ 45 kDa. It is a molecule with 320 amino-acid residues, and 157 amino acid extracellular domains. Fas is ubiquitously expressed on a variety of normal cells, including lymphocytes, hepatocytes, lung, and heart. The biological role of CD95 is best understood in the immune system, where it is implicated in peripheral clonal deletion of T-lymphocytes, activation-induced suicide of mature T-cells, cytotoxic response and induction of apoptosis in B-cells. Its expression can be increased by the activation of lymphocytes but also by cytokines such as interferon-g and TNF, but alternative splicing can result in a soluble form termed decoy receptor 3 (DcR3), that binds FasL and inhibits FasL-induced apoptosis.

Fas-mediated apoptosis is triggered by its ligand, FasL (Suda *et al.*, 1993). FasL is a TNF-related type II transmembrane of ~ 40 kDa molecular mass. It can be cleaved from the membrane by a metalloprotease. Both Fas and FasL probably form trimers, with extra-cellular cysteine-rich domain (CRD). CRD2 and CRD3 forms the major contact surfaces for FasL. FasL is expressed in more controlled manner than the receptor. Expression of FasL was also found to be associated with the immune privilege of certain sites in the body, such as, eye, testes, brain and placenta. These organs are protected from attack by the immune system. Constitutive cell surface expression of CD95L provides immunological privilege of these organs by killing infiltrating lymphocytes and inflammatory cells expressing CD95 receptor (Green and Ware, 1997). CD95L is predominantly expressed on activated lymphocytes, natural killer (NK) cells, erythroblasts, immune privileged tissues, and certain tumors.

All death receptors contain a 65 amino acid death domain (DD) in their cytoplasmic tail. The death domains mediate homo- and heterotypic protein-protein interactions to propagate signals. The DDs in the ligated TNF-R1/CD95 receptors recruit the C-terminal DD of the cytoplasmic adapter FADD/MORT-1. At its N-terminus, FADD possess a death effector domain (DED), which mediates interaction with the N-terminal prodomain of caspase-8. The recruitment of caspase-8 by FADD to the activated CD95 receptor generates a death inducing signaling complex (DISC) that leads to proteolytic autoactivation of caspase-8 (Kischkel *et al.*, 1995). Caspase-8 then activates other caspases, including caspase-1, and –3, which execute apoptotic dissolution of the cell (Enari *et al.*, 1996). This direct molecular interaction between external ligand and the basal effector apoptotic machinery occur only in type 1 cells (Scaffidi *et al.*, 1998). In type 2 Cells, CD95 leads to changes in mitochondria that activate downstream caspases by apoptosome formation. The amount of receptor-activated caspase-8 in type 2 cells is much lower than in type 1 cells (Scaffidi *et al.*, 1999) and probably insufficient to induce downstream effector caspase cleavage. However, the caspase-8 concentration is sufficient enough to cleave Bid, a Bcl-2 family member (Li *et al.*, 1998; Luo *et al.*, 1998). Truncated Bid then translocates to the mitochondria where it induces cytochrome-C release further leading to caspase activation and finally cell death. CD95-generated apoptotic signals activate acidic sphingomyelinase causing accumulation of ceramide, which probably may be a major determinant of the apoptotic decision (Fadok *et al.*, 1992).

The DDs of both CD95 and TNF-R1 also interact with C-terminal DD of a second receptor-interacting protein (RIP) (Stanger *et al.*, 1995). RIP N-terminus resembles a tyrosine kinase domain. DD of RIP binds to the C-terminal DD of another adapter protein called RAIDD (RIP associated ICH-1/ Ced-3 homologous protein with a death domain). At its N-terminus, RAIDD is homologous to caspase-2 (Ich-1). RAIDD oligomerizes with the prodomain of caspase-2. RIP is needed for TNF induced activation of transcription factor NF-κB. Cleavage of RIP by caspase-8 leads to blockage of NF-κB mediated anti-apoptotic signals (Lin *et al.*, 1999).

Daxx protein also binds to CD95 death domain but it does not have its own death domain (Yang *et al.*, 1997). Daxx is a nuclear protein that interacts and colocalizes with the tumor suppressive promyelocytic leukemia protein (PML) in nuclear bodies (Li *et al.*, 2000). DAXX is able to repress basal transcription. Overexpression of Daxx activates Jun N-terminal kinas (JNK), which leads to apoptosis.

FIST/HIPK3 (Fas-interacting serine/threonine kinase/homeodomain-interacting protein kinase) is another protein, which binds to the cytosolic domain of the CD95 receptor. FIST also interacts with FADD in a trimolecular complex of CD95, FADD and FIST. FIST kinase induces FADD phosphorylation and inhibits CD-95 mediated JNK (Jun N-terminal kinase) activation. FIST is localized in both cytoplasm as well as nucleus. It can also bind to Daxx.

TNF Receptors

Tumor necrosis factor (TNF) is a cytokine produced by activated T-cells and macrophages and plays a role in regulation of the host inflammatory response. It does so by influencing the proliferation, differentiation and apoptosis of cells involved in inflammation. TNF is the ligand for two receptors-TNF-R1, and –R2. TNF-R1 has a major role whereas, TNF-R2 provide an auxiliary function by helping in the binding of TNF to TNF-R1 (Tartaglia *et al.*, 1993). TNF-R1 signals caspase cascade by recruiting caspase-8 via the adaptor FADD. Although, FADD does not directly bind to TNF-R1, but it is recruited to the activated receptor via an intermediary cytoplasmic DD adapter called TRADD (TNF-R associated death domain). TRADD also binds RIP and link TNF-R1 to caspase-2 activation via RAIDD. TNF-R1 and –R2 also recruit another signaling adapter molecules, called TRAFs (TNFR-associated factors), which mediate activation of JNK or NF-κB (Lee and Choi, 1997) by interaction with the downstream signaling kinase NIK (NF-κB inducing kinase). NIK activates IκB kinases, which phosphorylate and inactivate IκB *i.e.* the endogenous cellular inhibitor of NF-κB (Malinin *et al.*, 1997). NF-κB could act as a powerful suppressor of apoptosis.

TRAIL Receptors

TRAIL (TNF-related apoptosis inducing ligand) is also called as apo-2L. It does not bind to either CD95 or TNF-R1. Like all members of TNF/CD95 family, TRAIL is synthesized as a membrane bound proprotein, which can be cleaved to generate a soluble ligand. TRAIL binding receptors (DR4/TRAIL-R1 and DR5/TRICK-2/TRAIL-R2/Killer) are expressed in most human tissues and some tumor cell lines. TRAIL induces apoptosis through these two-death domain-containing receptors requiring FADD and caspase-8 (Bodmer *et al.*, 2000). TRAIL affinity differences exist at physiological temperature, and the cell death inducing DR5 receptor seems to be the highest affinity receptor (Truneh *et al.*, 2000). Preclinical studies with mice and non-human primates have shown that TRAIL induced in vivo regression of tumors without severe side effects (Walczak *et al.*, 1999).

Inhibitors of CD95/TNF-R1 Death Signaling Pathways

FLIPs

FLIPs (Fas-associated death domain like ICE inhibitory proteins) interfere with recruitment of caspases to CD95/TNF-R1 signaling complexes. Many viruses were reported to encode FLIPs (v-FLIP) as part of their strategy to manipulate host cell suicide and viability. c-FLIP is a cellular homologue of v-FLIP, and is structurally similar to caspase-8 since it contains two death effector domains (DED) and an inactive caspase like domain lacking the conserved functional cysteine residue. c-FLIP is expressed in two isoforms (long and short), both of which are recruited to the DISC. c-FLIP blocks caspase-8 activation at the DISC and thereby inhibits CD-95 mediated apoptosis (Scaffidi *et al.*, 1999b).

Decoy Receptors

Other TRAIL receptors such as, decoy receptor 1 (DcR1/TRID/LIT/TRAIL-R3), decoy receptor 2 (DcR2/TRUNDD), and osteoprotegerin lack any functional death domain and hence are unable to tranduce a death signal. A soluble CD95 decoy receptor (DcR3) binds to CD95L and inhibit apoptosis (Bai *et al.*, 2000; Pitti *et al.*, 1998). Thus, they can therefore act as decoy receptors by abstracting TRAIL or CD95 ligand from productive interactions with DR4 or DR5. DcR1 and DcR2 were found to be expressed in many tissues but not in most cancer lines examined. DcR3 gene was reported to be amplified in around 50% of the primary lung and colon tumors.

IAP (Inhibitor of Apoptosis) Proteins

It was originally discovered in baculovirus and found to suppress the host cell apoptosis. IAP proteins contain BIR (baculovirus IAP repeats) motifs. These 65 amino acid repeats are typically found in the N-terminal region of the IAPs and mediate various types of protein-protein interactions. There are at least five human IAP family members, c-IAP1, c-IAP2, XIAP, NAIP, and survivin. All of the human IAP family members, with the exception of NAIP, have been shown to interact with specific caspases. C-IAP1, c-IAP2, XIAP (X-chromosome- linked IAP), and survivin have been reported to bind to and inhibit the active forms of the terminal caspases-3, -7, and –9 but do not interact with caspase-8. Human c-IAP1 and –2 have been purified and reported to be a part of TNF-R2-TRAF signaling complex (Rothe *et al.*, 1995). The gene encoding human NAIP (neuronal apoptosis inhibitory protein) was found to be partially deleted in individuals with spinal muscular atrophy, a disease involving apoptosis of motor neurons (Roy *et al.*, 1995).

Survivin, another member of IAP family was observed to be overexpressed in tumor cells (Ambrosini *et al.*, 1997). IAPs are found to be conserved during evolution and have been identified in *Drosophila*, *C. elegans*, and yeast. However, unlike IAPs of mammal and *Drosophila*, which are essential for cell survival, IAPs of *C. elegans*, and yeast do not seem to play a role in the inhibition of cell death but are rather involved in the regulation of cytokinesis and cell division, respectively (Fraser *et al.*, 1999; Goyal *et al.*, 2000; Uren *et al.*, 1999; Wang *et al.*, 1999). Mammalian survivin was also have shown these activities besides its anti-apoptotic function (Li *et al.*, 1999; Li *et al.*, 1998).

Aven is a conserved intacellular protein with a broad tissue distribution. Aven binds to Bcl-xl and Apaf-1. Thus, it interferes with the ability of Apaf-1 to self-associate leading to subsequent inhibition of Apaf-1 mediated activation of caspases. Livin, another novel member of the IAPs encodes a protein with a single BIR domain and a COOH-terminal RING domain. Expression of livin inhibited apoptosis by a number of stimuli. Livin can bind to caspases and could inhibit the proteolytic processing of caspase-9 in vitro. Livin was not detectable in most normal adult tissues with the exception of the placenta, but was present in fetal brain and in several cancer cell lines.

Caspases

Caspases are proteolytic enzymes. The name reflects the catalytic properties of these enzymes. The "c" denotes their cysteine protease mechanism *i.e.* they contain a cysteine residue within the active site and "aspase" refers to their ability to cleave after aspartic acid residues (Stennicke and Salvesen, 1998). The term "caspase' was adopted as a root name for all family members. Caspases are synthesized in cells undergoing apoptosis. They are first synthesized as inactive proenzymes (30-55 kDa) and than processed by self-proteolysis and/or cleavage by another protease. The processed forms consist of large and small subunits, which associate to form an active enzyme (Fig. 6.4).

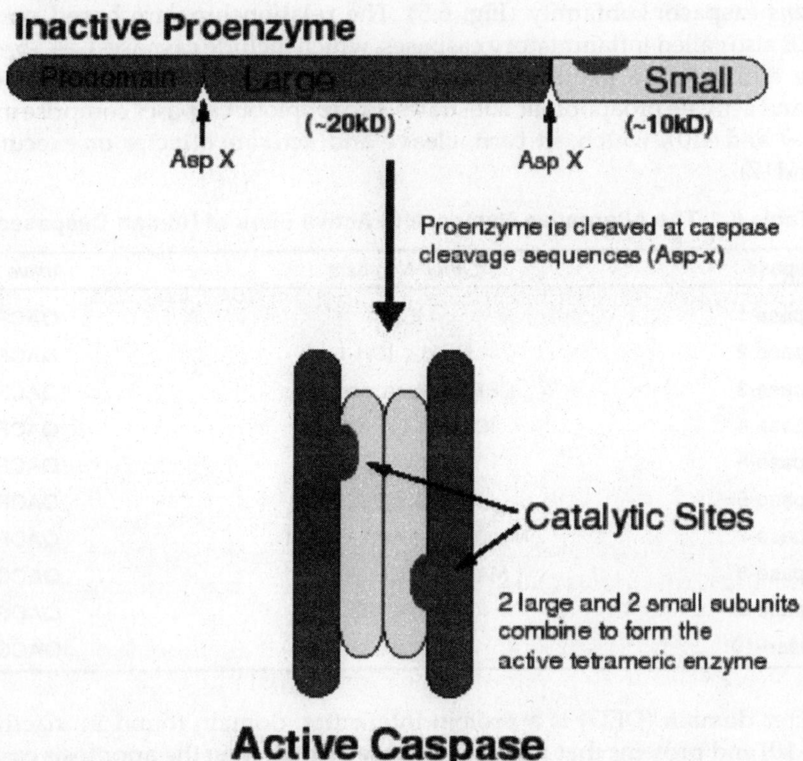

Fig. 6.4: Schematic Representation of Caspase Proenzyme and its Active Form [Reproduced from Chris McBride (www.cord.ubc.ca)]

Mammalian caspases are homologues of ced-3 of *C. elegans* (Earnshaw *et al.*, 1999; Thornberry and Lazebnik, 1998). The first mammalian homologue of Ced-3 was identified in the year 1993 and called as ICE (Interleukin converting enzyme). ICE is now called as caspase-1. In mammals altogether 14 caspases have been identified so far; 10 of which have long prodomains (Dorstyn, 1999a). All 10 long prodomain caspases contain either caspase recruitment domain (CARD) or death effecter domain (DED) sequences (Bertin *et al.*, 1997). Caspase-1 and –11 are mainly involved in proteolytic maturation of cytokines (*e.g.* interleukin-1β). They may have little to do with apoptosis. Caspase recruitment domains (CARD) are 90-100 amino acids long motifs. CARD mediates the association of adaptor proteins (*viz.* Apaf-1) and procaspases (-2 and -9), through heterodimerization of the respective CARD to upstream signaling complexes and allowing auto-activation.

In mammals three pathways are known that lead to caspase activation. In one pathway a serine protease, granzyme B, is directly delivered into the cytoplasm of target cells, where it activates executioner caspases (Yang *et al.*, 1998). In another two pathways, cytoplasmic adaptor proteins link a cell death signal transducer to a long prodomain caspase through a homophilic receptor. Adaptor-caspase interactions lead to effecter caspase activation.

Caspases share many common structural and catalytic features. All the caspases contain a conserved active site pentapeptide sequence QAC(R/Q/G)(G/E). The cysteine together with a distant histidine is directly involved in catalysis. Table 6.2 shows the active sites and the alternative names of the caspases. The human caspase sequences cluster into two subfamilies, the ICE (caspase-1) and the CED-3 (*C. elegans* caspase) subfamily (Fig. 6.5). The relationships are based on the full-length proenzymes. ICE also called inflammatory caspases, which include caspase-1, -4, -5, -11, -12, -13 and –14 and cleave and activate proinflammatory cytokines. CED-3 subfamilies (proapoptotic) caspases cleave and activate proapoptotic substrates. Proapoptotic caspases comprise initiator caspases (caspase-2, -8, -9 and -10), which, in turn, cleave and activate effecter or executioner caspases (caspase-3, -6 and -7).

Table 6.2: The Alternative Names, and Active Sites of Human Caspases

Caspase	*Other Names/s*	*Active site*
Caspase-1	ICE	QACRG
Caspase-2	Nedd2, ICH-1	QACRG
Caspase-3	CPP32, Yama, apopain	QACRG
Caspase-4	ICErelII, TX, ICH-2	QACRG
Caspase-5	ICErelIII, TY	QACRG
Caspase-6	Mch2	QACRG
Caspase-7	Mch3, ICE-LAP3, CMH-1	QACRG
Caspase-8	MACH, FLICE, Mch5	QACQG
Caspase-9	ICE-LAP6, Mch6	QACGG
Caspase-10	Mch4	QACQG

Death-effecter domain (DED) is a protein interaction domain found in inactive procaspases (caspase-8 and -10) and proteins that regulate caspase activation in the apoptosis cascade. The DED recruits procaspases into complexes with members of the TNF-receptor superfamily. This recruitment is mediated by a homotypic interaction between the procaspase DED and a second DED in a FADD adaptor molecule that is directly associated with activated members of the TNF-receptor superfamily.

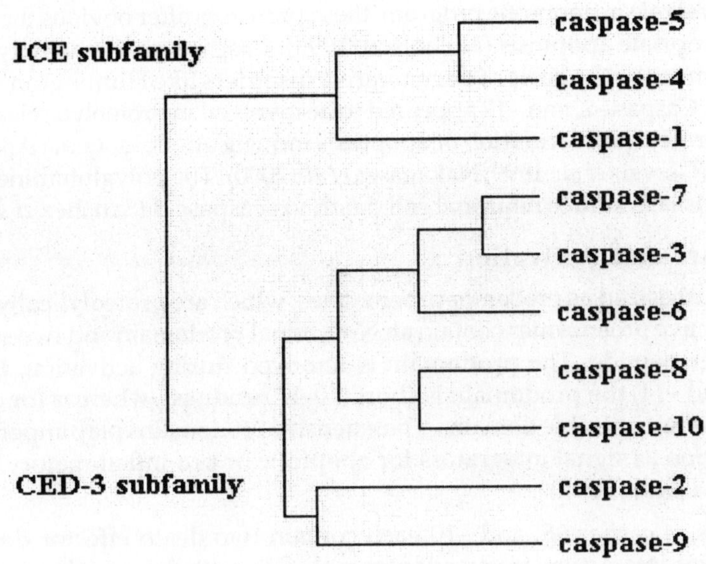

Fig. 6.5: Dendogram Showing Phylogenetic Relationship of Caspases

Complex formation allows transprocessing of procaspase to the active form. This, in turn, activates downstream caspases and initiates apoptosis (Armstrong *et al.*, 1997; Marti *et al.*, 2000).

Caspases form a proteolytic cascade, which leads to the cleavage and activation of various cytoplasmic and nuclear targets including other caspases, bcl-2 family proteins, lamins, poly ADP ribose polymerase (PARP), caspase activated DNAase (CAD) and others (Enari *et al.*, 1996; Nicholson, 2000; Sakahira *et al.*, 1998). The sequence of three amino acids before aspartate determines substrate specificity. Transactivation of procaspases by already activated caspases could generate sufficient proteolytic activity to overwhelm endogenous caspase inhibitors like inhibitor of apoptosis proteins (IAPs). Caspase-3 cleaves Bcl-2 and Bcl-xL, which destroys the anti-apoptotic function of these proteins and induces release of C-terminal fragment that are proapoptotic (Cheng *et al.*, 1997). Caspase-8 cleaves Bid, a pro-apoptotic Bcl-2 family member and generates a C-terminal fragment that induces release of mitochondrial cytochrome-C (Li *et al.*, 1998; Luo *et al.*, 1998).

Although caspases do not directly degrade DNA, they can degrade a protein called DFF45, which is an inhibitor of the nuclease responsible for DNA fragmentation that generally accompanies apoptosis. Caspase-3 cleaves ICAD/DFF45 allowing the ICAD (inhibitor of caspase activated DNAase) -bound nuclease CAD (caspase activated DNAase) to translocate to the nucleus and to cut DNA (Sakahira *et al.*, 1998). Caspase-3 also cleaves and activates gelsolin, a protein that regulates actin dynamics and promotes both cytoplasmic and nuclear apoptosis, including DNA fragmentation (Kothakota *et al.*, 1997). Caspase-6 cleaves lamins, a major structural protein within the nuclear envelope (Takahashi *et al.*, 1996). Cleavage of B-catenin and FAK by caspases leads to interruption of cell-cell contacts and cell-matrix adhesion. It helps in promoting cellular packaging and phagocytosis (Wen *et al.*, 1997). Cleavage of Poly ADP Ribose Polymerase (PARP) interrupts in repair of DNA double-strand break (Cohen, 1997).

Having the central role in apoptotic program, these proteases offer obvious therapeutic targets for the control of inappropriate apoptosis (Nicholson, 1996). Caspase over-reactivity promotes cellular suicide, and this seems to be the basis of degenerative disorders like Huntington's, Alzheimer's and Parkinson's diseases. Caspase-2, and –12 appeared to be involved in proteolytic cleavage of Alzheimer amyloid precursor protein and formation of apoptosis inducig amyloidogenic Aβ peptide leading to Alzheimer's disease (Gervais *et al.*, 1999; Nakagawa *et al.*, 2000). The polyglutamine repeats associated with Huntington's disease induce neuronal cell death via caspase-8 (Sanchez *et al.*, 1999).

Caspases: Structure and Activation

Caspases are synthesized as precursor proenzymes, which are proteolytically processed to their active forms. The inactive proenzymes contain an N-terminal prodomain and two subunits sometimes separated by a linker peptide. The prodomain is removed during activation. For some caspases (Caspase-3, -6, -7, and –14) the prodomain is short (10-40 residues), whereas for other caspases it is extensive and contains recognizable domains. The extensive prodomains play important role in caspase regulation and function as signal integrators for apoptotic or pro-inflammatory signals (Hofmann *et al.*, 1997; Fraser and Evan, 1997).

The prodomains of caspase-8, and –10 each contain two death effector domains (DED) that mediate the proenzyme interaction, via an adapter molecule, with the cytoplasmic tail of members of the TNF-R1/CD95 receptor family. This allows receptor-induced activation of caspases in response to ligand binding. Caspase-1, -2, -4, -9 and Ced-3 possess a distinct prodomain structure called CARD (caspase recruitment domain). CARD domains mediate specific intermolecular interactions that regulate caspase activation.

Active caspases are composed of a heterodimer comprising a large subunit (P20 for caspase-1, P17 for caspase-3) that contains the catalytic cysteine residue, and a smaller subunit (P10 for caspase-1, P12 for caspase-3) that contains determinants, which govern substrate specificity. During maturation the prodomain and the linker peptide are cleaved at specific Asp residues, resulting in the active caspase of two subunits. Family members are sequentially activated, and form a proteolytic cascade.

The structure of caspase-1 and caspase-3 has been determined by X-ray diffraction (Cohen, 1997). The overall structures are very similar, but major differences were found around the active site, especially in the substrate-binding pocket. The crystal structures of both caspase-1 and caspase-3 showed tetramer nature of the active enzyme with two small and two large subunits (Fig. 4). A small and a large subunit form a heterodimer, two of these heterodimers formed the tetramer. The inactive procaspase-1 is of 45 kDa molecular size. The proteolytic processing results into a 20 and a 10 kDa (p20 and p10) subunits, both of which are necessary for complete proteolytic activity (Fig. 4).

The catalytic machinery of caspases involves a diad composed of a cysteine sulphydryl group and a histidine imidazole ring. Both of them are present on the p20 subunit. The tetrahedral transition state of the cysteine protease is stabilized through hydrogen bonding with the backbone amide protons of Cys and Gly. The P1 Asp of the substrate seems to be stabilized by four residues (Arg, Gln from p20 and Arg, Ser from p10 subunits). The distinct specificities of caspase-1 and caspase-3 are observed from the important differences in the S4 subsites of the enzymes, which is the site accommodating the P4 residue of the substrate. The large hydrophobic depression in caspase-1 easily accommodates a tyrosyl side chain, present in the P4 position of pro-IL-1ß and the tetrapeptide inhibitor. In caspase-3 there is a surface loop spanning residues of Phe that does not have a counterpart in caspase-1. The S4 site is significantly smaller in caspase-3 and is capable of binding the carboxylated side chain of P4 Asp. This geometry explains the reason behind strict requirement for an Asp in the P4 position of caspase-3 substrates and inhibitors. Several of the sites responsible for distinct specificities of caspase-1 and caspase-3 are conserved in other family members, which suggested a similar substrate preference for these enzymes.

The members of caspase families undergo auto- and trans- cleavage during their activation. For instance, caspase-8 autoactivates itself upon recruitment to a death inducing signaling complex (DISC) and the active enzyme then in turn cleaves and activates the effector caspases-3, and –7 that execute the apoptotic program. On the other hand, caspase-9 does not necessarily require proteolytic processing, but instead requires binding to Apaf-1with which it forms an active holoenzyme (Rodriguez and Lazebnik, 1999; Stennicke *et al.*, 1999).

Bcl-2 Family Proteins

Bcl-2 is a human proto-oncogene (Hengartner, 2000; Humlova, 2002; Reed, 1998). Stanley Korsmeyer discovered it as the target gene present at the translocation site of the t (14; 18) chromosomal translocation breakpoint in the tumor cells of approximately 80% of patients with human B-cell follicular lymphoma and hence termed as bcl (Strasser *et al.*, 1997). The translocation places the bcl-2 gene under the control of 5′ immunoglobulin heavy chain gene enhancer (Eμ, chromosome 14), an element highly transcriptionally active in B-lymphoid cells. Bcl-2 is homologues to ced-9 of *C.elegans*. Bcl-2 family member proteins are structurally and functionally conserved throughout metazoan evolution. Bcl protein facilitates survival of cells in which it is expressed (Adams and Cory, 2001; Ashkenazai and Dixit, 1998; Dragovich *et al.*, 1998; Vaux *et al.*, 1988). Bcl-2 is expressed in a wide variety of fetal tissues but in the adult it tends to be more restricted to cells that are rapidly dividing and differentiating.

Bcl-2 family comprises of pro and anti-apoptotic members. Proapoptotic bcl-2 family members include, Bax, Bcl-x_s, Bad, Bak, Bid, Bik, Hrk, Bim, and Bok (Chong *et al.*, 2000; Duckett *et al.*, 1998; Srinivasan *et al.*, 1998). Antiapoptotic Bcl-2 family members, that protect the cells from apoptosis include, Bcl-2, Bcl-x_L, Bcl-w, Bfl-1, Brag-1, Mcl-1, A1, E1B19K, LMW5-HL, and EBV BHRF1 (Jurgensmeier *et al.*, 1997; Li *et al.*, 1997a).

Bcl-2 possesses a hydrophobic tail that allows it to associate with various cellular membranes, including mitochondrial outer membrane, endoplasmic reticulum and the nuclear envelope (Tsujimoto, 2003). Bcl-2 family members possess conserved amino acid sequence motifs, called bcl-2 homology (BH) domains (Ottilie *et al.*, 1997a; 1997b). There are at least 4 different BH domains (BH 1-4). Some of these BH domains confer pro-apoptotic activity, while others confer anti-apoptotic activity to the proteins in which they are present. BH1, BH2 and BH4, together with the carboxy-terminal hydrophobic transmembrane region found in most family members, are important for functional activity. BH3 domain seems to be dispensable for the death inhibiting function of anti-apoptotic Bcl-2 homologues. BH3 is also essential for cytotoxic activity of the pro-apoptotic family members (Kelekar and Thompson, 1998). BH3 molecules induce the activation of the multidomain proapoptotic Bax and Bak, which results in the permeabilization of the outer mitochondrial membrane and the efflux of cytochrome (Scorrano and Korsmeyer, 2003).

The bcl-2 family members can form hetero- and homo-dimers, suggesting a model of regulation where the cellular susceptibility or threshold for apoptosis is partly influenced by the repertoire and level of expression (Diaz *et al.*, 2000). Their dimerization reinforces or cancels out cell death or cell survival promoting activities (Diaz *et al.*, 1997). Bcl-2 knockout mice appear almost normal at birth but later in life develop hair hypopigmentation due to death of melanocytes, distortion of their small intestines and polycystic kidney disease (Veis *et al.*, 1993). Regulatory nature of bcl-2 family proteins has shown the promise to control the diseases due to imbalanced apoptosis. Recently, bcl-2 antisense oligonucleotide such as G3139 has been explored as a potential therapeutic agent of cancer (Stephens and Rivers, 2003).

Bcl-x

Similar to Bcl-2, Bcl-x also acts as a death suppressor protein (Boise *et al.*, 1993). 2.7 kb bcl-x transcript encodes a protein with 44% sequence identity to human and mouse bcl-2. However, unlike bcl-2, the bcl-x gene encodes two polypeptides arising from alternative splicing. The longer mRNA encodes a death suppressing functional Bcl-2 homologue, called as Bcl-x_L. It is of around 28 kDa comprising 233 amino acids. The shorter mRNA encodes the 18 kDA Bcl-x_s protein having 170 amino acids. Compared to Bcl-x_L, Bcl-x_s lacks a 63 amino acid stretch, which includes the BH1 and BH2 domains, and hence, Bcl-x_s may act as a dominant interfering regulator of Bcl-x_L or Bcl-2. Bcl-x is widely expressed and is particularly high in brain, kidney and adult thymus. Bcl-x_L is transcriptionally activated by NF-κB (Lee *et al.*, 1999) or by STAT 5 (Silva *et al.*, 1999). Bcl-2 as well as Bcl-x_L both heterodimerize with pro-apoptotic Bcl-2 family members such as Bax and Bak and antagonize PCD.

Bax

It is a proapoptotic Bcl-2 family member. Bax and Bak (another proapoptotic Bcl-2 family member) both possess BH1, BH2, and BH3 regions. However, their BH4 is poorly preserved. Bax and Bak cytotoxic activity is dependent upon a small region of each protein that comprises the BH3 domain. The same BH3 region mediates dimerization between Bax/Bak and Bcl-2, Bcl-$x_{L,}$ E1B19K, Bid or itself (Sattler *et al.*, 1997). As per the rheostst model for Bcl-2/Bax function, propensity to undergo PCD depends upon the relative ratios of Bax/Bax homodimers. An excess of Bax homodimers promotes cell death, whereas Bax complexed with Bcl-2 favors survival. Bax is expressed in a wide variety of tissues including lymphoid organs, lung, stomach, and kidney. In general, absence of Bax greatly diminishes the ability of p53 to retard tumor growth (Yin *et al.*, 1997).

Various forms of cellular stress that promote apoptosis result in activation of pro-apoptotic bcl-2 family members such as bax. Bax moves from the cytoplasm to the mitochondria and punctures holes in the outer mitochondrial membrane. This results in the release of cytochrome-C from mitochondria and subsequently the binding of cytochrome-C to Apaf1 to create an active complex, which in turn activates caspase-9 and finally caspase-3 leading to apoptosis (Cecconi *et al.*, 1998; Hakem *et al.*, 1998; Kuida *et al.*, 1998).

Bad and Other Bcl-2 Proteins (BH3 Domain Only)

Mammalian proteins such as Bad, Bik, Blk, Hrk, Bid, Bim, Noxa and Egl-1 of C. *elegans* share only the short (9 to 16 residues) BH3 domain with the Bcl-2 family. This domain allows them to bind to the anti-apoptotic Bcl-2 molecules and neutralize their function. Bim is required for hematopoietic homeostasis and act as a barrier to autoimmunity. Bad contains 204 amino acid residues and its molecular size is ~22 kDa. It binds with Bcl-2/Bcl-x_L and promotes apoptosis (Yang *et al.*, 1995). However, unlike many other Bcl-2 family proteins, Bad and Bid do not possess any C-terminal membrane anchor sequence and probably they are not an integral membrane protein. Bad protein heterodimerizes with both Bcl-2 and Bcl- x_L but not with Bcl-$x_{s,}$ Bax, Mcl-1, A1 or itself. Bad counters the death repressor activity of Bcl- x_L, and probably exerts its pro-apoptotic function by competing with Bax for Bcl- x_L binding. Bad has been reported to be regulated by phosphorylation (Franke and Cantley, 1997). Bad is phosphorylated at three serine sites. Phosphorylated Bad binds to the cytosolic

14-3-3 protein and becomes sequestered and functionally inactivated (Datta *et al.*, 2000; Virdee *et al.*, 2000).

Bid, another BH-3 only domain protein, is regulated by proteolytic cleavage rather than phosphorylation. Bid is cleaved by caspase-8 and the truncated form then translocates to mitochondria, where it activates Bax and Bak by inducing their oligomerization (Eskes *et al.*, 2000; Gross *et al.*, 1999). Noxa is another BH-3 only protein. It is regulated by p53 (Spector *et al.*, 1997). Noxa also gets localized to the mitochondria after an apoptotic stimulus and interact with other Bcl-2 family proteins.

Functions of Bcl-2 Family Proteins

Bcl-2 is localized to the outer mitochondrial and nuclear membranes and the endoplasmic reticulum. No direct link was observed between the anti-apoptotic activity of Bcl-2 and oxidative phosphorylation taking place in mitochondria. Levels of cellular ATP or oxygen consumption were not found to be affected by Bcl-2 expression (Zhong *et al.*, 1993). However, Bcl-2 expression affected the intracellular Ca^{2+} homeostasis (Moroy and Zoring, 1997). It seems that Bcl-2 either directly modulates calcium channels or acts to protect lipid membranes from damage by peroxide radicals, which is known to disrupt Ca^{2+} homeostasis.

At present three possible models are used to explain Bcl-2 function. These models describe Bcl-2 proteins as ion channels, or as proteins that modulate activation of caspases, or as inhibitors of cytochrome-C export from mitochondria. The possibility of role of Bcl-2 in formation of membrane channels emerged from elucidation of the detailed three-dimensional structure of Bcl-x_L, which resembles the membrane insertion domain of bacterial toxins such as diphtheria toxin and colicin (Muchmore *et al.*, 1996). Ion channel activity has been detected with Bcl-2, Bcl-x_L, and Bax proteins in isolated lipid bilayers in vitro (Reed *et al.*, 1997). Bax is able to interact with voltage dependent ion channel (VDAC) and adenine nucleotide translocator (ANT) to release cytochrome-C (Shimizu *et al.*, 2000). After oligomerization it may also form cytochrome-C releasing pores by itself (Antonsson *et al.*, 2000). Bcl-2 family members having BH4 domain such as Bcl-2, and Bcl-xL can block the cytochrome-C release.

Bcl-2 and other family member not only inhibit or promote apoptosis but also have an influence on cell cycle progression. Over expression of anti-apoptotic Bcl-2 family members slows down the entry of cells into cell cycle progression (Fujise *et al.*, 2000). On the other hand, over expression of pro-apoptotic Bax protein accelerate cell cycle entry (Gil-Gomez *et al.*, 1998). The cell cycle regulatory function of Mcl-1 is mediated through its binding to proliferating cell nuclear antigen (PCNA). This interaction seems to inhibit PCNA'a ability to promote progression through S phase (Fujise *et al.*, 2000).

Apoptotic Protease Activating Factors (APAFS)

Biochemical analysis has revealed two cytosolic apoptotic protease activating factors, Apaf-1 and Apaf-3. Apaf-1 shares partial homology with the product of the *C. elegans* ced-4 gene product (Soengas *et al.*, 2001; Zou *et al.*, 1997). Apaf-3 is homologous to caspase-9 (Li *et al.*, 1998). The Ced-4 like domain in Apaf-1 is flanked on one side by a region with strong homology to the CARD motif within the prodomains of Ced-3 and mammalian caspases-2 and –9 and on the other side by several WD-40 repeats, which are presumed to mediate protein-protein interactions. The CARD domain in Apaf-1 and the prodomain of caspase-9 interact and, in the presence of cytochrome C and either ATP or ADP, form a complex called apoptosome, which induces autocatalytic activation of the caspase-9. Activated caspase-9 then activates the downstream effector caspases -3, -6, and –7 (Cecconi *et al.*, 1998; Hakem *et al.*, 1998; Kuida *et al.*, 1998).

Studies on human Apaf1 gene revealed that it maps to a region of human chromosome 12, which has been implicated in a genetic syndrome called Noonan syndrome. Individuals that show this syndrome have many of the characteristic defects, including craniofacial, limb, and retinal abnormalities. The Apaf-1 deficient mice also showed severe craniofacial malformations, brain overgrowth, the persistence of the interdigital webs, and dramatic alterations in the development of the lens and retina (Inohara and Nunez; 2003; Yoshida *et al.*, 1998). Transgenic mice lacking Apaf-1 exhibited significant developmental defects associated with alterations in apoptosis during development and did not survive more than 16 days (Cecconi *et al.*, 1998; Yoshida *et al.*, 1998).

PCD and Mitochondria

Permeabilization of mitichondrial membranes leads to release of cytochrome-C, which is normally located in the inter membrane space of mitochondria and takes part in oxidative phosphorylation. Upon exposure of cells to apoptotic stimuli, cytochrome-C is released from mitochondria into the cytoplasm (Slee *et al.*, 1999). An apical caspase (caspase-8), which is activated at the death-inducing complex (DISC) of cell surface receptor, cleaves the pro-apoptotic Bcl-2 family member Bid (Li *et al.*, 1998; Kothakota *et al.*, 1997). Cleaved or truncated Bid (t-Bid) then binds to Bax. It leads to Bax oligomerization and integration into the outer mitochondrial membrane where it triggers cytochrome c release (Eskes *et al.*, 2000). Similarly, t-Bid binds to and oligomerizes another pro-apoptotic Bcl-2 homologue, Bak, resulting in cytochrome-C release (Wei *et al.*, 2000). Recently it has been shown that the release of cytochrome-C always precedes exposure of phosphatidylserine at the cell surface and the loss of plasma membrane integrity (Goldstein *et al.*, 2000). It has also been observed that the drop in the mitochondrial membrane potential typically seen in apoptotic cells occurs later than cytochrome-C release from mitochondria (Goldstein *et al.*, 2000). This seems to be dependent on caspase activation.

There are various competing models exist to explain the depolarization of mitochondrial membrane during the process of apoptotic cell death (Martinou *et al.*, 2000). In general outer mitochondrial membrane could rupture due to swelling of the mitochondrial matrix. Different models explain the possible mechanism/s of swelling of matrix leading to release of cytochrome-C. Permeability transition pore (PTP) openers such as the proapoptotic Bcl-2 family member Bax, cause permeabilization of the inner membrane and mitochondrial depolarization by binding to the adenine nucleotide translocator (ANT), which is located in the inner mitochondrial membrane (Marzo *et al.*, 1998). This process allows entry of water and solutes into the matrix and leads to mitochondrial swelling. Another component needed for the PTP formation is the voltage-dependent anion channel (VDAC), which is located in the outer mitochondrial membane. ANT and VDAC span both the inner and outer mitochondrial membranes at sites at the two membranes are opposed. Swelling also happens due to defect in mitochondrial ATP/ADP exchange, which leads to the closure of VDAC resulting in hyperpolarization of the inner mitochondrial membrane and subsequent matrix swelling.

An alternative model proposes that the Bax channel itself can mediate release of cytochrome-C independent of PTP and loss of membrane potential. Bax oligomers can form large conductance channels in lipid planer bilayers (Martinou *et al.*, 2000), which allow the passage of cytochrome-C in to the cytosol. Bax cooperate with the VDAC to form a cytochrome-C conducting channel (Shimizu *et al.*, 1999).

Some other proteins are also released from mitochondria during apoptosis such as Smac/Diablo, and apoptosis inducing factor (AIF). Smac/Diablo bind and inactivate inhibitors of apoptosis (IAPs), which inhibit cell death by binding to procaspases and active caspases, and blocking their processing and activity (Du *et al.*, 2000). Smac/Diablo and the pro-apoptotic *Drosophila* proteins Reaper, Grim and Hid seem to function in a similar way by inhibiting IAP activity. A short stretch sequence similarity of N-terminal 14 amino acids had been observed in these proteins indicating their structural and functional homology (Silke *et al.*, 2000). AIF (apoptosis inducing factor) redistributes from mitochondria and induces some of the nuclear morphology associated with apoptosis in a caspase independent manner (Susin *et al.*, 1999).

PCD and Oncogenes

Many oncogene products induce apoptosis upon overexpression (Harrington *et al.*, 1994a). Some of these include transcription factors such as c-Myc, EIA (adenovirus protein), C-Jun and c-Fos (Bossy-Wetzel *et al.*, 1997). The dual signal hypothesis provides an explanation for apoptosis induced by the overexpression of oncogenes (Harrington *et al.*, 1994b). According to this hypothesis oncogene such as myc promotes both pathways proliferation and apoptosis at the same time. The apoptotic pathway is suppressed as long as appropriate survival factors deliver anti-apoptotic signals.

c-Myc induces apoptosis through its action as a transcription factor-presumably by modulation of appropriate target genes. Some of the candidate Myc-regulated genes include, ornithine decarboxylase (ODC), Cdc25A, TRAP-1 (TNF-R associated protein), and Bcl-2 family member Bax (Coller *et al.*, 2000). ODC is a rate-limiting enzyme in polyamine biosynthesis, necessary for DNA synthesis, while Cdc25A encodes a tyrosine phosphatase involved in activation of the key inducers of mitosis, the cyclin-dependent kinase (CDK) complexes. Both these candidate Myc targets, ODC and cdc25A, when over-expressed, induce apoptosis in cells lacking survival factors. Thus, it seems that the cell proliferation and cell death pathways bifurcate downstream of c-Myc. Other targets of c-Myc and E1A are p53, which gets activated via p19ARF, a protein encoded by the alternate reading frame of the p16 tumor suppressor gene (Zindy *et al.*, 1998). E2F-1, a component of G1 cell cycle progression machinery seems to induce apoptosis by upregulation of the p53 homologue p73 (Stiewe and Putzer, 2000).

Recently, it has been observed that c-Myc induced apoptosis requires interaction on the cell surface between CD95 and CD95L (Hueber *et al.*, 1997). In general c-Myc acts upstream of CD95/CD95L by inducing expression of their cognate genes. C-Myc promotes apoptosis by causing release of cytochrome-C by upregulation of Bax (Mitchell *et al.*, 2000). Cytochrome-C release is caspase independent and is blocked by the survival factors such as insulin-like growth factor-1 (IGF-1). c-Myc also sensitizes target cells for the cytotoxic action of activated natural killer (NK) cells (Klefstrom *et al.*, 1999).

p53 and PCD

p53 is a latent and short-lived transcription factor which is induced by various conditions such as microtubules and growth factor depletion, ribonucleotide depletion, cytokines, hypoxia, viral infection, oncogenes activation and DNA damage. Oxidative stress and radiations like X-rays, gamma-rays, UV-rays lead to the damage of DNA. In general, p53 protein is present in all normal cells and is expressed at a relatively constant level, but it does not accumulate. The half-life of p53 is around 20 min and it is virtually undetectable in normal cells. p53 in normal cells remains in an inactive conformation. The process of activation requires changes in its conformation and accumulation.

p53 is encoded by a gene on chromosome 17p. It is a relatively small protein of 393 amino acids (Prokocimer and Rotter, 1994). p53 gene consists of 11 exons and encodes a 2.8 kb mRNA. Normal p53 is an oligomeric DNA-binding protein, which functions to trigger cellular responses to DNA damage, both by protein-protein interactions and regulation of transcription (Miyashita and Reed, 1995). For these reasons, it was also termed as the 'guardian of the genome' (Attardi and Jacks, 1999; Oren and Rotter, 1999; Soussi, 2000). p53 controls the G1-S and G2-M checkpoints of the cell cycle to slow down the growth and induce DNA repair, or, if the damage is too severe, to induce apoptosis. In the cell cycle, p53 activates a protein p21(WAF1/CIP-1) and this leads to cell cycle arrest in G1 and probably also in the G2/M interface (Cox and Lane, 1995). Transactivation of p21 leads to inactivation of cyclin dependent kinases (CDK)-cyclin complexes, which block cell cycle and result in growth arrest.

In general, p53 is not necessary for normal cell growth or differentiation as p53-null mice could develop normally. However, p53 is required for its tumour suppressor activity and hence also called as tumor suppressor protein (TP53). The p53 is functionally inactivated in 70% of human tumors (Evan *et al.*, 1995). The upstream effectors such as the atm gene product which is defective in ataxia telangiectasia and which is part of a pathway responding to DNA damage as a result of ionizing radiation selectively activate p53 to regulate specific downstream pathways (Barlow *et al.*, 1997).

Several possible mechanisms have been proposed to explain the trigger of PCD by p53. The pro-apoptotic Bcl-2 family member Bax appears to be transcriptionally induced by p53 after DNA damage (Miyashita and Reed, 1995). p53 also causes transcriptional induction of Trail receptor DR5 through an intronic sequence specific DNA binding site (Takimoto and Ei-Deiry, 2000). CD95 receptor has also been suggested as a target gene for transcriptional activation by p53 (Owen-Schaub *et al.*, 1995). C-fos proto-oncogene transcription was found to be stimulated by p53 in cells undergoing apoptosis (Elkeles *et al.*, 1999). Mcg10 has been observed as a novel p53 target gene encoding a KH domain RNA binding protein capable of inducing apoptosis and cell cycle arrest in the G_2-M phase (Zhu and Chen, 2000). Pidd, a death domain containing protein was found to be induced by p53 and to promote apoptosis (Lin *et al.*, 2000). Human p53-regulated apoptosis inducing protein-1 (p53AIP1) gets induced by wild-type p53 (Oda *et al.*, 2000). p53AIP1 is located within the mitochondria and its over expression leads to disruption of the mitochondrial membrane potential and apoptotic cell death.

Ser-46 phosphorylation within p53 regulates the transcriptional activation of this apoptosis-inducing gene. p53 induces apoptosis also through non-transcriptional mechanism. p53 activation was found to transiently increase surface receptor expression by resulting its transport from the golgi complex (Bennett *et al.*, 1998). A small fraction of stress induced p53 traffics to mitochondria leading to release of cytochrome c and hence caspase-3 activation. Recently, few other proteins with the same structure and similar biological properties as p53 have also been observed. Two of them have been identified as p73 and p63 (Benard *et al.*, 2003).

p53 protein expression was found to be regulated by at least two mechanisms. One is transcriptionally controlled, giving rise to a short living protein, and the other is a post-transnational controlled, leading to the stabilization of the p53 protein. It appears that when a cell has to address an urgent need, such as a response to DNA damage, it will use the protein accumulation mechanism, which is rapid and gives rise to excess protein levels leading to PCD. On the other hand, the transcriptional control mechanism is used in a more programmed way to address the normal physiological needs of the cell, such as cell cycle control, and exits from the cell cycle towards cell differentiation. In this case the accumulation of the protein is much slower and requires lower levels of p53 protein, which is most likely to be more economical for the cell.

Oncoprotein Murine double minute 2 (Mdm2) is a negative modulator of p53 activity. It binds to N-terminus of p53 and prevents its interaction with transcriptional machinery. Mdm2 binding also targets p53 for ubiquitin-mediated degradation. Mdm2 regulation takes place by the small ubiquitin-like modifier protein-1 (SUMO-1). Conjugation of Mdm2 with SUMO-1 excludes its self-ubiquitination and increases Mdm's ability to ubiquitinate p53 (Buschmann *et al.*, 2000). DNA damage leads to reduction in Mdm2 sumolyation ultimately resulting in increased p53 stability.

Caspase Independent PCD

Recently, the existence of a Ras-mediated caspase independent programmed cell death such in human cancer cells has been reported. It was characterized morphologically by formation of autophagic vacuoles and paucity of nuclear and mitochondrial changes. This cell death was negative for TUNEL staining and was neither accompanied by caspase activation nor blocked by caspase inhibitors such as z-VAD-fmk and z-Asp-CH2-DCB (Kitanaka and Kuchino, 1999). Ras-induced cell death was not inhibited even by the overexpresion of Bcl-2 (Kitanaka and Kuchino, 1999). Moreover, the activation of phosphatidyl inositol 3-OH kinase (PI3K), a physiological effector of Ras, was required for induction of caspase-independent cell death. Caspase-independent cell death pathway involved apoptosis inducing factor (AIF), which is localized in the mitochondria. AIF is translocated from mitochondria to nuclei in response to death stimuli and leads to large scale DNA fragmentation and peripheral chromatin condensation (Susin *et al.*, 2000). Detailed analysis has indicated that AIF seems to be a probable oxidoreductase comprising around 613 amino acid residues and of around 67-kDa molecular sizes. There are also other reports of caspase-independent cell death (Borner and Monney, 1999; Chautan *et al.*, 1999; Xiang *et al.*, 1996) such as Fas receptor is able to kill activated primary T-cells in the absence of active caspases. It involves necrotic morphological changes and depends on Rip kinase (Holler *et al.*, 2000).

Conclusion

The forgoing text provides an overview of the state-of-the-art developments in the understanding of the mechanism of programmed cell death from prokaryotes to mammals. This evolutionarily conserved phenomenon is essential for the normal development and existence of organisms. The existing gaps in our understanding of several facts of apoptotic cell death warrant extensive research in this field. A complete understanding of the mechanism of apoptosis may one day provide clues to control life threatening human diseases.

REFERENCES

1. Adams, J. M., Cory, S. (2001) Trends in Biochemical. Sci., 26: 61-6.
2. Alnemri, E.S., Livingston, D. J., Nicholson, D. W., Salvesen, G., Thornberry, N. A., Wong, W. W., Yuan, J. (1996) Cell, 87: 171.
3. Antonsson, B., Montessuit, S., Lauper, S., Eskes, R., Martinou, J. C. (2000) Biochem. J., 345: 271-8.
4. Armstrong, R., Aja, T., Hoang, K., Gaur, S., Bai, X., Alnemri, E., Litwack, G., Karanewsky, D., Fritz, L., Tomaselli, K. (1997) J. Neuroscience, 17: 553-62.
5. Armstrong, R., Aja, T., Xiang, J., Gaur, S., Krebs, J., Hoang, K., Bai, X., Korsmeyer, S., Karanewsky, D., Fritz, L., Tomaselli, K. (1996) J. Biol. Chem., 271: 16850-5.
6. Ambrosini, G., Adida, C., Altieri, D. C. (1997) Nature Med., 3: 917–21.
7. Ashkenazai, A., Dixit, V. M. (1998) Science, 281: 1305-8.
8. Attardi, L. D., Jacks, T. (1999) Cell. Mol. Life. Sci., 55: 48-63.

9. Bai, C., Connolly, B., Metzker, M. L., Hilliard, C. A., Liu, X., Sandig, V., Soderman, A., Galloway, S. M., Liu, Q., Austin, C. P., Caskey, C. T. (2000) PNAS (USA), 97: 1230–5.

10. Baker, S. J., Reddy, E. P. (1996) Oncogene, 12: 1–9.

11. Barlow, C., Brown, K. D., Deng, C. X., Tagle, D. A., Wynshaw-Boris, A. (1997) Nature Genet., 17: 453-6.

12. Bassler, B. L. (2002) Cell, 109: 421-4.

13. Bayles, K. W. (2003) Trends in Microbiology, 11: 306-11.

14. Benard, J., Douc-Rasy, S., Ahomadegbe, J. C. (2003) Hum. Mutat., 21: 182-91.

15. Bennett, M., Macdonald, K., Chan, S. W., Luzio, J. P., Simari, R., Weissberg, P. (1998) Science, 282: 290-3.

16. Bernard, P., Couturier, M. (1992) J. Mol. Biol., 226: 735-45.

17. Bertin, J., Armstrong, R. C., Ottilie, S., Martin, D. A., Wang, Y., Banks, S., Wang, G. H., Senkevich, T. G., Alnemri, E. S., Moss, B., Lenardo, M. J., Tomaselli, K. J., Cohen, J. I. (1997) PNAS (USA), 94: 1172-6.

18. Bodmer, J. L., Holler, N., Reynard, S., Vinciguerra, P., Schneider, P., Juo, P., Blenis, J., Tschopp, J. (2000) Nature Cell Biol., 2: 241–3.

19. Boise, L. H., Gonzalez-Garcia, M., Postema, C. E., Ding, L., Lindsten, T., Turka, L. A., Mao, X., Nunez, G., Thompson, C. B. (1993) Cell, 74: 597-608.

20. Borner, C., Monney, M. (1999) Cell Death Differ., 6: 497–507.

21. Bossy-Wetzel, E., Bakiri, L., Yaniv, M. (1997) EMBO J., 16: 1695–1709.

22. Brachmann, C. B., Cagan, R. L. (2003)

23. Trends Genet., 19: 91-6.

24. Brown, D. A., Rose, J. K. (1992) Cell 68: 533–44.

25. Buschmann, T., Fuchs, S. Y., Lee, C. G., Pan, Z., Q., Ronai, Z. (2000) Cell, 101: 753-62.

26. Cecconi, F., Alverez-Bolado, G., Meyer, B. I., Roth, K. A., Gruss, P. (1998) Cell, 94: 727-37.

27. Chan, F.K., Chun, H. J., Zheng, L., Siegel, R. M., Bui, K. L., Lenardo, M. J. (2000) Science, 288: 2351–4.

28. Chautan, M., Chazal, G., Cecconi, F., Gruss, P., Golstein, P. (1999) Curr. Biol., 9: 967–70.

29. Cheng, E.H., Kirsch, D.G., Clem, R.J., Ravi, R., Kastan, M.B., Bedi, A., Ueno, K., Hardwick, J.M. (1997) Science, 278: 1966–68.

30. Chichkova, N. V., Kim, S. H., Titova, E. S., Kalkum, M., Morozov, V. S., Rubtsov, Y. P., Kalinina, N. O., Taliansky, M. E., Vartapetian, A. B. (2003) The Plant Cell, 16: 157-71.

31. Chinnaiyan, A. M., O'Rourke, K., Lane, B. R., Dixit, V. M. (1997) Science, 275: 1122-26.

32. Chong, M. J., Murray, M. R., Gosink, E. C., Russell, H. R. C., Srinivasan, A., Kapsetaki, M., Korsmeyer, S. J., McKinnon, P. J. (2000) PNAS (USA), 97: 889-94.

33. Cohen, G. M. (1997) Biochem. J., 326: 1-16.

34. Coller, H. A., Grandori, C., Tamayo, P., Colbert, T., Lander, E. S., Eisenman, R. N., Golub, T. R. (2000) PNAS (USA), 97: 3260–5.

35. Conradt, B., Horvitz, H. R. (1998) Cell, 93: 519-29.

36. Cory, S., Adams, J. M. (1998) Biochim. Biophys. Acta, 1377: R25-44.

37. Cox, L. S., Lane, D. P. (1995) Bioessays, 17: 501-8.

38. Darzynkiewicz, Z., Bruno, S., Del Bino, G., Gorczyca, W., Holz, M. A., Lassota, P., Traganos, F. (1992) Cytometry, 13: 795-808.

39. Datta, S. R., Katsov, A., Hu, L., Petros, A., Fesik, S. W., Yaffe M. B., Greenberg, M. E. (2000) Mol. Cell, 6: 41-51.

40. Diaz, J. L., Oltersdorf, T., Horne, W., McConnell, M., Wilson, G., Weeks, S., Garcia, T., Fritz, L. (1997) J. Biol. Chem., 272: 11350-5

41. Diaz, J. L., Oltersdorf, T., Fritz, L. C. (2000) Methods Enzymol., 322: 255-66.

42. Dickman, M. B., Park, Y. K., Oltersdorf, T., Li, W., Clemente, T., French, R. (2001) PNAS, USA, 98: 6957-62.

43. Dorstyn, L., Colussi, P. A., Quinn, L. M., Richardson, H., Kumar, S. (1999a) PNAS (USA) 96: 4307-12.

44. Dorstyn, L., Read, S. H., Quinn, L. M., Richardson, H., Kumar, S. (1999b) J. Biol. Chem., 274: 30778-83.

45. Dragovich, T., Rudin, C. M., Thompson, C. B. (1998) Oncogene, 17: 3207-13.

46. Du, C., Fang, M., Li, Y., Li, L., Wang, X. (2000) Cell, 102: 33–42.

47. Duckett, C., Li, F., Wang, Y., Tomaselli, K., Thompson, C., Armstrong, R. (1998) Mol. and Cell. Biol., 18: 608-15.

48. Earnshaw, W. C., Martins, L. M., Kaufmann, S. H. (1999) Annul. Rev. Biochem., 68: 383-424.

49. Elkeles, A., Juven-Gershon, T., Israeli, D., Wilder, S., Zalcenstein, A., Oren, M. (1999) Mol. Cell. Biol., 19: 2594-600.

50. Ellis, H. M., Horvitz, H. R. (1986) Cell, 44: 817-29.

51. Enari, M., Talanian, R. V., Wong, W. W., Nagata, S. (1996) Nature, 380: 723–6.

52. Engidawork, E., Gulesserian, T., Yoo, B. C., Cairns, N., Lubec, G. (2001) Biochem. Biophys. Res. Commun., 281: 84-93.

53. Eskes, R., Desagher, S., Antonsson, B., Martinou, J. C. (2000) Mol. Cell. Biol., 20: 929–35.

54. Evan, G. E., Brown, L., Whyte, M., Harrington, E. (1995) Curr. Opin. Cell Biol., 7: 825-34.

55. Fadok, V. A., savill, J. S., Haslett, C., Bratton, D. L., Doherty, D. E., Campbell, P. A., Henson, P. M. (1992) J. Immunol., 149: 4029–35.

56. Franke, T. F., Cantley, L. C. (1997) Nature, 390: 116–117.

57. Fraser, A., Evan, G. (1997) Cell, 85: 781–4.

58. Fraser, A. G., James, C., Evan, G. I., Hengartner, M. O. (1999) Curr. Biol., 9: 292–301.

59. Fraser, A. G., McCarthy, N. J., Evan, G. I. (1997) EMBO J., 16: 6192-9.

60. Fujise, K., Zhang, D., Liu, J. L., Yeh, E. T. (2000) J. Biol. Chem., 275: 39458-65.
61. Gautam, S. (2003) Ph. D. (Science) Thesis, University of Mumbai.
62. Gautam, S. and Sharma, A. (2002a) Molecular Microbiology 44(2): 393-401.
63. Gautam, S. and Sharma, A. (2002b) Journal of General and Applied Microbiology 48: 67-76.
64. Gautam, S., Misra, H. S., Khairnar, N. P., Mukhopadhyay, R., Mahajan, S. K., and Sharma, A. (2002) GenBank accession AY 125331.
65. Gerdes, K., Poulsen, L. K., Thisted, T., Nielsen, A. K., Martinussen, J., Andreasen, P. H. (1990) New Biol., 2: 946-56.
66. Gervais, F. G., Xu, D., Robertson, G. S., Vaillancourt, J. P., Zhu, Y., Huang, J., LeBlanc, A., Smith, D., Rigby, M., Shearman, M. S., Clarke, E. E., Zheng, H., Van Der Ploeg, L. H., Ruffolo, S. C., Thornberry, N. A., Xanthoudakis, S., Zamboni, R. J., Roy, S., Nicholson, D. W. (1999) Cell, 97: 395–406.
67. Gil-Gomez, G., Berns, A., Brady, H. J. (1998) EMBO J., 17: 7209-18.
68. Goldstein, J. C., Waterhouse, N. J., Juin, P., Evan, G. I., Green, D. R. (2000) Nature Cell Biol., 2: 156–62.
69. Goyal, L., McCall, K., Agapite, J., Hartwieg, E., Steller, H. (2000) EMBO J., 19: 589–97.
70. Green, D. R., Ware, C. F. (1997) Proc. Natl. Acad. Sci. USA, 94: 5986–90.
71. Gross, A., Yin, X. M., Wang, K., Wei, M. C., Jockel, J., Milliman, C., Erdjument-Bromage, H., Tempst, P., Korsmeyer, S. J. (1999) J. Biol. Chem., 274: 1156-63.
72. Guohua, P., O'Rourke, K., Chinnaiyan, A. M., Gentz, R., Ebner, R., Ni, J., *et al.* (1997) Science, 276: 111-3.
73. Hakem, R., Hakem, A., Duncan, G. S., Henderson, J. T., Woo, M., Soengas, M. S., Elia, A., de la Pompa, J. L., Kagi, D., Khoo, W., Potter, J., Yoshida, R., Kaufman, S. A., Lowe, S. W., Penninger, J. M., Mak, T. W. (1998) Cell, 94: 339-52.
74. Harrington, E. A., Bennett, M. R., Fanidi, A., Evan, G. I. (1994a) EMBO J., 13: 3286–95.
75. Harrington, E. A., Fanidi, A., Evan, G. (1994b) Curr. Opin. Genet. Dev., 4: 120–9.
76. Hawkins, C. J., Yoo, S. J., Peterson, E. P., Wang, S. L., Vernooy, S. Y., Hay, B. A. (2000) J. Biol. Chem., 275: 27084-93.
77. Hengartner, M. O. (2000) Nature, 407: 770-6.
78. Hengartner, M. O., Ellis, R.E., Horvitz, H. R. (1992) Nature, 356: 494–99.
79. Hengartner, M. O., Horvitz, H. R. (1994a) Cell, 76: 665-76.
80. Hengartner, M. O., Horvitz, H. R. (1994b) Nature, 369: 318-20.
81. Hengartner, M. O., Horvitz, H. R. (1994c) Philos. Trans. R. Soc. Lond B. Biol. Sci., 345: 243-6.
82. Hiroo, F. (2000) Plant Mol. Biol., 44: 245-53.
83. Hofmann, E. R., Milstein, S., Hengartner, M. O. (2000) Cold Spring Harb. Symp. Quant. Biol., 65: 467-73.
84. Hofmann, K., Bucher, P., Tschopp, J. (1997) Trends Biochem. Sci., 22: 155–6.

85. Holler, N., Zaru, R., Micheau, O., Thome, M., Attinger, A., Valitutti, S., Bodmer, J. L., Schneider, P., Seed, B., Tschopp, J. (2000) Nature Immunol., 1: 489–95.

86. Hueber, A. O., Zornig, M., Lyon, D., Suda, T., Nagata, S., Evan. G. I. (1997) Science, 278: 1305-8.

87. Hughes, D. P. M., Crispe, I. N. (1995) J. Exp. Med., 182: 1395–1401.

88. Humlova, Z. (2002) Sb. Lek., 103: 419-25.

89. Hussein, M. R., Haemel, A. K., Wood, G. S. (2003) J. Pathol., 199: 275-88.

90. Inohara, N., Nunez, G. (2003) Nat. Rev. Immunol., 3: 371-82.

91. Irmler, M., Hofmann, K., Vaux, D., Tschopp, J. (1997) FEBS Letter, 406: 189-90.

92. Jaffe, A., Ogura, T., Hiraga, S. (1985) J. Bacteriol., 163: 841-9.

93. James, C., Gschmeissner, S., Fraser, A., Evan, G. I. (1997) Curr. Biol., 7: 246-52.

94. Jones, B. E., Lo, C. R., Liu, H., Srinivasan, A., Streetz, K., Valentio, K. L., Czaja, M. J. (2000) J. Biol. Chem., 275: 705-12.

95. Jurgensmeier, J., Krajewski, S., Armstrong, R., Wilson, G. M., Oltersdorf, T., Fritz, L., Reed, J. C., Ottilie, S. (1997) Mol. Biol. of the Cell, 8: 325-39.

96. Katoch, B., Begum, R. (2003) J. Biosci., 28: 581-8.

97. Katoch, B., Sebastian, S., Sahdev, S., Padh, H., Hasnain, S. E., Begum, R. (2002) Indian J. Exp. Biol., 40: 513-24.

98. Kelekar, A., Thompson, C. B. (1998) Trends Cell Biol., 8: 324–30.

99. Kerr, J. F., Wyllie, A. H., Currie, A. R. (1972) Br. J. Cancer, 26: 239-57.

100. Kischkel, F.C., Hellbardt, S., Behrmann, I., Germer, M., Pawlita, M., Krammer, P. H., Peter, M. E. (1995) EMBO J., 14: 5579–88.

101. Kitanaka, C., Kuchino, Y. (1999) Cell Death Differ., 6: 508-15.

102. Klefstrom, J., Kovanen, P. E., Somersalo, K., Hueber, A. O., Littlewood, T., Evan, G. I., Greenberg, A. H., Saksela, E., Timonen, T., Alitalo, K. (1999) Oncogene, 18: 2181-8.

103. Knapp, B. (2003) Ophthalmologe, 100: 67-8.

104. Koopman, G., Reutelingsperger, C. P., Kuijten, G. A., Keehnen, R. M., Pals, S. T., van Oers, M. H. (1994) Blood, 84: 1415-20.

105. Kothakota, S., Azuma, T., Reinhard, C., Klippel, A., Tang, J., Chu, K., McGarry, T. J., Kirschner, M. W., Koths, K., Kwiatkowski, D. J., Williams, L. T. (1997) Science, 278: 294–8.

106. Krammer, P. H. (2000) Nature, 407: 789-95.

107. Kroemer, G., Petit, P., Zamzami, N., Vayssiere, J. L., Mignotte, B. (1995) FASEB J., 9: 1277-87.

108. Kuida, K., Haydar, T. F., Kuan, C. Y., Gu, Y., Taya, C., Karasuyama, H., Su, M. S. S., Rakic, P., Flavell, R. A. (1998) Cell, 94: 325-37.

109. Lee, H. H., Dadgostar, H., Cheng, Q., Shu, J., Cheng, G. (1999) Proc. Natl. Acad. Sci., USA 96: 9136-41.

110. Lee, S. Y., Choi, Y. (1997) J. Exp. Med., 185: 1275–85.

111. Lehnherr, H., Maguin, E., Jafri, S., Yarmolinsky, M. B. (1993) J. Mol. Biol., 233: 414-28.

112. Li, F., Ackermann, E. J., Bennett, C. F., Rothermel, A. L., Plescia, J., Tognin, S., Villa, A., Marchisio, P. C., Altieri, D. V. (1999) Nature Cell Biol., 1: 461–6.

113. Li, F., Ambrosini, G., Chu, E. Y., Plescia, J., Tognin, S., Marchisio, P. C., Altieri, D. C. (1998) Nature, 396: 580–4.

114. Li, F., Srinivasan, A., Wang, Y., Armstrong, R., Tomaselli, K., Fritz, L. (1997a) J. Biol. Chem., 272: 30299-305.

115. Li, H., Leo, C., Zhu, J., Wu, X., O'Neil, J., Park, E. J., Chen, J. D. (2000) Mol. Cell. Biol., 20: 1784–96.

116. Li, H., Zhu, H., Xu, C. J., Yuan, J. (1998) Cell, 94: 491–501.

117. Li, P., Nijhawan, D., Budihardjo, I., Srinivasula, S. M., Ahmad, M., Alnemri, E. S., Wang, X. (1997b) Cell, 91: 479–89.

118. Lin, Y., Devin, A., Rodriguez, Y., Liu, Z. G. (1999) Genes Dev., 13: 2514–26.

119. Lin, Y., Ma, W., Benchimol, S. (2000) Nature Genet., 26: 122-7.

120. Liu, Q. A., Hengartner, M. O. (1999) Curr. Biol., 9: 1347-50.

121. Lohmann, I. (2003) Gene Expr. Patterns, 3: 159-63.

122. Losick, R., Stragier, P. (1992) Nature, 355: 601-4.

123. Maccarrone, M., Zadelhoff, G. V., Veldink, G. A., Vliegenthart, J. F. G., Finazzi-Agro, A. (2000) Eur. J. Biochem., 267: 5078-84.

124. Luo, X., Budihardjo, I., Zou, H., Slaughter, C., Wang, X. (1998) Cell, 94: 481–90.

125. MacFarlane, M. (2003) Toxicol. Lett., 139: 89-97.

126. Madeo, F., Herker, E., Maldener, C., Wissing, S., Lachelt, S., Herlan, M., Fehr, M., Lauber, K., Sigrist, S. J., Wesselborg, S., Frohlich, K. U. (2002) Mol. Cell, 9: 911-7.

127. Malinin, N.L., Boldin, M. P., Kovalenko, A. V., Wallach, D. (1997) Nature, 385: 540–4.

128. Marti, A., Graber, H., Lazar, H., Ritter, P., Baltzer, A., Srinivasan, A., Jaggi, R. (2000) Adv. Exp. Med. Biol., 480: 195-201.

129. Martinou, J. C., Desagher, S., Antonsson, B. (2000) Nature Cell Biol., 2: E41–E43.

130. Marzo., Brenner, C., Zamzami, N., Jurgensmeier, J. M., Susin, S. A., Vieira, H. L., Prevost, M. C., Xie, Z., Matsuyama, S., Reed, J. C., Kroemer, G. (1998) Science, 281: 2027–31.

131. Meier, P., Finch, A., Evan, G. (2000) Nature, 407: 796-801.

132. Metzstein, M. M., Stanfield, G. M., Horvitz, H. R. (1998) Trends Genet., 14: 410-6.

133. Mitchell, K. O., Ricci, M. S., Miyashita, T., Dicker, D. T., Jin, Z., Reed, J. C., El-Deiry, W. S. (2000) Cancer Res., 60: 6318–25.

134. Miyashita, T., Reed, J. C. (1995) Cell, 80: 293-9.

135. Moroy, T., Zornig, M. (1997) J. Cell Physiol. Biochem., 6: 312–36.

136. Muchmore, S. W., Sattler, Liang, H., Meadows, R. P., Harlan, J. E., Yoon, H. S., Nettesheim, D., Chang, B. S., Thompson, C. B., Wong, S. L., Ng, s. L., Fesik. S. W. M. (1996) Nature, 381: 335-41.

137. Naito, T., Kusano, K., Kobayashi, I. (1995) Science, 267: 897-9.

138. Naito, Y., Naito, T, Kobayashi, I. (1998) Biol. Chem., 379: 429-36.

139. Nakagawa, T., Zhu, H., Morishima, N., Li, E., Xu, J., Yankner, B. A., Yuan, J. (2000) Nature, 403: 98–103.

140. Nicholson, D. W. (2000) Nature, 407: 810-6.

141. Nicholson, D. W. (1996) Nature Biotechnol., 14: 297–301

142. Nystrom, T. (2003a) Bioessays, 25: 204-11.

143. Nystrom, T. (2003b) Mol. Microbiol., 48: 17-23.

144. Oberhammer, F., Wilson, J., Dive, C., Morris, I., Hickman, J., Wakeling, A., Walker, P., Sikorska, M. (1993) EMBO J., 12: 3679–84.

145. Oda, K., Arakawa, H., Tanaka, T., Matsuda, K., Tanikawa, C., Mori, T., Nishimori, H., Tamai, K., Tokino, T., Nakamura, Y., Taya, Y. (2000) Cell, 102: 849-62.

146. Olson, M. R., Holley, C. L., Yoo, S. J., Huh, J. R., Hay, B. A., Kornbluth, S. (2003) J. Biol. Chem., 278: 4028-34.

147. Opferman, J. T., Korsmeyer, S. J. (2003) Nat. Immunol., 4: 410-5.

148. Oren, M., Rotter, V. (1999) Cell. Mol. Life. Sci., 55: 9-11.

149. Ottilie, S., Diaz, J. L., Horne, W., Chang, J., Wang, Y., Wilson, G., Chang, S., Weeks, S., Fritz, L., Oltersdorf, T. (1997a) J. Biol. Chem., 272: 30866-72.

150. Ottilie, S., Wang, Y., Banks, S., Chang, J., Vigna, N. J., Weeks, S., Armstrong, R., Fritz, L., Oltersdorf, T. (1997b) Cell Death and Differentiation, 4: 526-33.

151. Owen-Schaub, L. B., Zhang, W., Cusack, J. C., Angelo, L. S., Santee, S. M., Fujiwara, T., Roth, J. A., Deisseroth, A. B., Zhang, W. W., Kruzel, E. (1995) Mol. Cell. Biol., 15: 3032-40.

152. Peter, M. E., Heufelder, A. E., Hengartner, M. O. (1997) PNAS, USA, 94: 12736-37.

153. Pitti, C., Marsters, S. A., Lawrence, D. A., Roy, M., Kischkel, F. C., Dowd, P., Huang, A., Donahue, C. J., Sherwood, S. W., Baldwin, D. T., Godowski, P. J., Wood, W. I., Gurney, A. L., Hillan, K. J., Cohen, R. L., Goddard, A. D., Botstein, D., Ashkenazi, A. (1998) Nature, 396: 699–703.

154. Prokocimer, M, Rotter, V. (1994) Blood, 84: 2391-411.

155. Proskuryakov, S. Y., Konoplyannikov, A. G., Gabai, V. L. (2003) Exp. Cell Res., 283: 1-16.

156. Raff, M. C. (1992) Nature, 356: 397-400.

157. Reed, J. C. (1998) Oncogene, 17: 3225-36.

158. Reed, J. C. (1997) Nature, 387: 773-6.

159. Rice, K. C. and Bayles, K. W. (2003) Molecular Microbiology, 50: 729-38

160. Rich, T., Allen, R. L., Wyllie, A. H. (2000) Nature, 407: 777-83.

161. Riess, O; Berg, D., Kruger, R., Schulz, J. B. (2003) J. Neurol., 250: 3-10.

162. Roberts, R. C., Strom, A. R., Helinski, D. R. (1994) J. Mol. Biol., 237: 35-51.

163. Rodriguez, J., Lazebnik, Y. (1999) Genes Dev., 13: 3179-84.

164. Rossi, D., Gaidano, G. (2003) Haematologica, 88: 212-8.

165. Rothe, M., Pan, M. G., Henzel, W. J., Ayres, T. M., Goeddel, D. V. (1995) Cell, 83: 1243–52.

166. Roy, N., Mahadevan, M. S., McLean, M., Shutler, G., Yaraghi, Z., Farahani, R., Baird, S., Besner-Johnston, A., Lefebvre, C., Kang, X. (1995) Cell, 80: 167–78.

167. Ruiz-Echevarria, M. J., Berzal-Herranz, A., Gerdes, K., Diaz-Orejas, R. (1991) Mol. Microbiol., 5: 2685-93.

168. Sakahira, H., Enari, M., Nagata, S. (1998) Nature, 391: 96–9.

169. Sanchez, C., Xu, J., Juo, P., Kakizaka, A., Blenis, J., Yuan, J. (1999) Neuron, 22: 623–33.

170. Sanchez, P., Zabala, M. D. T., Grant, M. (2000) The Plant Journal, 21: 393.

171. Sattler, M., Liang, H., Nettesheim, D., Meadows, R. P., Harlan, J. E., Eberstadt, M., Yoon, H. S., Shuker, S. B., Chang, B. S., Minn, A. J., Thompson, C. B., Fesik, S. W. (1997) Science, 275: 983-6.

172. Scaffidi, C., Fulda, S., Srinivasan, A., Friesen, C., Li, F., Tomaselli, K., Debatin, K. M., Krammer, P., Peter, M. (1998) EMBO J., 6: 1675-87.

173. Scaffidi, C., Schmitz, I., Krammer, P. H., Peter, M. E. (1999a) J. Biol. Chem., 274: 1541–8.

174. Scaffidi, C., Schmitz, I., Zha, J., Korsmeyer, S. J., Krammer, P. H., Peter, M. E. (1999b) J. Biol. Chem., 274: 22532–8.

175. Schmitt, C. A. (2003) Nat. Rev. Cancer, 3: 286-95.

176. Schreader, B. A., Wang, Y., Nambu, J. R. (2003) Apoptosis, 8: 129-39.

177. Scorrano, L., Korsmeyer, S. J. (2003) Biochem. Biophys. Res. Commun., 304: 437-44.

178. Sharma, A., Nair, P. M., Pawar, S. E. (1993) Euphytica, 67: 95-9.

179. Sharma, A., Syed, A. N., Nair, P. M. (1994) J. Phytopathol., 141: 53-8.

180. Sharma, A. (1999) Encyclopedia of food microbiology, London: Academic Press. pp.2323-9.

181. Shimizu, S., Ide, T., Yanagida, T., Tsujimoto, Y. (2000) J. Biol. Chem., 275:12321-5.

182. Shimizu, S., Narita, M., Tsujimoto, Y. (1999) Nature, 399: 483–7.

183. Silke, J., Verhagen, A. M., Ekert, P. G., Vaux, D. L. (2000) Cell Death Differ., 7: 1275.

184. Silva, M., Benito, A., Sanz, C., Prosper, F., Ekhterae, D., Nunez, G., Fernandez-Luna, J. L. (1999) J. Biol. Chem., 274: 22165–9.

185. Skulachev, V. P. (2002) FEBS Lett., 528: 23-6.

186. Slee, E. A., Harte, M. T., Kluck, R. M., Wolf, B. B., Casiano, C. A., Newmeyer, D. D., Wang, H. G., Reed, J. C., Nicholson, D. W., Alnemri, E. S., Green, D. R., Martin, S. J. (1999) J. Cell. Biol., 144: 281–92.

187. Smith, C. A., Farrah, T., Goodwin, R. G. (1994) Cell, 76: 959-62.

188. Smits, E., Criekinge, W. van., Plaetinck, G., Bogaert, T. (1999) Curr. Biol., 9: 1351-4.

189. Soengas, M. S., Capodieci, P., Polsky, D., Mora, J., Esteller, M., Opitz-Araya, X., McCombie, R., Herman, J. G., Gerald, W. L., Lazebnik, Y. A., Cordon-Cardo, C., Lowe, S. W. (2001) Nature, 409: 207-11.

190. Sogame, N., Kim, M., Abrams, J. M. (2003) PNAS (USA) 100: 4696-701.

191. Soussi, T. (2000) Ann. NY Acad. Sci., 910: 121-39.

192. Spector, M. S., Desnoyers, S., Hoeppner, D. J., Hengartner, M. (1997) Nature, 385: 653-6.

193. Srinivasan, A., Li, F., Wong, A., Kodandapani, L., Smidt, R. J., Krebs, J., Fritz, L., Wu, J., Tomaselli, K. (1998) J. Biol. Chem., 273: 4523-9.

194. Srinivasula, S. M., Ahmad, M., Fernandes-Alnemri, T., Alnemri, E. S. (1998) Mol. Cell, 1: 949–57.

195. Stadelmann, C., Deckwerth, T. L., Srinivasan, A., Bancher, C., Bruck, W., Jellinger, K., Lassmann, H. (1999) Am. J. Pathol., 155: 1459-66.

196. Stanger, B. Z., Leder, P., Lee, T. H., Kim, E., Seed, B. (1995) Cell, 81: 513–23.

197. Steller, H. (1995) Science, 267: 1445-9.

198. Stennicke, H. R., Deveraux, Q. L., Humke, E. W., Reed, J. C., Dixit, V. M., Salvesen, G. S. (1999) J. Biol. Chem., 274: 8359–62.

199. Stennicke, H. R., Salvesen, G. S. (1998) Biochim. Biophys. Acta, 1387: 17-31.

200. Stephens, A. C., Rivers, R. P. (2003) Curr. Opin. Mol. Ther., 5: 118-22.

201. Stiewe, N., Putzer, B. M. (2000) Nature Genet., 26: 464–9.

202. Strasser, A., Huang, D. C., Vaux, D. L. (1997) Biochim. Biophys. Acta 1333: F151–F178.

203. Suda, T., Takahashi, T., Golstein, P., Nagata, S. (1993) Cell, 75: 1169-78.

204. Susin, S. A., Daugas, E., Ravagnan, L., Samejima, K., Zamzami, N., Loeffler, M., Costantini, P., Ferii, K. F., Irinopoulou, T., Prevost, M. C., Brothers, G., Mak, T. W., Penninger, J., Earnshaw, W. C., Kroemer, G. (2000) J. Exp. Med., 192: 571-80.

205. Susin, S. A., Lorenzo, H. K., Zamzami, N., Marzo, I., Snow, B. E., Brothers, G. M., Mangion, J., Jacotot, E., Costantini, P., Loeffler, M., Larochette, N., Goodlett, D. R., Aebersold, R., Siderovski, D. P., Penninger, J. M., Kroemer, G. (1999) Nature, 397: 441–6.

206. Takahashi, A., Alnemri, E. S., Lazebnik, Y. A., Fernandes-Alnemri, T., Litwack, G., Moir, R. D., Goldman, R. D., Poirier, G, G., Kaufmann, S. H., Earnshaw, W. C. (1996) Proc. Natl. Acad. Sci. USA, 93: 8395–400.

207. Takimoto, R., El-Deiry, W. S. (2000) Oncogene, 19: 1735-43.

208. Tartaglia, L. A., Pennica, D., Goeddel, D. V. (1993) J. Biol. Chem., 268: 18542–8.

209. Thisted, T., Neilsen, A. K., Gerdes, K. (1994) EMBO J., 13: 1950-9.

210. Thornberry, N. A., Lazebnik, Y. (1998) Science, 281: 1312-6.

211. Tibbetts, M. D., Zheng, L., Lenardo, M. J. (2003) Nat. Immunol., 4: 404-9.

212. Truneh, A., Sharma, S., Silverman, C., Khandekar, S., Redy, M. P., Deen, K. C., McLaughlin, M. M., Srinivasula, S. M., Livi, G. P., Marshall, L. A., Alnemri, E. S., Williams, W. V., Doyle, M. L. (2000) J. Biol. Chem., 197: 23319–25.

213. Tsuchimoto, S., Nishimura, Y., Ohtsubo, E. (1992) J. Bacteriol., 174: 4205-11.

214. Tsujimoto, Y. (2003) J. Cell. Physiol., 195: 158-67.

215. Tsuruo, T., Naito, M., Tomida, A., Fujita, N., Mashima, T., Sakamoto, H. (2003) Cancer Sci., 94: 15-21.

216. Ueda, N., and Shah, S. V. (1994) J. Lab. Clin. Med., 124: 169-77.

217. Uren, A. G., Beilharz, T., O'Connell, M. J., Bugg, S. J., van Driel, R., Vaux, D. L., Lithgow, T. (1999) Proc. Natl. Acad. Sci. USA, 96: 10170–5.

218. Vaux, D. L., Aguila, H. L., Weissman, I. L. (1992) Int. Immunol., 4: 821-4.

219. Vaux, D. L., Cory, S., Adams, J. M. (1988) Nature, 335: 440-2.

220. Veis, D. J., Sorenson, C. M., Shutter, J. R., Korsmeyer, S. J. (1993) Cell, 75: 229-40.

221. Vermes I., Haanen, C., Steffens-Nakken, H., Reutelingsperger, C. (1995) J. Immunol. Methods., 184: 39-51.

222. Virdee, K., Parone, P. A., Tolkovsky, A. M. (2000) Curr. Biol., 10: 1151-4.

223. Walczak, H., Miller, R. E., Ariail, K., Gliniak, B., Griffith, T. S., Kubin, M., Chin, W., Jones, J., Woodward, A., Le, T., Smith, C., Smolak, P., Goodwin, R. G., Rauch, C. T., Schuh, J. C., Lynch, D. H. (1999) Nature Med., 5: 157–63.

224. Wallach, D. (1997) Trends Biochem. Sci., 1997: 107–9.

225. Wang, S. L., Hawkins, C. J., Yoo, S. J., Muller, H. A., Hay, B. A. (1999) Cell, 98: 453–63.

226. Ware, C. F., VanArsdale, S., VanArsdale, T. L. (1996) J. Cell Biochem., 60: 47–55.

227. Wei, M. C., Lindsten, T., Mootha, V. K., Weiler, S., Gross, A., Ashiya, M., Thompson, C. B., Korsmeyer, S. J. (2000) Genes Dev., 14: 2060–71.

228. Wen, L.P., Fahrni, J.A., Troie, S., Guan, J. L., Orth, K., Rosen. G.D. (1997) J. Biol. Chem., 272: 26056–61.

229. Wu, D., Wallen, H. D., Nunez, G. (1997) Science, 275: 1126-9.

230. Wyllie, A. H. (1993) Br. J. Cancer, 67: 205-8.

231. Wyllie, A. H. (1980) Nature, 284: 555–6.

232. Wyllie, A. H., Kerr, J. F. R., Currie, A. R. (1980) Int. Rev. Cytol., 68: 251-306.

233. Xiang, J., Chao, D. T., Korsmeyer, S. J. (1996) Proc. Natl. Acad. Sci. USA, 93: 14559–63.

234. Xue, D., Horvitz, H. R. (1997) Nature, 390: 305-8.

235. Yang, E., Zha, J., Jockel, J., Boise, L. H., Thompson, C. B., Korsmeyer, S. J. (1995) Cell, 80: 285-91.

236. Yang, X., Khosravi-Far, R., Chang, H., Baltimore, D. (1997) Cell, 89: 1067–76.

237. Yang, X., Stennicke, H. R., Wang, B., Green, D. R., Janicke, R. U., Srinivasan, A., Seth, P., Salvesen, G. S., Froelich, C. J. (1998) J. Biol. Chem., 273: 34278-83.

238. Yarmolinsky, M. B. (1995) Science, 267: 836-7.

239. Yin, C., Knudson, C. M., Korsmeyer, S. J., Van Dyke, T. (1997) Nature, 385: 637-40.

240. Yorgey, P., Lee, J., Kordel, J., Vivas, E., Warner, P., Jebaratnam, D., Kolter, R. (1994) PNAS (USA) 91: 4519-23.

241. Yoshida H., Kong, Y. Y., Yoshida, R., Elia, A. J., Hakem, A., Hakem, R., Penninger, J. M., Mak, T. W. (1998) Cell, 94: 739-50.

242. Young, R. Y. (1992) Microbiol. Rev., 56: 430-81.

243. Yuan, J., Yankner, B. A. (2000) Nature, 407: 802-9.

244. Zhang, Y. *et al.* (2003) Mol. Cell., 12: 913-23.

245. Zhong, L. T., Sarafian, T., Kane, D. J., Charles, A. C., Mah, S. P., Edwards, R. H., Bredesen, D. E. (1993) Proc. Natl. Acad. Sci. USA, 90: 4533-7.

246. Zhu, J., Chen, X. (2000) Mol. Cell. Biol., 20: 5602-18.

247. Zindy, F., Eischen, C. M., Randle, D. H., Kamijo, T., Cleveland, J. L., Sherr, C. J., Roussel, M. F. (1998) Genes Dev., 12: 2424–33.

248. Zou, H., Henzel, W. J., Liu, X., Lutschg, A., Wang, X. (1997) Cell, 90: 405-13.

Advances in Biochemistry and Biotechnology Volume 1 ***Pages 158-182***
***Edited by:* Dr. Chiranjib Chakraborty**
***Published by:* DAYA PUBLISHING HOUSE**

7

Biocatalysis: An Emerging Field in the Synthesis of Eenantio-Pure Compounds

*Mani Shankar Bhattacharyya, Sawraj Singh, Anirban Banerjee, Praveen Kaul, Pankaj Soni, Brajesh Barse, Rohit Sharma & U.C. Banerjee**
**Department of Pharmaceutical Technology, National Institute of Pharmaceutical Education and Research, Sector 67. S.A.S. Nagar-160062, Punjab*
E-mail: ucbanerjee@niper.ac.in

ABSTRACT

A majority of the research work in drug design time is focused towards the understanding of interaction of small molecules with the biological macromolecules. Understanding of the mechanism of the drug interaction at the molecular level has led to the realization of the importance of chirality as the key factor. Therefore, optically active chiral entities have become a major field of interest not only for the organic synthetic chemists, but also for the biotechnologists and pharmacologists. Organic synthesis of chiral molecules has enormous potential for practical use. Application of enzymes as biocatalyst in the synthesis of chiral entities has established its root not only in the drug and pharmaceutical sectors but also in the in the synthesis of agrochemical, polymers and many other classes of functional molecules. Here we have described the application of various enzymes as biocatalysts, for the large-scale production of chiral entities and chemicals, various advantages and disadvantages with a forward look.

* Correspondent Author

Introduction

Chirality (handedness: left or right) is an intrinsic universal feature of various levels of matters. Molecular chirality plays an important role in science and technology. In particular, life depends on chirality. In the biological system most physiological phenomena is the result of highly precise molecular interactions, in which host chiral molecule interacts with the two different enantiomeric guest molecules in different ways. Chiral receptor sites of the human body interact with the chiral molecules, mostly drug molecules, which are having proper absolute configuration to interact with the receptors, showing a marked difference in the pharmacological activities of the enantiomers. To appreciate the relationship between the pharmacological activity and the molecular chirality, the tragic case of administration of thalidomide to pregnant women in 1960s can be exemplified. (R)-Thalidomide has desirable sedative properties. On the other hand the *S* enantiomer is teratogenic and induces fatal malformations. Nevertheless, in the recent time with the advent of modern analytical methodologies and with the increasing understanding of mechanisms of drug interactions in the biological system demand for the enantiomerically pure drugs are increasing. Still in the 1990s about 90% of the synthetic chiral drugs were racemic (1).

To day approximately, 80% of the chiral intermediates and related products go to the pharmaceutical market. As the preparation of chiral, enantiomerically pure pharmaceutical agents has become more important, the need for new synthetic methods for their efficient preparation has also increased (2). In 2001, single-enantiomer pharmaceuticals accounted for 36% of the US $410 billion drug industry (3). Marketing estimates for drug manufacturing suggest that the need for chiral materials in pharmaceutical preparations will continue to rise with a 9.4% annual growth between 2000 and 2005. The introduction of novel asymmetric methods can offer effective new avenues for the preparation of the enantiomerically pure drugs. But though variety of methods is available in the recent time for the synthesis of enantiomerically compounds, most of the methodologies are tedious; require expensive and toxic or hazardous reagents causing environmental problems. On the other hand increased regulatory pressure to market homochiral drugs by the U.S. Food and Drug Administration and other regulatory agencies created an atmosphere leading to the use of alternative approaches for the synthesis of the chiral compounds. Biocatalysis gave a new dimension, with new innovative approaches and with enormous opportunity to prepare industrially useful chiral compounds. The advantages with the biocatalysts over chemical catalysts are that the enzyme- catalysed reactions are stereoselective and regioselective, and can be carried out at an ambient temperature and pressure, and under environment-friendly conditions. The process also minimizes the problem of isomerization, racemization, and rearrangement of molecules that may occur during chemical processes.

In this chapter we will first try to give a brief account of various enzymes used in the synthesis of enantiomerically pure compounds and then we will gradually look towards the various processes that have been developed using enzyme systems like lipases, nitrilases and oxidoreductases. In case of oxidoreductase-mediated biotransformation we will concentrate mainly on the reduction of ketones both by the conventional and nonconventional yeasts.

Enzymes in the Synthesis of Enantiomerically Pure Compounds

Microorganisms are present in a variety of ecological niche and posses' a variability in their enzymatic system, within the same group, even within the same species. By virtue of these phenomena microbial enzymes can be utilized for catalyzing a variety of reactions in the organic synthesis. As a consequence of these realizations, hydrolytic enzymes such as lipases, esterases, proteases, acylases, amidases, nitrilases, lyases, hydantoinases and dehaloginases are used in the resolution of racemic compounds and in the asymmetric synthesis of optically active compounds. oxidoreductase are used in the synthesis of chiral alcohols, aminoalcohols and amines; aldolases and decarboxylases catalyse aldol-condentation and acyloin-condensation reactions; oxygenases such as monooxygenases and dioxygenases are used in the stereoselective hydroxylation and epoxidation reactions; epoxide hydrolases are used in the stereoselective synthesis of chiral diols. Various other complex type of reactions catalysed by the microbial cells and the enzymes there in are Baeyer-Villiger reactions; enantioselective formation of chiral cyanohydrins, production of β-hydroxy acids etc. A discussion of each and every enzymatic system and their application in the synthesis of enantiopure compounds is beyond the scope of this chapter. Therefore, we will try to give an over view on the some of the mostly used enzymes on which work has now been focused in our laboratory; *viz.* lipases, nitrilases, and oxidoretases along with some of the examples of processes developed using them.

Lipases

Lipases (EC 3.1.1.3) are a versatile group of biocatalysts. Lipases (triacylglycerol acylhydrolases) belong to the class of serine hydrolases and therefore do not require any cofactor. their natural function is the hydrolysis of triglycerides to partial glycerides and fatty acids during digestion. Unlike carboxyl esterases (EC 3.1.1.1), lipase catalysis occurs at the lipid–water interface and show the phenomenon of interfacial activation–highly active in the presence of a hydrophobic phase like lipid droplet dispersed in water or an organic solvent. This phenomenon has been related to the presence of a hydrophobic oligopeptide covering the entrance to the active site. In a hydrophobic environment, the lid moves aside and the substrate can enter the binding pocket. The tetrahedral intermediate formed during the transition state is also stabilized by an oxyanion hole (4). Enzymes find use in many fields such as food, dairy, pharmaceutical, detergent, textile, cosmetic industries and enantioselective synthesis. Lipases are active in organic solvents that can dissolve apolar compounds and possess regio-and stereoselctivity hence, extensively used in enantioselective synthesis of chirally pure molecules. In absence of water, lipase is capable of reversing the reaction to esterification and formation of glycerides from fatty acids and glycerol. Lipases have been well reported in the synthesis of enatiomerically pure compounds and their intermediates.

Lipases in Pharmaceuticals and Agrochemicals

Calcium Channel Blocker

An impressive example of the commercial application of lipases in the resolution of racemic mixtures is the hydrolysis of epoxyester alcohols. The reaction products, (*R*)-glycidyl esters and (*R*)-glycidol are readily converted to (*R*)- and (*S*)-glycidyltosylates, which is attractive, intermediates for the preparation of optically active *ß*-blockers and a wide range of other products. Diltiazem is a typical calcium channel blocker and has been clinically used as an effective antianginal and antihypertensive agent. The production of (2S, 3S)-acetoxy-5-[2-(dimethylamino)ethyl]-2,3-dihydro-2-(4-methoxyphenyl)-1,5-benzothiazepin-4(5H)-one, a hydrochloride salt, (Diltiazem), has been carried out conventionally by a chemical synthetic process through nine steps from *p*-anisaldehyde and chloroacetic acid methyl ester (5,6). The desired pharmacological effect resides in one of the two enantiomers of diltiazem hydrochloride. However, in this process, a large amount of loss in the raw material occurs, as the optical resolution was carried out at the later stage of the process using the large molecular compound *i.e. threo*-2-hydroxy-3-(4-methoxyphenyl)-3-(2-nitrophenylthio)propanoic acid. Using lipase, the number of steps for the synthesis of diltiazem from the raw materials (anisaldehyde and chloroacetic acid) is reduced to five, which not only reduces the cost of production but also gives chirally pure drug (Fig. 7.1).

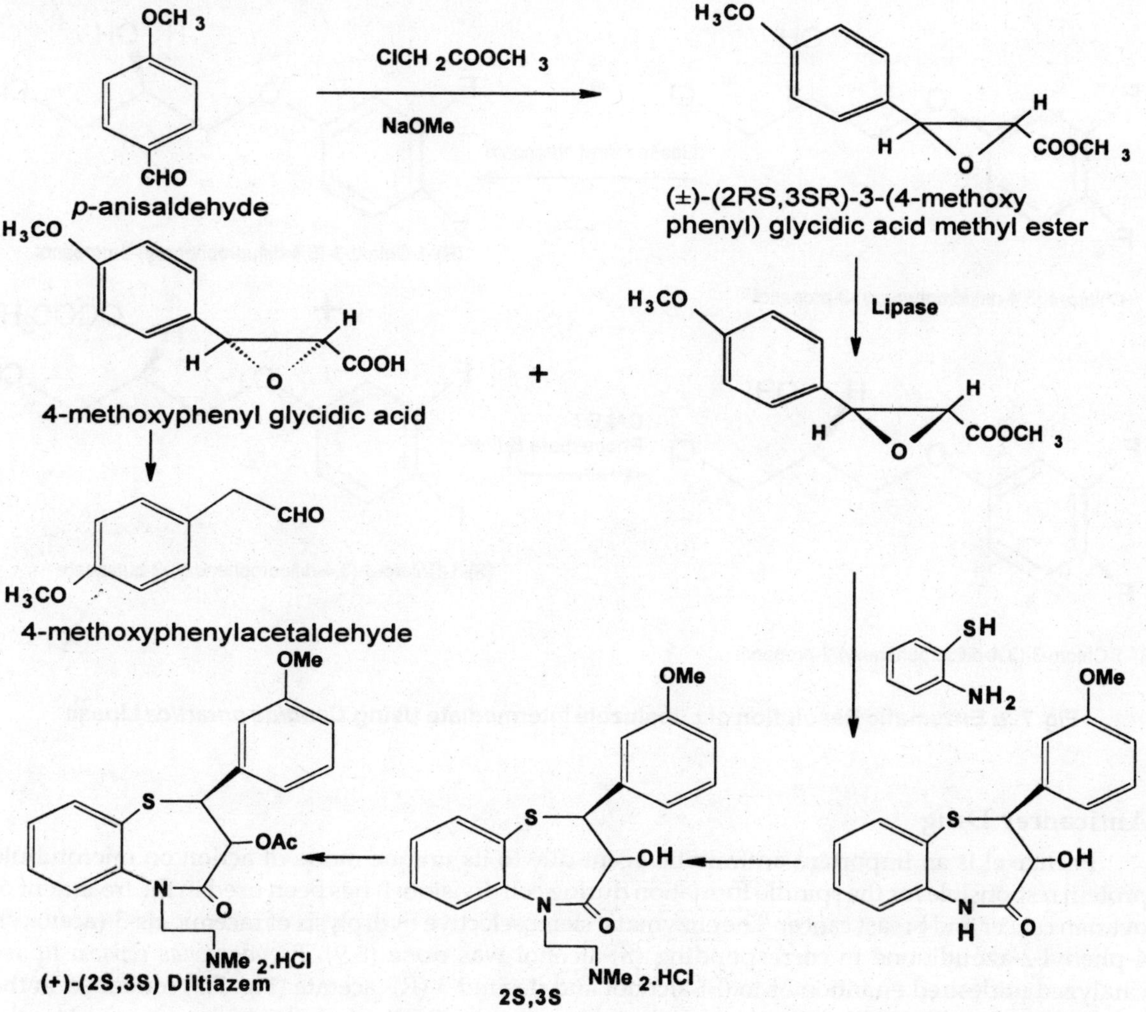

OCH_3

$ClCH_2COOCH_3$

NaOMe

H_3CO

H

H COOCH$_3$

O

CHO

p-anisaldehyde

(±)-(2RS,3SR)-3-(4-methoxy phenyl) glycidic acid methyl ester

H_3CO

H

H COOH

O

4-methoxyphenyl glycidic acid

+

H_3CO

Lipase

H

H COOCH$_3$

O

CHO

H_3CO

4-methoxyphenylacetaldehyde

SH

NH_2

OMe

S

OAc

N

O

NMe $_2$.HCl

(+)-(2S,3S) Diltiazem

OMe

S

OH

N

O

NMe $_2$.HCl

2S,3S

OMe

S

OH

N

H

O

Fig. 7.1: Synthesis of Diltiazem Using Lipase

Anti-ischemic Drug

Lubeluzole is the (S) enantiomer of benzothiazole used in treatment of ischemic stroke. It reduces infarct size, accelerates recovery of tissue in the penumbra and protects neurologic function. Lubeluzole is synthesised from intermediates 1-choloro-3 (3,4-difluorophenoxy)-2-propanol and N-methyl–N-(4-piperidinyl-2-benzothiazolamine. Racemic 1-choloro-3 (3,4-difluorophenoxy)-2-propanol is lipase catalysed to get (S)-enantiomer of 1-choloro-3(3,4-difluorophenoxy)-2-propanol which is used in the synthesis of (S)- lubeluzole. *Candida antartica* lipase B enantioselectively converts the (S)-alcohol to (S)-ester (Fig. 7.2) in vinyl butanoate in organic media with enantiomeric excess of 98% (7).

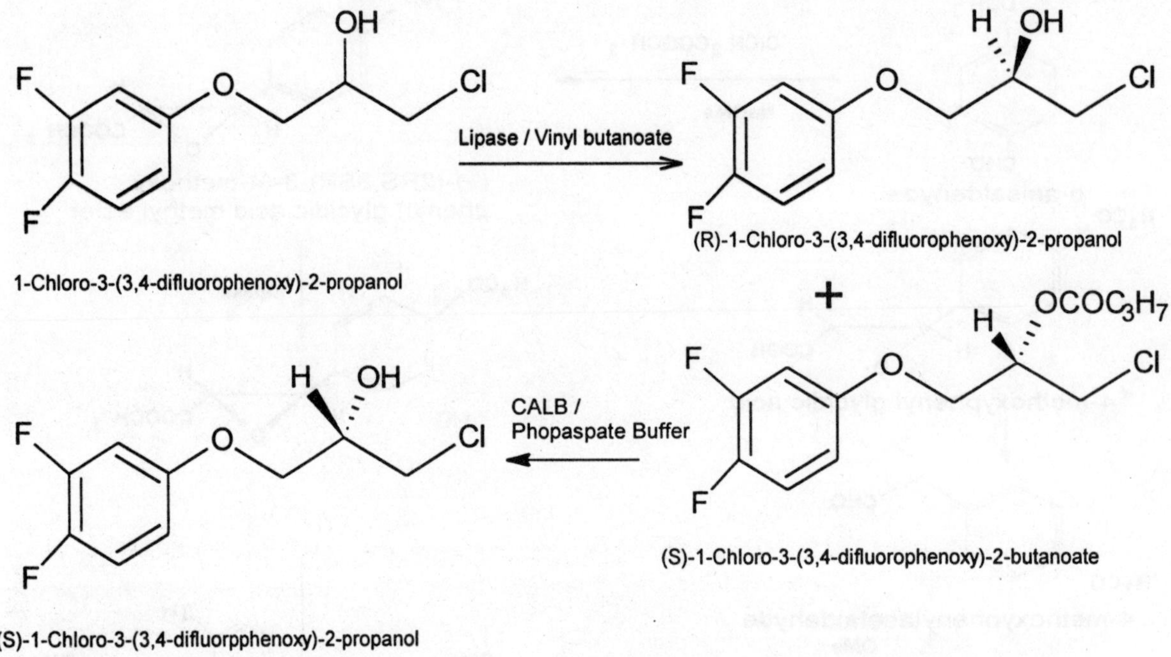

1-Chloro-3-(3,4-difluorophenoxy)-2-propanol

Lipase / Vinyl butanoate

(R)-1-Chloro-3-(3,4-difluorophenoxy)-2-propanol

+

(S)-1-Chloro-3-(3,4-difluorophenoxy)-2-butanoate

CALB / Phopaspate Buffer

(S)-1-Chloro-3-(3,4-difluorpphenoxy)-2-propanol

Fig. 7.2: Enzymatic Resolution of Lubeluzole Intermediate Using *Candida antartica* Lipase

Anticancer Drug

Paclitaxel is an important antimitotic agent due to its unique mode of action on microtubule protein responsible for the spindle formation during cell division. It has been used in the treatment of ovarian cancer and breast cancer. The enzymatic stereoselective hydrolysis of racemic cis-3-(acetoxy)-4-phenyl-2-azetidinone to corresponding (S)-alcohol was done (8,9). *Pseudomonas cepacia* lipase catalyzed undesired enantiomer to (S)-alcohol and desired 3-(R)- acetate (Fig. 7.3) remained in the reaction mixture. The 3-(R)- acetate is chemically converted to 3-(R)- alcohol, which is, required for the synthesis of paclitaxol.

Anticholestrol Drugs

It is well known that the three major risk factors for ischemic heart disease are hypercholesterolemia, hypertension and smoking. Hypercholesterolemia has long been considered to be the most important of these factors (10). Cholesterol is accumulated endogenously by both, absorption from diet and biosynthesis. The biosynthesis of cholesterol constitutes an extensive biochemical process exceeding more than 20 steps starting from acetyl-coenzyme A. The rate limiting enzyme of this pathway is 3-hydroxy-3-methylglutaryl (HMG)-CoA reductase, which catalyses the reduction of HMG-CoA to mevalonate. A lipase catalysed stereoselective acetylation of racemic [4-[4α, 6β(E)]]-6[4,4-bis[(4-fluorophenyl)-3-(1-methyl-1H-tetrazol-5-yl)-1,3-butadienyl]-tetrahydro-4-hydroxy-2H-phyrene-2-one in organic solvent gave chiral alcohol (Fig. 7.4). For this a number of lipases were evaluated among which lipase PS-30 (Amano resolution) and BMS lipase efficiently catalyzed the acetylation of the

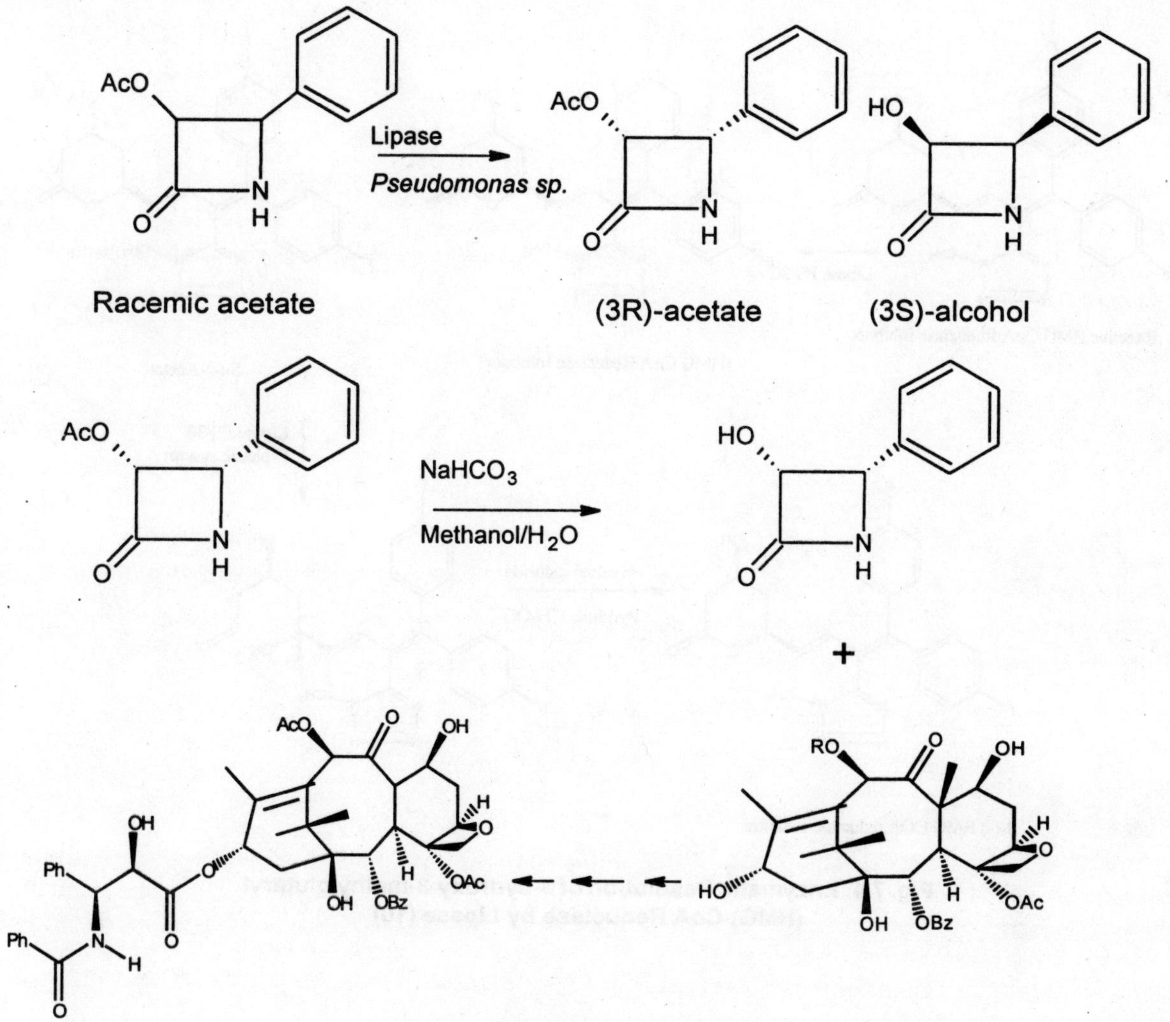

Racemic acetate — Lipase, *Pseudomonas sp.* → (3R)-acetate + (3S)-alcohol

$NaHCO_3$, Methanol/H_2O

Fig. 7.3: Synthesis of Paclitaxel

undesired enantiomer of the racemic substrate to give S-(-)-acetylated product. The reaction yield and enantiomeric excess was found to be 49 and 98.5%, respectively.

Anti-inflammatory Drugs

Naproxen, (S)-2-(6-methoxy-2-napthyl) propionic acid is a nonsteroidal antinflammatory and analgesic agent. S-naproxen, the eutomer has been obtained unreacted in *Candida cylinderacea* lipase catalysed reaction in isooctane and trimethylsilane (Fig. 7.5). In this reaction the undesired isomer was esterified (11).

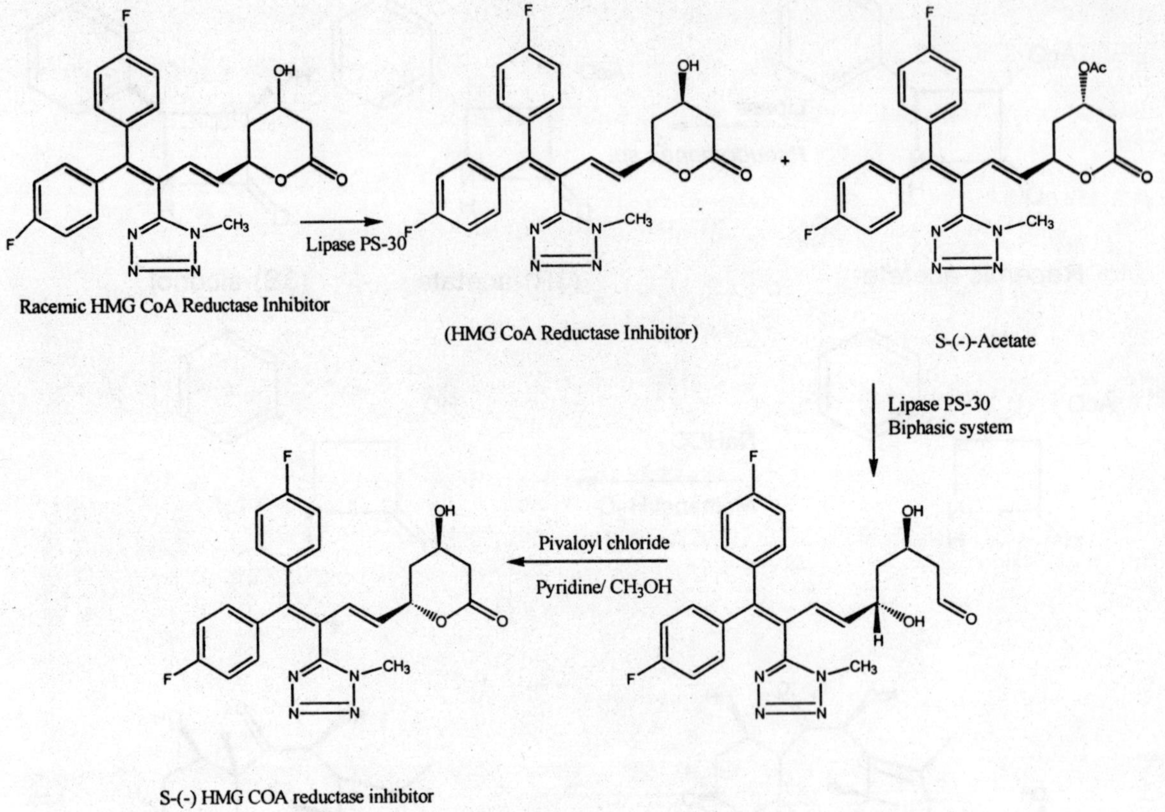

Racemic HMG CoA Reductase Inhibitor

Lipase PS-30

(HMG CoA Reductase Inhibitor)

S-(-)-Acetate

Lipase PS-30
Biphasic system

Pivaloyl chloride
Pyridine/ CH_3OH

S-(-) HMG COA reductase inhibitor

Fig. 7.4: Enzymatic Resolution of 3-hydroxy-3-methylglutaryl (HMG)-CoA Reductase by Lipase (10)

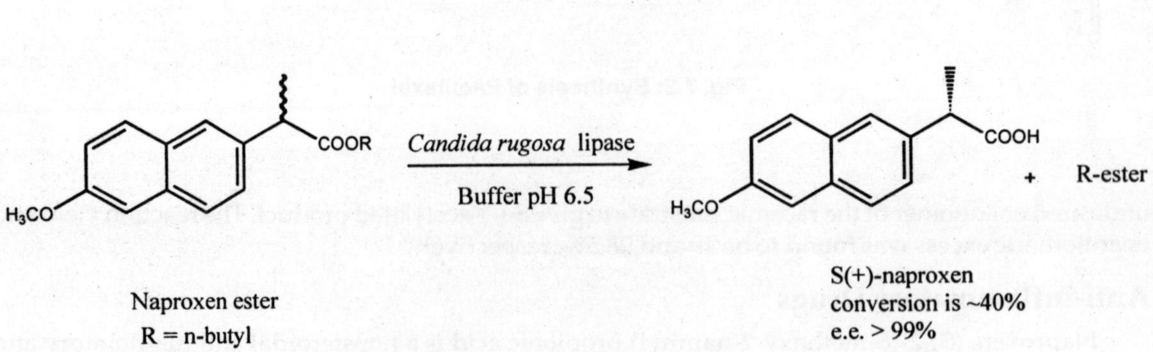

Candida rugosa lipase
Buffer pH 6.5

Naproxen ester
R = n-butyl

S(+)-naproxen
conversion is ~40%
e.e. > 99%

+ R-ester

Fig. 7.5: Preparation of (S)-naproxen and R-ester in Lipase Catalysed Synthesis

Ibuprofen (Fig. 7.6) is a well-known analgesic, antiinflammatory drug. This drug has been synthesised at gram scale. The enzymatic synthesis if S-ibuprofen has been done by enantioselective ammonolysis of ibuprofen 2-choloroethylester by *Candida antartica* lipase giving 44 and 96% e.e., The enantioselective enzymatic esterification of racemic ibuprofen has been done using *Candida cylinderacea* lipase at water in oil emulsion in which (S)-enantiomer of ibuprofen was esterified and (R)- isomer remain unreacted with a yield of 35% (12).

Antiviral Drugs

Lamivudine is an effective anti-viral agent used against HIV2 and HIV3 infection. It has two chiral acetal centres sharing oxygen atom. Cytidine deaminase from *E. coli* has been used to deaminate 2′-deoxy-3′-thiacytidine enantioselectively to prepare optically pure (2′R-cis)-2′-deoxy-3-thiacydine (Fig. 7.7). In enzymatic resolution of α-acetoxysulfides by *Pseudomonas fluorescens* lipase, lamivudine has been synthesised with e.e. of 97% (13).

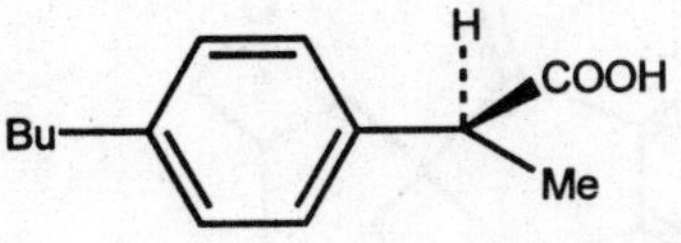

Fig. 7.6: (S)-Ibuprofen

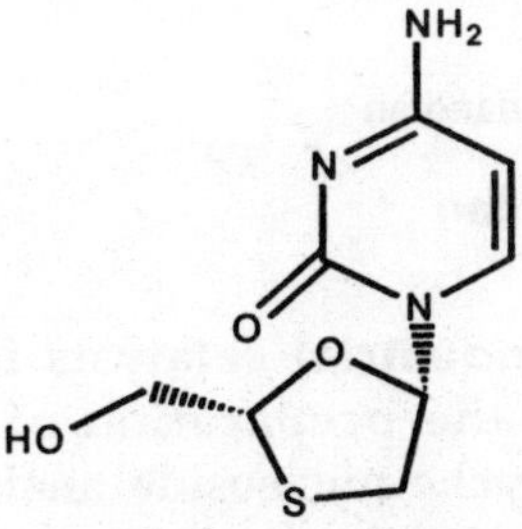

There are very few lipase-catalysed molecules, which have been commercialized. Zopiclone (Fig. 7.8) has been commercialized for the treatment of insomnia. Racemic zopiclone is a hypnotic agent of the cyclopyrrolone class; the (*S*)-configured enantiomer is more active 14). An important intermediate (±)-6-(5-chloropyridin-2-yl)-7-chloromethyl-oxycarbonyloxy-6,7-dihydro-5*H*-pyrrolo[3,4-*b*]pyrazin-5-one has been catalyzed by *Candida antarctica* B lipase enantioselectively in the synthesis of (*S*)-(+)-zopiclone. This enzyme also catalyzes the resolution of the corresponding 2-chloroethylcarbonate derivative.

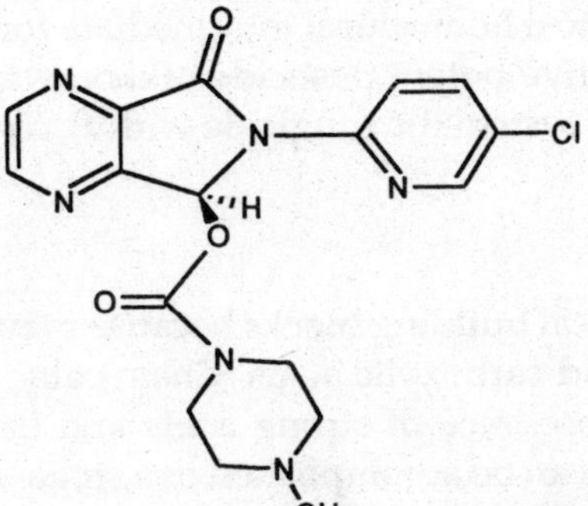

Fig. 7.7: Lamuvidine

Herbicide

Lipase has been used in the multikilogram synthesis of herbicide indanofan well known herbicide. This is syntheised from racemic indanofan, which is kinetically resolved using lipase (Fig. 7.9) and chemically inversed (15). The o-xylene complex of a hemiketal used as precursor, is converted to (S)-indanofan and recovered from o-xylene.

Another optically active compound (1*S*,5*R*)-(-)-frontalin (Fig. 7.10) is synthesized by the lipase catalyzed reaction (16). Frontalin (1,5-dimethyl-6,8-dioxabicyclo[3,2,1]octane) is the aggregation pheromone of pine beetles of the Dendroctonus family. 1,2 Frontalin has recently been identified in the bark of several angiosperm trees. Frontalin from the pine beetle *Dendroctonus frontalis* is an 85:15 mixture of (1*S*,5*R*)- and (1*R*,5*S*)-enantiomers. Bioassays have shown that the (1*S*,5*R*)-(-) enantiomer of frontalin is much more active than the (1*R*,5*S*)-(+)-enantiomer. 2,5 Pheromones have been used to control the progression of pine beetle infestations in forests.

Fig. 7.8: Zopiclone

Racemic Indanofan

(S)-Indanofan

Fig. 7.9: Lipase Catalysed Resolution of Indanofan

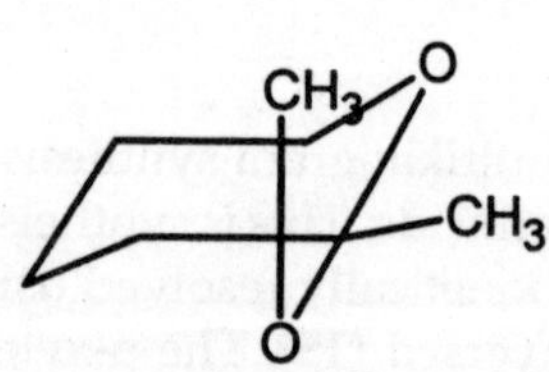

Fig. 7.10: Frontalin

Lipases have applications as industrial catalysts for the resolution of racemic alcohols in the preparation of some prostaglandins, steroids, and carbocyclic nucleoside analogues. Lipases are also useful in the synthesis of the artificial sweetener sucralose by regioselective hydrolysis of octa-acetylsucrose (17). Akita *et al.*, in 1995 (18) described a highly stereospecific synthesis of the versatile chiral synthon possessing two stereogenic centres, which was subsequently converted into a homochiral intermediate for the synthesis of the biologically active potent pesticide nikkomycin-B. Production of enantiomerically pure (*S*)-(-)- fenpropimorph, a systematic fungicide with *P. cepacia* lipase mediated kinetic resolution has been reported (19).

Nitrilases

Nitriles are convenient and synthetically important chemical building blocks because they are readily generated and can be converted into useful amides and carboxylic acids. Chemically, this conversion requires hydrolysis at high temperatures in the presence of strong acids and bases. Enzymatically, the conversion of a nitrile into a carboxylic acid can be accomplished in a single step by the action of a nitrilase with the incorporation of two mole equivalents of water and the liberation of a mole equivalent of ammonia, or in a two step process by the sequential actions of a nitrile hydratase to generate an amide and its subsequent hydrolysis by an amidase to yield the carboxylic acid (Fig. 7.11) (20).

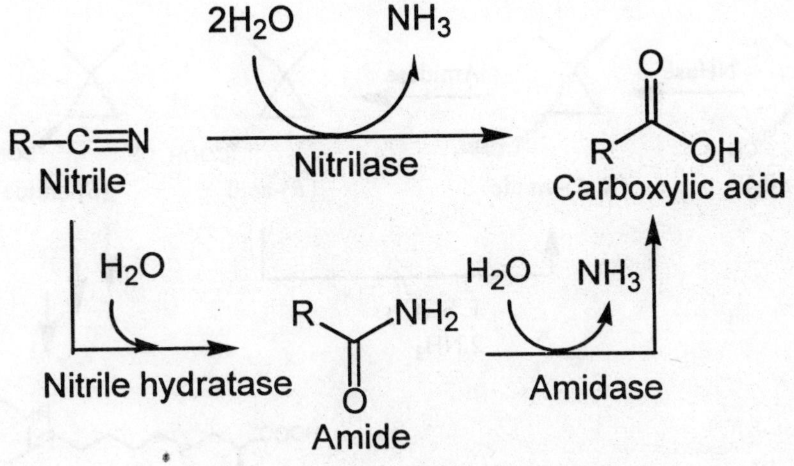

Fig. 7.11: Enzymology of Nitrile Metaboism

From the synthetic viewpoint, nitriles represent a widely applicable chiral synthon, which can be employed for the homologation of a carbon framework; on the other hand, further transformations of the nitriles thus obtained are impeded due to the harsh reaction conditions required for its hydrolysis. In this context, the *chemo*-selective biocatalytic hydrolysis of nitriles represents a valuable alternative because it occurs at ambient temperature and near physiological pH (21). Many reactions catalyzed by the nitrile metabolizing enzymes are either already operated in large scales or has the potential to become so. The success of nitrile-converting organisms such as *Rhodococcus* N-774, *Pseudomonas chlororaphis* B23 and *R. rhodochrous* J1 in the kiloton production of acrylamide and synthesis of nicotinamide and nicotinic acid from 3-cyanopyridine have demonstrated the commercial viability of these enzymes (22). In recent years the use of these nitrile-converting enzymes, especially nitrilase and NHases, has increased tremendously.

Hydrolases, in particular have shown to accept a wide range of substrates. An example is a process of Lonza for the synthesis of (S)-2,2-dimethylcyclopropanecarboxamide which is an intermediate for the production Cilastatin, a β-lactamase inhibitor (Fig. 7.12). The product is produced at 15 m^3 scale the yield is 35% with an ee value of > 98%. The (R)-acid can be recycled chemically into the substrate for amidase, which leads to higher yields and minimises waste (23).

An example of application of an amidases for production of a chiral building block is the production of (S)-piperazine-2-carboxylic acid by Lonza, an intermediate in the synthesis of anti-HIV drug Crixivan, whereby a *Klebsiella* amidases is used to resolve the racemic (RS)-piperazine -2-carboxamide (Fig. 7.13) (24).

α-hydroxy acids are versatile chiral building blocks, *e.g.* (R)-(-)-mandelic acid is a key intermediate for the production of semi-synthetic cephalosporins and penicillins (25). It is also used as a chiral resolving agent and chiral synthon for the synthesis of anti-tumor agents (26) anti-obesity agents (27) etc.

The arylacetonitrilases of several organisms including *Alcaligenes faecalis* ATCC 8750 (28) and *Pseudomonas putida* MTCC 5110 (29) have been reported to bring about the enantioselective hydrolysis

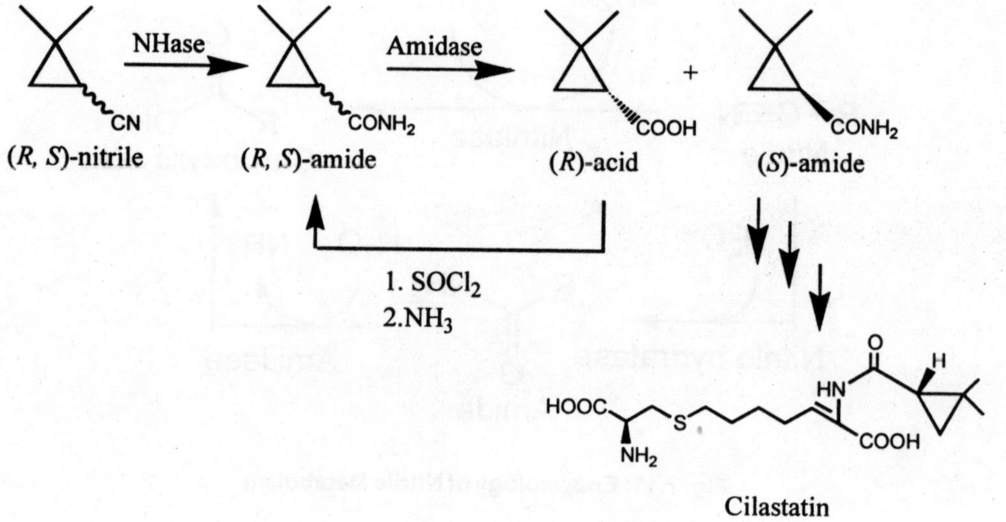

Fig. 7.12: Route for Production of Cilastatin Intermediate

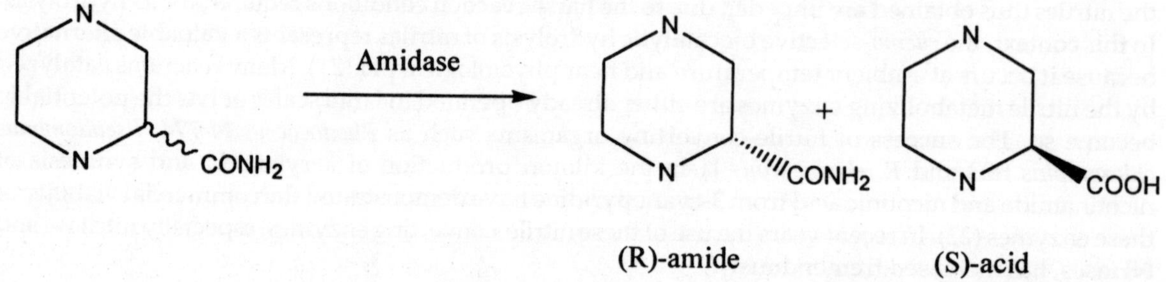

Fig. 7.13: Selective Conversion of Racemic Piperazine-2-carboxamide into the (S)-acid

of mandelonitrile to (R)-(-)-Mandelic acid. *P. putida* has been proposed to be a good candidate for the biotransformation taking into account its higher growth rate, faster reaction rate and higher stability (29). At slightly alkaline pH, mandelonitrile undergoes spontaneous degradation to benzaldehyde and hydrocyanic acid. This can constitute an efficient dynamic resolution, where the unreacted (S)-mandelonitrile undergoes racemization to give theoretical 100% conversion to (R)-(-)-mandelic acid by nitrile hydrolysis as opposed to 50% theoretical maximum in kinetic resolution. BASF and Mitsubisih Rayon use the nitrilase process for the production (R)-mandelic acid and its derivatives on a multi ton scale. An advantage of the nitrilase process is that it does not require any organic solvent. Moreover, the synthesis of nitrile is avoided as the cyanohydrin is formed from HCN and benzaldehyde under the reaction conditions.

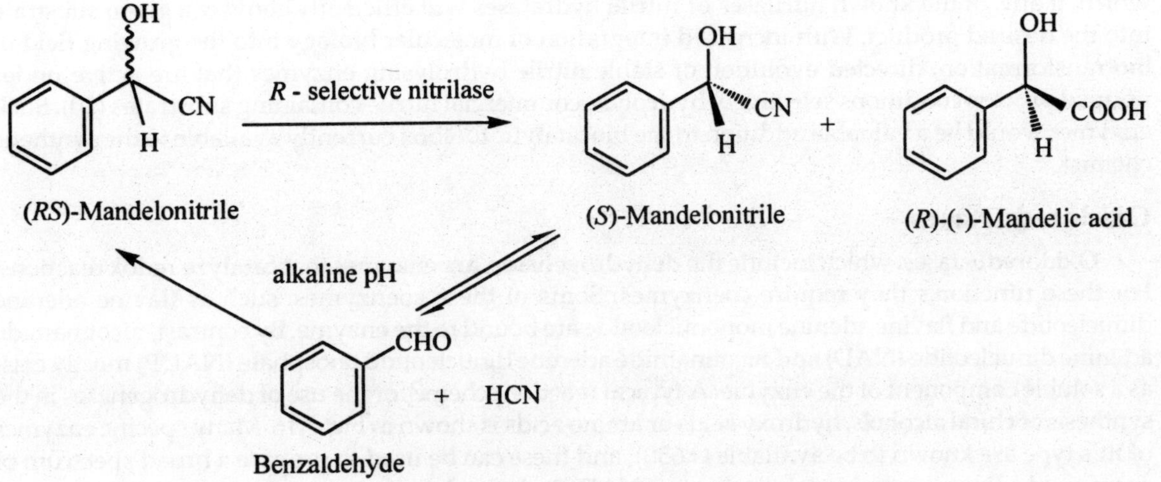

OH
CN
H
(RS)-Mandelonitrile
R - selective nitrilase
OH
CN
H
(S)-Mandelonitrile
+
OH
COOH
H
(R)-(-)-Mandelic acid
alkaline pH
CHO
+ HCN
Benzaldehyde

Fig. 7.14: Scheme for Dynamic Resolution of Nitrilase Mediated Hydrolysis of Mandelonitrile

By using nitrilases and starting with epichlorohydrin, Diversa Corp. has a adopted a three step route to the intermediate of multi billion dollar drug, Lipitor which initially required six long steps. The drug is recommended for hyperlipidemic patients. After the first chemical step, next, the symmetric prochiral substrate is desymmetrized by the action of a nitrilase to a (R)-hydroxy acid (Fig. 7.15) (30).

All the above mentioned nitrile conversions are carried out by whole cells, mainly because nitrile hydrolases are intracellular in nature and partly because the purified enzymes are too unstable. Although systems that involve whole cells avoid expensive enzyme purifications, they are in general undesirable, as contaminating side-products can result from the reactions of other enzymes within the cell. Recently only five commercially available nitrilases have come up. For reactions in which *regio-* or *stereo-* selectivity is desired, it has been found that the degree, extent and even absolute

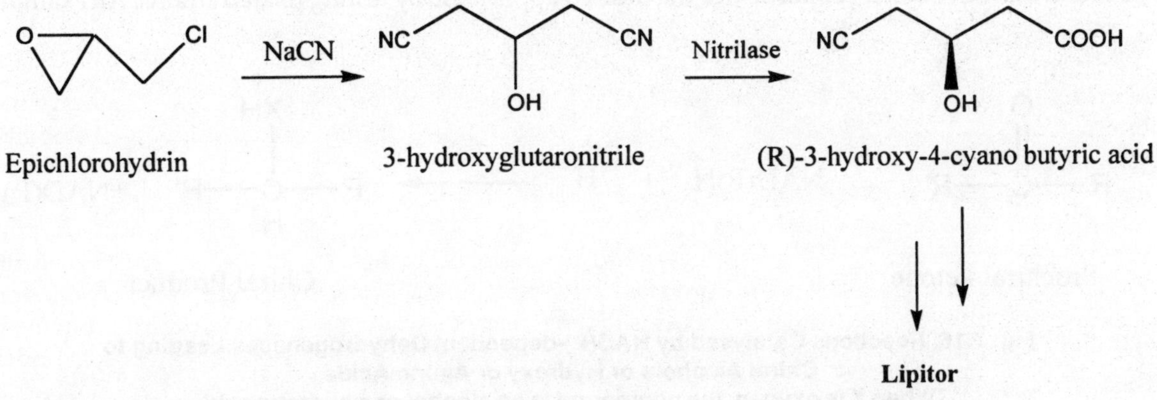

O
Cl
Epichlorohydrin
NaCN
NC
CN
OH
3-hydroxyglutaronitrile
Nitrilase
NC
COOH
OH
(R)-3-hydroxy-4-cyano butyric acid
Lipitor

Fig. 7.15: Diversa Short-cut Route to the Lipitor Intermediate

configuration of the final product are highly dependent on the substrate. Thus it is difficult to predict which, if any, of the known nitrilases or nitrile hydratases will efficiently convert a given substrate into the desired product. With increased integration of molecular biology into the growing field of biotransformation, directed evolution of stable nitrile hydrolysing enzymes that are active under optimal process conditions selectively hydrolyse commercial nitrile containing substrates (30). Such enzymes would be a valuable addition to the biocatalytic toolbox currently available to the synthetic chemist.

Oxidoreductases

Oxidoreductases, which include the dehydrogenases, are enzymes that catalyze redox reactions. For these function,s they require coenzymes. Some of these coenzymes, such as flavine adenine dinucleotide and flavine adenine mononucleotide are bound to the enzyme. By contrast, nicotinamide adenine dinucleotide (NAD) and nicotinamide adenine dinucleotide phosphate (NADP) mostly exist as a soluble component of the enzyme. A typical reaction scheme for the use of dehydrogenases in the synthesis of chiral alcohols, hydroxy acids or amino acids is shown in Fig. 7.16. Many specific enzymes of this type are known to be available (<650), and these can be used to prepare a broad spectrum of compounds. For preparative applications of NAD(P)-dependent enzymes, at least three components need to be available: an appropriate enzyme, an efficient coenzyme regenerating step and a suitable reaction-engineering technique.

Sources of Oxidoreductases

Oxidoreductase producing microorganisms have been isolated from various sources. These organisms include bacteria, fungi, yeasts and also some alga. Isolated enzyme from various plant and animal sources have also been employed for the reduction of various types of ketones and even examples are available in literature showing application of whole cell from plant sources for the dehyrdogenase mediated bioreduction with high enantioselectivity. NAD(P)-dependent enzymes have been isolated by a variety of screening procedures. Bacteria with an NAD-dependent L-phenylalanine dehydrogenase were isolated from soil samples by adding L-phenylalanine to the culture medium. An NAD-independent trimethylamine dehydrogenase was obtained using enrichment cultivation techniques in a chemostat (1). Strains of *Thermoanaerobium brockii* that produce a thermostable NADP-dependent alcohol dehydrogenase (ADH) were isolated from a hot-spring site in Yellowstone Park (WY, USA). Once such producer strain has been isolated and identified, other candidate enzymes may be found that have better properties for the desired application by testing related strains. An example

$$\underset{\text{Prochiral ketone}}{R{-}\overset{\overset{\displaystyle O}{\|}}{C}{-}R'} \;+\; \text{NAD(P)H} \;+\; H^{+} \longrightarrow \underset{\text{Chiral Product}}{R{-}\overset{\overset{\displaystyle XH}{|}}{\underset{\underset{\displaystyle H}{|}}{C}}{-}R'} \;+\text{NAD(P)}$$

Fig. 7.16: Reactions Catalysed by NAD(P)-dependent Dehydrogenases Leading to Chiral Alcohols or Hydroxy or Amino Acids (When X is oxygen, the compound is an alcohol or a hydroxy acid; when X is NH, the compound is an amino acid)

of this is (*R*)-specific ADH, significantly more thermostable variants that were detected by an antibody-supported screening method.

Strategies for the Bioreduction

The inter-conversion of a ketone to the corresponding alcohol and vice versa represents one of the most common redox-reactions in organic chemistry. Whereas traditional synthetic methods predominantly use toxic metals and expensive complex hydrides, biotransformations offer some significant advantages. The vast majority of dehydrogenases and reductases used for ketone reduction and alcohol oxidation require nicotinamide cofactors, such as NADH and NADPH. For their efficient recycling, two strategies are feasible (Fig. 7.17).

The most elegant and simple approach makes use of a single enzyme, which simultaneously transforms substrate plus co-substrate (Fig. 7.17a)The latter, however, must be applied in large excess to drive the equilibrium into the desired direction. As a consequence, these so called 'coupled-substrate' systems are commonly impeded by cosubstrate-inhibition. The precondition for efficient, large-scale coupled-substrate systems is the availability of chemo-stable dehydrogenases. These drawbacks can be employing cofactor recycling process by using a second (and preferably irreversible) enzymatic reactions, which renders a more complex coupled enzyme approach. The most advanced regeneration system makes use of innocuous hydrogen and oxygen as reductant and oxidant, respectively (Fig. 7.17b). An additional benefit of the coupled-substrate system is that it may be used in the oxidation and reduction mode, whereas separate cofactor-recycling systems are required for the coupled-enzyme approach (32).

Reduction of Carbonyl Compounds by Yeast Cells

In 1914 Neuberg and Nord showed that a growing yeast culture reduced n-Pentanal to the corresponding alcohol in 68% yield on a more than 20 g scale (33). Neuberg also provided the first demonstration that yeast reduction of the prochiral ketones could yield an optically pure alcohol (33) Since that time, prochiral ketone reduction have been the most common application of bakers yeastand by other conconventional yeasts, and bacteria in organic synthesis (34). These reductions are simple

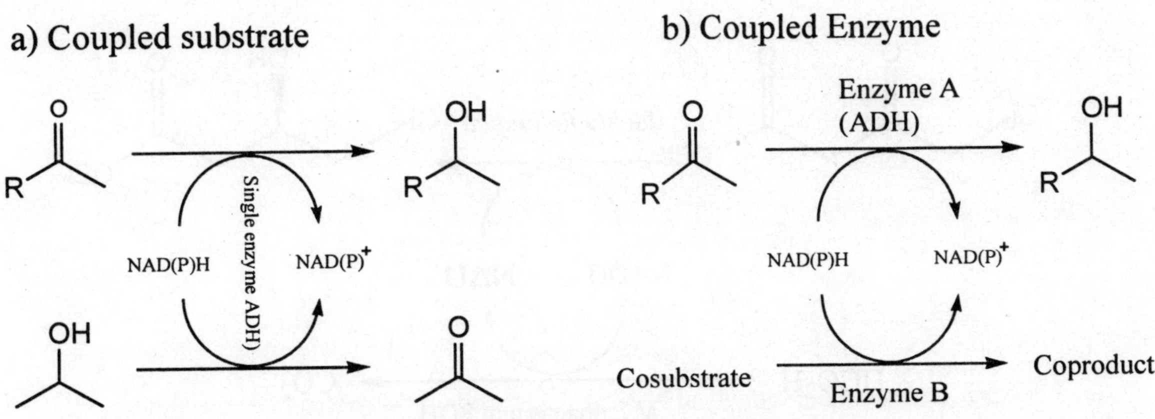

Fig. 7.17: Different Types of Nicotinamide Cofactor Recycling

to perform, the cells are inexpensive and readily available, the technique is available to a very broad array of carbonyl compounds (35). In addition yeast reduction help to minimise the environmental impact of organic synthesis: reductions are generally carried out in aqueous solution, and the spent catalyst is entirely biodegradable. These advantages have led the use of yeast mediated ketone reduction on the industrial scale. In the reduction of a particular carbonyl compound, when a single yeast enzyme dominates the process, the use of whole yeast cells çan provide the corresponding chiral alcohol in very high optical purity. However the optical purity of the product diminishes when multiple yeast reductase enzymes with different enantioselectivities are involved. (36,37).

The use of purified yeast enzymes for the carbonyl reduction avoids problems associated with competing catalysts with different stereoselectivities. A variety of reductase enzymes have been purified from baker's yeast. But unfortunately yeast alcohol dehydrogenase, the sole commercially available enzyme acceptt a limited range of ketone as substrate (38). Moreover, reductase enzymes also require reduced nicotinamide co-factors, which must be provided in stoichiometric amounts or via regeneration system. These experimental complications have inspired the search for methods to improve the stereoselectivities of carbonyl reduction using intact yeast cells since the cells provide a source of both the enzyme and co-factor in a single package. The key to improve the stereoselectivities of whole cell mediated reduction is to arrange conditions so that only few yeast enzymes reduce the added carbonyl substrate. The techniques include the modification of substrate structure (39,40) carefully controlling the substrate concentration (39) changing the growth conditions, use of organic solvent or two-phase system (41) and the inclusion of the enzyme inhibitor in the culture medium (42). While these methodologies favour one catalyst over competitors, they are based on empirical findings and is difficult to predict their effect a priori.

Recent Developments in the Bioreduction of Ketones

A 'classic' regeneration system for NADH is the use of formate dehydrogenase (FDH) and formate as hydrogen donor (Fig. 7.18), with the formation of innocuous CO_2 as co-product, which ensures irreversibility. Instead of using two isolated enzymes–for instance, alcohol dehydrogenase (ADH) from *Rhodococcus erythropolis* (43) plus FDH from *Candida boidinii*–a more simple and elegant approach was chosen by over expressing both enzymes in a single host (*e.g. Escherichia coli*), which was

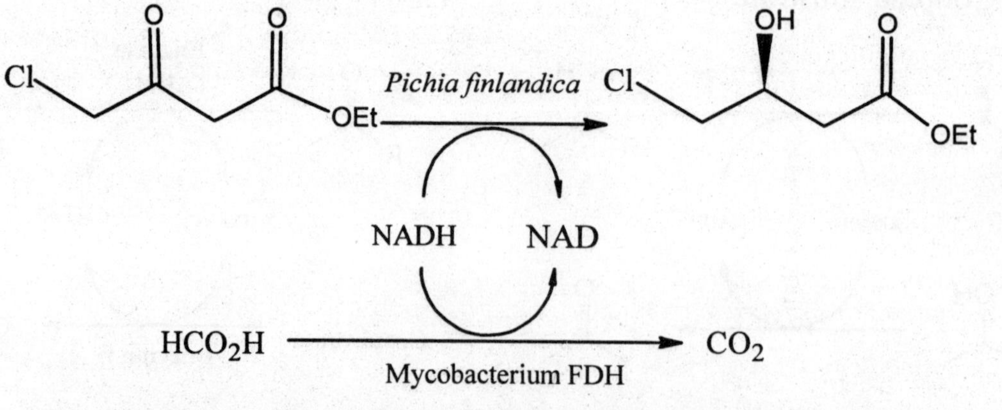

Cl O O OEt
Pichia finlandica
Cl OH O OEt
NADH NAD
HCO_2H CO_2
Mycobacterium FDH

Fig. 7.18: β-keto Ester Bioreduction Using Coupled Enzyme Approachusing ADH and FDH Co-expressed in *E. coli*

demonstrated for *Pichia finlandica* ADH and Mycobacterium FDH. Thus, reduction of ethyl-4-chloroacetoacetate (32 g/l) gave (S)-ethyl-4-chloro-3-hydroxybutanoate in 98.5% yield and 99% enantiomeric excess (ee) (Fig. 7.19) (44).

Out of simplicity, there is a strong trend for 'designerbugs' (*i.e.* whole-cell organisms co-expressing all enzymes needed), as shown for the co-expression of a carbonyl reductase from *Candida magnoliae* and glucose dehydrogenase from *Bacillus megaterium* in *E. coli* (45,46). An entirely new approach for separate NAD(P)H-recycling makes use of hydrogenases with molecular hydrogen as reducing agent. For instance, hydrogenase I from *Pyrococcus furiosus* was employed for NADPH recycling, showing a turnover frequency up to 44 h_1 (Fig. 7.19) (47,48). Unfortunately, the enhanced activity of the hyperthermophilic enzyme at elevated temperature (maximum 80°C) could not be used due to the thermal instability of NADPH and, as a compromise, the reactions had to be performed at 40°C. Despite the oxygen-sensitivity of hydrogenase I, this system represents an elegant and novel system with significant potential.

In view of the simplicity, single-enzyme system performing the substrate transformation and cofactor recycling simultaneously are preferable ('coupled substrate', (Figure17a). In view of the stepwise reaction mechanism, coupled substrate-system bear a strong resemblance to transition-metal catalyzed hydrogen-transfer reactions rather than to the Meerwein-Ponndorf-Verley reduction, which acts in a concerted fashion. To push the productivity limits of coupled-substrate system further, the maximum tolerable concentration of iso-propanol applied for cofactor regeneration with different ADHs (or ADH-preparations) was constantly increased by using even more chemo-stable enzyme preparations.

With few exceptions, bacterial ADHs dominate the stability ranking, and top-values were reached by using an ADH from *Rhodococcus ruber* DSM 44541 (49). Elevated concentrations of iso-propanol up to 50% (v/v) using whole cells (80% v/v for the partially purified ADH) not only shift the equilibrium towards the product-side, but also enhance the solubility of lipophilic substrates in the monophasic aqueous/organic medium. The ADH from *Rhodococcus ruber* employed as whole cell preparation tolerates substrate concentrations of approximately one to two mol/l. For example, reduction of 6-methyl-hept-5-en-2-one (sulcatone) at a concentration of 95 g/l was achieved in a batch process at a

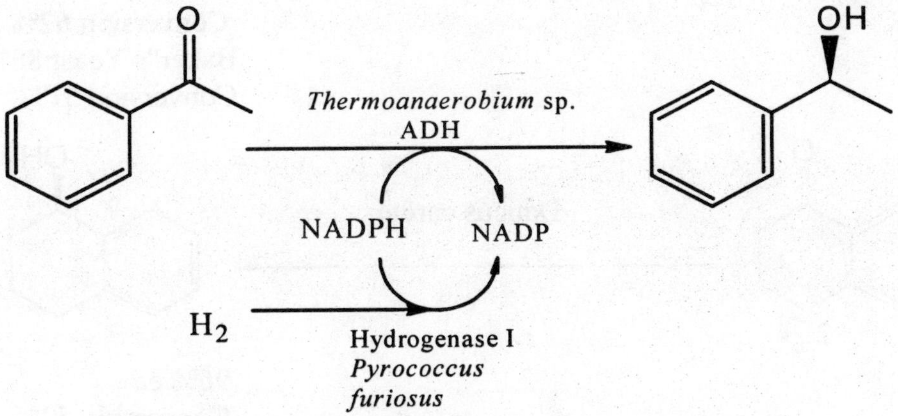

O
OH
Thermoanaerobium sp.
ADH
NADPH
NADP
H_2
Hydrogenase I
Pyrococcus furiosus

Fig. 7.19: Cofactor Recycling Using Dehydrogenase I System of *Pyrococcus furiosus*

space-time yield of 282 g/l (50). For large-scale applications, the acetone formed can be removed by organophilic pervaporation or stripping, thereby canceling all equilibrium-limitations and ensuring complete conversion.

Further Novel ADHS

Substrates possessing sterically demanding groups on each side of the carbinol moiety are, for most alcoholdehydrogenases, very bad substrates or non-substrates. Only very few ADHs are able to transform sterically demanding ketones/alcohols. For example, bulky ketones, such as 1 or 2 (Fig. 7.20) were reduced by a carbonyl reductase purified from bakers' yeast (51) or by whole cells of carrot (Daucus carota root) to furnish the corresponding (S)-alcohols in reasonable ee, albeit at low substrate concentration (1.5 g/l) (52). In a related fashion, different species of liverworts were able to reduce the sterically hindered terpenoid (+)-n-dihydrocarvone 5 in a diasteroselective fashion to yield neo-dihydrocarveol 6 (Fig. 7.21) (53). One disadvantage of using whole cells is the presence of competing dehydrogenase enzymes, which reduces the enantioselectivity of the reaction or leads to unexpected reduction products. This behavior is most pronounced for α, β-unsaturated ketones, which are often reduced to the corresponding saturated carbonyl compounds by enoate reductases, before the desired carbonyl reduction takes place. For instance, reduction of (-)-carvone 4 by whole cells of liverwort gave (+)-n-dihydrocarvone 5, before reduction of the carbonyl moiety occurred in a second step. Related bottlenecks exist with respect to the availability of alcohol dehydrogenases showing opposite 'anti-Prelog' stereospecificity. Besides the limited number of 'anti- Prelog' ADHs, novel activities were

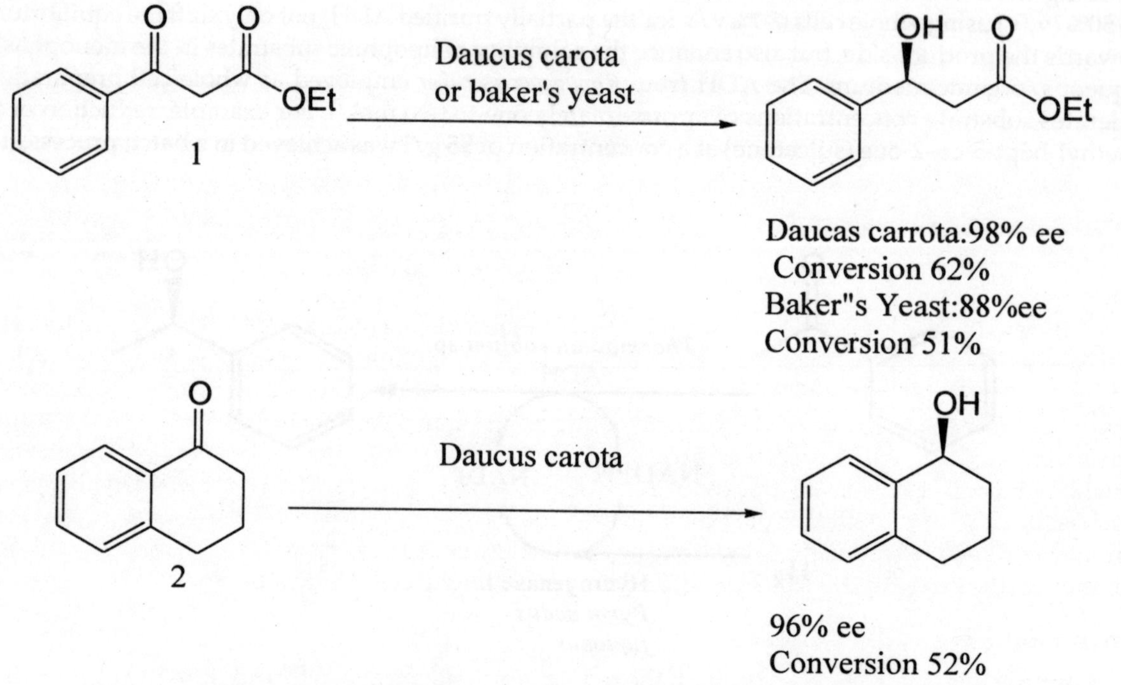

O O
OEt
1
Daucus carota
or baker's yeast
OH O
OEt
Daucas carrota:98% ee
Conversion 62%
Baker"s Yeast:88%ee
Conversion 51%

O
2
Daucus carota
OH
96% ee
Conversion 52%

Fig. 7.20: Sterically Demanding Ketone Reduction by Carrot Cells or Baker's Yeast

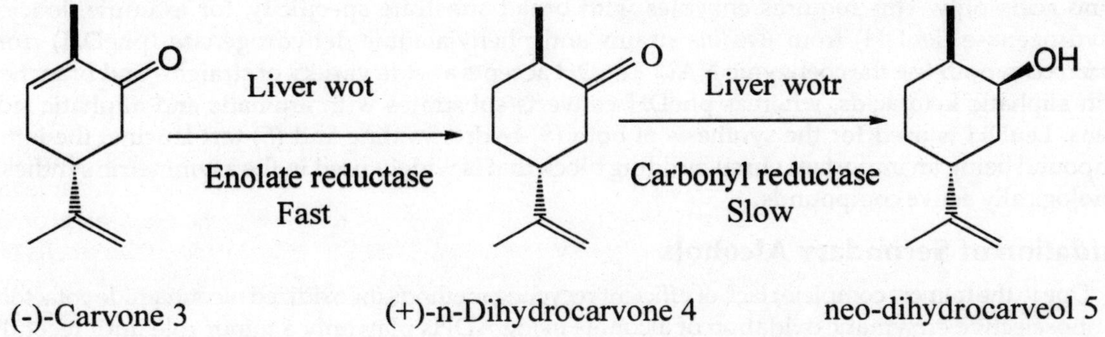

Liver wot
Enolate reductase
Fast
Liver wotr
Carbonyl reductase
Slow

(-)-Carvone 3 (+)-n-Dihydrocarvone 4 neo-dihydrocarveol 5

Fig. 7.21: Sterically Demanding Ketone Reduction by Liver Wort

reported for *Yarrowia lipolytica* (54). An interesting stereochemical switch was induced by varying the reaction conditions and the physiological state of yeast cells to provide either the (S)- or the (R)-enantiomer using the same organism (55). Novel activities for the bioreduction of various carbonyl compounds are recently described in *Pseudomonas putida* UV4 (pyridyl carbinols) (56), *Aspergillus terreus, Rhizopus oryzae* (fluoro-acetophenones) (57), *Mucor rouxii* (for various β-ketoesters) (58) and ***Saccharomyces montanus*** CBS 6772 (2-oxocycloalkanecarbonitriles) (59).

Hydroxy-acid Dehydrogeneses

Hydroxy-acid dehydrogenases are useful catalysts for obtaining chiral hydroxy acids from keto acids. In general, all these enzymes use the nonphosphorylated coenzyme NAD. LDHs are well-known representatives of this enzyme group. L-LDH from several animal tissues or *Bacillus stearothermophilus* and D-LDH from *Leuconostoc mesenteroides* are commercially available, inexpensive and fairly stable, and exhibit a high specific activity. Unfortunately, most LDHs only accept pyruvate as a substrate, 2-oxobutyrate and some pyruvate derivatives (*e.g.* the chloro compound). For the synthesis of other chiral 2-hydroxy acids, new NAD-dependent hydroxy-acid dehydrogenases isolated from *lactobacillus* sp. are also available.

Chiral hydroxy acids are valuable synthons, and several compounds of this type have already been used as precursors for useful chiral products. For example, enantiomerically pure epoxyacrylic acid can be obtained from (*S*)-chlorolactic acid, which is synthesized by L-LDH. D-Hydroxyisocaproate dehydrogenases (from *Lactobacillus casei* or related strains) and D-mandelate dehydrogenases (from *Lactobacillus curvatus* or *Streptococcus faecalis*) convert a broad variety of 2-oxo acids [*e.g.* aliphatic straight- or branched-chain substrates and aromatic keto acids (*e.g.* phenylpyruvate or benzoylformate, which is the prochiral precursor of D-mandelate)]. Similarly, L-hydroxyisocaproate dehydrogenase from *Lactobacillus confuses* yields (*S*)-hydroxy acids. All these substrates are converted with high specific activities (in the range of 200–2000 U mg) and very high enantioselectivity (60-62).

Amino acid Dehydrogeneses

Amino acid dehydrogenases perform the reductive amination of 2-keto acids into L-amino acids. Owing to the competition provided by the fermentative production processes of naturally occurring L-amino acids, the enzyme-catalysed process should be considered for the synthesis of unnatural L-amino acids only. This requires enzymes with broad substrate specificity, for example, leucine dehydrogenase (leuDH) from *Bacillus* strains and phenylalanine dehydrogenase (pheDH) from *Rhodococcus,* both use the coenzyme NAD. LeuDH accepts a wide variety of straight- and branched chain aliphatic keto acids, whereas pheDH converts substrates with aromatic and aliphatic side chains. LeuDH is used for the synthesis of both (*S*)-hydroxyvaline and (*S*)-tert-leucine, the latter compound being an important chiral building block that is widely used in the asymmetric synthesis of biologically active compounds.

Oxidation of Secondary Alcohols

Due to the (almost complete) lack of efficient recycling methods for oxidized nicotinamide cofactors, enantioselective enzymatic oxidation of alcohols using ADHs play only a minor role until recently. Now, impressive progress has been achieved in enzyme-coupled alcohol oxidation using NADH- and NADPH-oxidases for cofactor regeneration. Thus, NADH or NADPH is oxidized at the expense of the most innocuous and cheapest oxidant, molecular oxygen, by producing hydrogen peroxide or H_2O as by-product (Fig. 7.22). Furthermore, the high redox-potential of the O_2/H_2O or O_2/H_2O_2 couple results in a strong thermodynamic driving force. Among them, four-electron transfer to form H_2O is strongly preferred over two-electron reduction yielding H_2O_2, which frequently causes enzyme deactivation even at low concentration. Numerous water-forming NADH oxidases were identified recently in different microorganisms, such as *Pyrococcus furiosus* (63), *Lactobacillus brevis* (64), *Borrelia burgdorferi* and *Lactobacillus sanfranciscensis* (65). The nicotinamide- oxidase from Lactobacillus sanfranciscensis is able to utilize not only NADH but also NADPH as substrate with an activity ratio of approximately 3:1(65). The water-forming NADH oxidases from *Lactobacillus brevis* and *Lactobacillus sanfranciscensis* were recently employed for the oxidative kinetic resolution of racemic 1-phenylethanol (Fig. 7.22) (66,67).

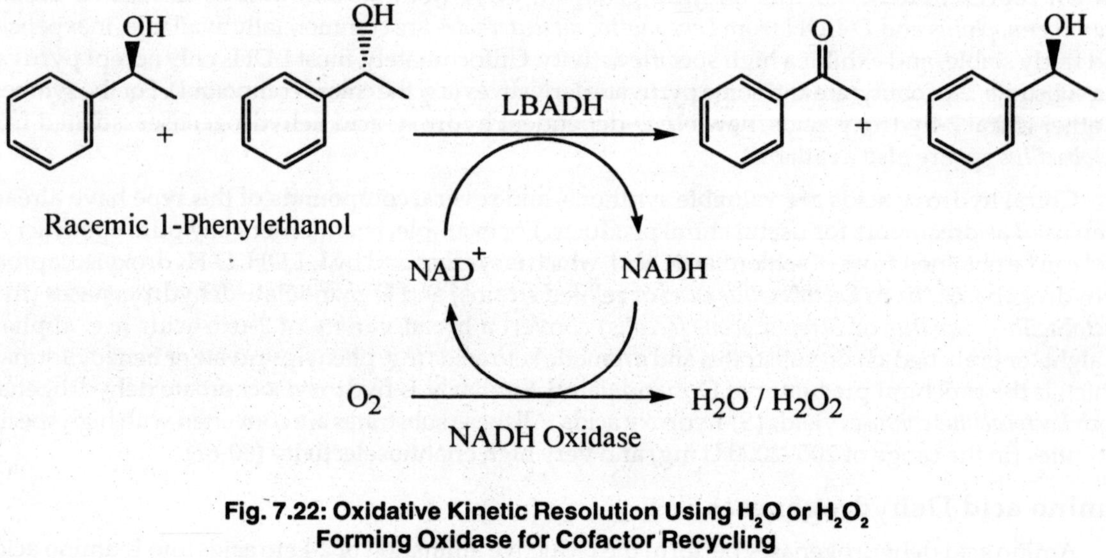

OH + OH — LBADH → O + OH

Racemic 1-Phenylethanol

NAD^+ NADH

O_2 → H_2O / H_2O_2

NADH Oxidase

Fig. 7.22: Oxidative Kinetic Resolution Using H_2O or H_2O_2 Forming Oxidase for Cofactor Recycling

Medium Engineering

To increase the productivity of (isolated) ADHs, such as HLADH, LBADH and TBADH, biphasic solvent systems were evaluated. It was shown that the commonly used logP value is not a satisfying parameter to assess the compatibility of an organic phase with ADHs, but the functional groups within the organic solvent play a more important role (68). For instance, using tert-butyl methyl ether as the organic solvent allowed a 10-fold increase of the substrate concentration for LBADH using iso-propanol for cofactor regeneration. On the other hand, FDH-based cofactor regeneration was enhanced in biphasic systems containing n-hexane or n-heptane(69).

X-ray Structures

The availability of the three-dimensional structure of an enzyme is a crucial prerequisite not only for a deeper understanding of its catalytic mechanism of action, but also of its stability during operational conditions. The crystal structure of the apo-form of the biocatalytically important LBADH, a member of the short-chain dehydrogenase/reductase superfamily, showed that although the activity of the enzyme strongly depends on the presence of Mg^{2+}, the metal is not directly involved in the catalytic mechanism. However, the binding of Mg^{2+} causes structural changes by activating the enzyme, which provides a plausible explanation for the metal dependency of this dehydrogenase(70).

Future Outlook

During the past two years, impressive results have been achieved towards the increased applicability of alcohol dehydrogenases/reductases, most notably in the development of novel system for cofactor-recycling based on hydrogenases and nicotinamide-oxidases for coupled-enzyme system. Alternatively, the availability of chemo-stable alcohol dehydrogenases has allowed their application in highly efficient single-enzyme system. Despite these advances, there is still a striking lack of studies for the ADHs that can tolerate high substrate/cosubstrate concentration. Furthermore, the search for novel ADHs that can accept sterically demanding ketones or alcohols and enzymes showing 'anti-Prelog' stereo-specificity is an ongoing challenge.

Conclusion

Amongst the various enzymes that have potentiality for industrial application, lipases have got highest number of successful stories. The enzyme has got some inherent characteristics that have made it more suitable over the others for its large-scale application. The extracellular, comparatably stable enzyme can be produced and harvested easily from a vast array of microorganisms, can be successfully immobilized on a broad array of matrices, and applied for hydrolysis of esters, esterification reaction, transesterification reaction, acylation- deacylation reactions and resolution of a great variety of racemic alcoholic compounds. Alcoholic compounds being one of the most important class of chiral compounds and speciality chemicals has got a high market value. They are chiral intermediates for a variety of pharma-chemicals too. Biochemically, optically avtive chiral alcohols can be made by lipase mediated stereoselective hydrolysis of fatty acid esters or by oxidoreductase mediated reduction of prochiral ketones. However, since the NAD or $NAD(P)^+$ dependent oxidoreductases directly introduce chirality into the target molecule yielding an optically active product, a theoretical yield up to 100% can be achieved as compared to the 50% maximum yield that can be achieved via kinetic resolution with a lipase lipase mediated process to yield chiral alcohols. In spite of that advantage, intracellular localization and the cofactor dependency has made oxidoreductases, still to date, largely unsuitable for industrial application. On the other hand, nitrilases being intracellular in nature often show very high activity and substrate specificity, which are considered to be the prime reason for their development as potential indistrial biocatalyst.

The potentiality of enzymes, particularly, as biocatalytic agent for a host range of organic reactions has now been realized, a practical difficulty for its broad scale application is the lack of a host range of easily applicable, highly stable, off-the shelf, and robust biocatalyst. To achieve this there is a need for an efficient screening procedure guided by high-throughput assay system for generating an enzyme-mediated process for a particular compound. Rational application of molecular biology techniques has now directed towards generation of new biocatalysts for industrial application. Techniques like directed evolution combines site directed mutagenesis, random mutagenesis, multiple cassette gene mutagenesis, and obviously guided by a high-throughput assay system for the development new generation of biocatalysts.

REFERENCES

1. Noyori, R. (2002). Asymmetric catalysis: science and opportunities (Noble Lecture). **Angew. Chem. Int. Ed.** (41) 2008-2022.
2. Stinson, S.C. (2001). Chiral pharmaceuticals. **Chem. Eng. News** (40) 79–97
3. Rouhi, A.M. (2002). Chiral roundup. **Chem. Eng. News** (23) 43–50
4. Bornscheuer, U., T., Bessler, C., Srinivas. and Krishna, S., H. (2002). Optimizing lipases and related enzymes for efficient application. **Trends. Biotechnol.** (20) 433-437.
5. Senuma, M., Shibazaki, M., Nishimoto, S.,Shibata, K., Okamura, K. and Date, T. (1989). The practical resolution of (2R,3S)-2-hydroxy-3-(4-methoxyphenyl)-3-(2-nitrophenylthio)propionic acid, a key intermediate for diltiazem, with L-lysine. **Chem. Pharm. Bull.** (37) 3204-3208.
6. Matsumae, H., Furmi, M. and Shibatani, T. (1993). Lipase catalyzed asymmetric hydrolysis of 3-Phenylglycidic acid ester, the key intermediate in the synthesis of diltiazem hydrochloride. **J. Ferment. Bioeng.** (73) 93-98.
7. Lui, H., Hoff, B.H., Berg, T.C., and Anthonsen, T. (2001). Synthesis of anti stroke drug lubeluzole and its enantiomer. Lipase catalysed resolution of chiral building block. **Chirality**. (13) 135-139.
8. Patel, R.N., Banerjee, A., Ko, R.Y., Howell, J.M., Li, W.S., Comezoglu, F.T., Partyka, R.A., Szarka, F.T. (1994). Enzymic preparation of (3R-cis)-3-(acetyloxy)-4-phenyl-2-azetidinone: a taxol side-chain synthon. **Biotechnol. Appl. Biochem.** (20) 23–33.
9. Patel, R.N., F.T., Partyka, R.A., Szarka, F.T. (1996) Enzymatic processes for resolution of enantiomeric mixtures of compounds useful as intermediates in the preparation of taxanes. **United States Patent** 5,567,614.
10. Page, H.I., Berrettoni, J.N., Butkus, A., Sones, Jr F.M. (1970). Prediction of coronary heart disease based on clinical suspicion, age, total cholesterol, and triglyceride. **Circulation**. (42) 625-645
11. Tsai S., Wei, H. (1994). Enantioselective esterification of racemic naproxen by lipases in organic solvent. **Enzyme Microbiol. Tech.** (16) 328-333.
12. Hedstrom,G., Backlund, M. and Slotte, J., P. (1993). **Biotechnol. Bioeng.** (42) 618-624.
13. Milton J., Brand S., Jones M., F., Rayner C., M. (1995). Enantioselective Enzymatic Synthesis of the Anti-Viral Agent Lamivudine. **Tetrahedron Letters** (36) 6961-6964.

14. Solares, L.F., Dý´az, M., Brieva, R., Sa´nchez, V.M., Miguel Bayod, M. and Gotora, V. (2002) Enzymatic resolution of new carbonate intermediates for the synthesis of (*S*)-(+)-zopiclone. **Tetrahedron: Asymmetry** (13) 2577–2582.

15. Tanaka, K., Yoshida, K., Sasaki, C. and Osano, Y., T. (2002) Practical asymmetric synthesis of the herbicide (S)- indanofan via lipase catalysed kinetic resolution of a diol and stereoselective acid-catalyzed hydrolysis of a chiral epoxide. **J. Org. Chem.** (67) 3131-3133.

16. Che^ nevert, R., Caron, D. (2002) Chemoenzymatic enantioselective synthesis of (1*S*,5*R*)-(-)-frontalin. **Tetrahedron: Asymmetry** (13) 339–342.

17. Bornemann, S., Casselle, J. M., Combs, C. L., Dordick, J. S. and Hacking, A. J. (1989) **UK Patent No. 2 224 504.**

18. Avdagic, A., Cotarca, L., Ruzic, K., S., Gelo, M., Sunjic, V. (1994). An Efficient Chemoenzymatic Synthesis of S-(-)-Fenpropimorph, **Biocatalysis** (9) 49-60.

19. Akita, H., Chen, C. Y. and Uchida, K. A. (1995). A Formal Total Synthesis of Nikkomycin B Based on Enzymatic Resolution of a Primary Alcohol Possessing Two Stereogenic Centers,**Tetrahedron:Asymmetry**. (6) 2131–2134.

20. Kobayashi, M. and Shimizu, S. (2000). Nitrile hydrolases, **Curr. Opin. Chem. Biol.** 4 95–102.

21. Banerjee, A., Sharma, R. and Banerjee U. C. (2002). The nitrile degrading enzymes: current status and future prospects. **Appl. Microbiol. Biotechnol.** (60) 33-44.

22. Mathew, C., Nagasawa, T., Kobayashi, M. and Yamada H. (1988). Nitrilase catalyzed production of nicotinic acid from 3-cyanopyridine in *Rhodococcus* ***rhodochrous*** **J1. Appl. Environ. Microbiol. (54)** 1030–1032.

23. Birnbaum, J., Kahan, F. M., Kropp, H. and Macdonald, J. S. (1985). Carbapenems, a new class of beta-lactam antibiotics, **Am. J. Med.** (78) (Suppl. 6), 3-21.

24. Kiener, A., Roduit, J. –P, Kohr, J. and Shaw, N. (Lonza AG). (1995). **EP 686698.**

25. Terreni, M., Pegani, G., Ubiali, D., Fernandez, R., Mateo, C. and Gussian, J. M. (2001). Modulation of penicillin acylase properties via immobilization techniques: one pot chemo-enzymatic synthesis of cephamandole from cephalosporin C. **Bioorg. Med. Chem. Lett.** (11) 2429-2432.

26. Surivet, J. P. and Valete, J. M. (1999). Total synthesis of antitumor Goniothalamus styryllactones. **Tetrahedron.** (55) 13011-13028.

27. Mills, J., Schmiegel, K. K. and Shaw, W. N. (1983). Phenethanolamines, compositions containing the same and method for effecting weight control. **U. S. 4391826.**

28. Yamamoto, K., Oishi, K., Fujimatsu, I. and Komatsu, K. I. (1991) Production of (R)-(-)-mandelic acid from mandelonitrile by *Alcaligenes faecalis* ATCC 8750. **Appl. Environ. Microbiol.** (57) 3028-3032.

29. Kaul, P., Banerjee, A., Mayilraj, S. and Banerjee U. C. (2004) Screening for enantioselective nitrilases: kinetic resolution of racemic mandelonitrile to (R)-(-)-mandelic acid by new bacterial isolates. **Tetrahedron: Asymmetry.** (15) 207-211.

30. De Santis, G., Wong, K., Farwell, B., Chatman, K., Zhu, Z., Tomlinson, G., Huang, H., Tan, X., Bibbs, L., Chen, P., Kretz, K., Burk, M. J. (2003). Creation of a productive, highly enantioselective nitrilase through gene site saturation mutagenesis (GSSM). **J. Am. Chem. Soc.** (125) 11476-11477.

31. Hummel, W. (1999). Large-scale applications of NAD(P)-dependent oxidoreductases: recent developments. **Trends Biotechnol.** (17) 487-492.

32. Wolfgang, K., Harald.M.Klaus, E., Faber, K. (2004) Recent advances in the biocatalytic reduction of ketones and oxidation of sec-alcohols. **Current Opion Chem Biolo.** (8) 120–126

33. Stewart, J. D., and Rodriguez, S. (2001) Highly stereoselective reagents for b-keto ester reduction by genetic engineering of baker's yeast. **J. Am. Chem. Soc.** (123) 1547-1555.

34. Roy, A., Bhattacharyya, M. S., Ravi Kumar, L., Chawla, H. P. S., Banerjee, U. C. (2003). Microbial reduction of 1-acetonapthone: a highly efficient process for multigram synthesis of S (-)-1-(1¢-napthyl) ethanol. **Enzyme Microbiol Technol.** (33) 576-580.

35. Nakamura, K., Yamanaka, R., Matsuda, T., Harada, T. (2003). Recent developments in asymmetric reduction of ketones with biocatalysts. **Tetrahedron: Asymmetry.** (14) 2659-2681.

36. Shieh, W. R., Gopalin, A. S., and Sih, C. J. (1985). Stereochemical control of yeast reduction. 5. Characterization of oxodoreductases involved in the reduction of b-keto ester. **J. Am. Chem. Soc.** (107) 2993-2994.

37. Matsuda, T., Harada, T., Nakajima, N., Itoh, T., Nakamura, K. (2000). Two classes of enzymes of opposite stereochemistry in an organism: 0ne for fluorinated and another for nonfluorinated substrates. **J Org Chem.** (65) 157-163.

38. Zhou, B. N., Gopalin, A. S., VanMiddlesworth, F., Shieh, W. R., and Sih, C. J. (1983). Stereochemical control of yeast reduction. 1. asymmetric synthesus of l_carnetine **J. Am. Chem. Soc.** (105) 5925-5926

39. Nakamura, K., Miyai, T., Nozaki, K., Ushio, K., Oka, S., and Ohno, A. (1986). Diasterio and enantioselective reduction of 2-methyl-3-oxobutanoateby bakers' yeast **Tetrahedron Lett**. 3155-3156.

40. D'Arrigo, P., Fantoni, G. P., Serci, S., and Strini, A. (1997). The effect of absorbing resins on substrate concentration and enantiomeric excessin yeast reduction **Tetrahedron: Asymmetry.** (8) 2375-2379

41. Medson, C., Smallridge, A. J., and Trewhella, M. A. (1997). The stereoelective preparation of b-hydroxy esters using a yeast reduction in an organic solvent. **Tetrahedron: Asymmetry.** (8) 1049-1054

42. Guo, Z.-W., and Sih, C. J. (1989). Enantioselective inhibition: astretagy for improving the enantioselectivity of Biocatalytic system. **J. Am. Chem. Soc.** (111) 6836-6841

43. Hummel,W., Abokitse, K., Drauz, K., Rollmann, C., Gro¨ger, H. (2003). Towards alarge-scale asymmmetric reduction process with isolated enzymes: expression of an (S)-alcohol dehydrogenase in E. coli and studies on the synthetic potential of this biocatalyst. **Adv Synth Catal**, (345) 153-159.

44. Matsuyama, A., Yamamoto, H., Kobayashi, Y. (2002). Practical application of recombinant whole-cell biocatalysts for the manufacturing of pharmaceutical intermediates such as chiral alcohols. **Org Process Res Dev.** (6) 558-561.

45. Kizaki, N., Yasohara, Y., Hasegawa, J., Wada, M., Kataoka, M., Shimizu, S. (2001). Synthesis of optically pure ethyl (S)-4-chloro-3-hydroxybutanoate by Escherichia coli transformant cells coexpressing the carbonyl reductase and glucose dehydrogenase genes. **Appl Microbiol Biotechnol.** (55) 590-595.

46. Yasohara, Y., Kizaki N, Hasegawa J, Wada M, Kataoka M, Shimizu S (2001). Stereoselective reduction of alkyl 3-oxobutanoate by carbonyl reductase from Candida magnoliae. **Tetrahedron Asymmetry** (12) 1713-1718.

47. Mertens, R., Greiner, L., van den Ban, E.C.D., Haaker, H.B.C.M., Liese, A. (2003). Practical applications of hydrogenase I from Pyrococcus furiosus for NADPH generation and regeneration. **J Mol Catal B Enzym** (24-25) 39-52.

48. Greiner, L., Mu¨ ller, D.H., van den Ban, E.C.D., Wo¨ ltinger, J., Wandrey, C., Liese, A. (2003) Membrane aerated hydrogenation: enzymatic and chemical homogeneous catalysis. **Adv Synth Catal** (345) 679-683.

49. Stampfer, W., Kosjek, B., Moitzi, C., Kroutil, W., Faber K. (2002). Biocatalytic asymmetric hydrogen transfer. **Angew Chem Int Ed Engl.** (41) 1014-1017.

50. Stampfer, W., Kosjek, B., Kroutil, W., Faber, K. (2003). On the organic solvent and thermostability of the biocatalytic redox system of Rhodococcus ruber DSM 44541. **Biotechnol Bioeng**. (81) 865-869.

51. Ema, T., Moriya, H., Kofukuda, T., Ishida, T., Maehara, K., Utaka, M., Sakai, T. (2001). High enanatioselectivity and broad substrate specificity of a carbonyl reductase: toward a versatile biocatalyst. **J Org Chem**. (66). 8682-8684.

52. Yadav, J.S., Nanda, S., Thirupathi, Reddy, P., Bhaskar Rao, A. (2002). Efficient enantioselective reduction of ketones with Daucus carota root. **J Org Chem** (67) 3900-3903.

53. Speicher A, Roeser H, Heisel R. (2003). Stereoselective oxidoreductase type bioconversion of exogenous substrates by cell suspension cultures of bryophytes. **J Mol Catal, B Enzym.** (22) 71-77.

54. Lagos, F.M., Del Campo, C., Llama, E.F., Sinisterra, J.V. (2002). New yeast strains for enantioselective production of halohydrin precursor of (S)-propranolol. **Enzyme Microb Technol** (30) 895-901.

55. Bertau, M. (2002). How cell physiology affects enantioselectivity of the biotransformation of ethyl 4-chloro-acetoacetate with Saccharomyces cerevisiae. **Biocatal Biotransform**. (20) 363-367.

56. Garret, M.D., Scott, R., Sheldrake, G.N. (2002). Enantiocomplementary preparation of (S)- and (R)-1-pyridylalkanols via ketone reduction and alkane hydroxylation using whole cells of Pseudomonas putida UV4. **Tetrahedron Asymmetry**. (13) 2201-2204.

57. Comasseto, J.V., Omori, A.T., Andrade, L.H., Porto, A.L.M. (2003). Bioreduction of fluoroacetophenones by the fungi Aspergillus terreus and Rhizopus oryzae. **Tetrahedron Asymmetry.** (14) 711-715.

58. Mangone, C.P., Pereyra, E.N., Argimon, S., Moreno. S., Baldessari, A. (2002). Chemo- and stereoselective reduction of b-keto esters by spores and various morphological forms of Mucor rouxii. **Enzyme Microb Technol**. (30) 596-601.

59. Dehli, J., Gotor, V. (2002). Dynamic kinetic resolution of 2- oxocycloalkanecarbonitriles chemoenzymatic syntheses of optically active cyclic b- and c-amino alcohols. **J Org Chem**. (67) 6816-6819.

60. Yamazaki. Y., Maeda. H. (1986). Enzymatic synthesis of optically pure (R)-(-) mandelic acid and other 2-hydroxy carboxylic acids: Screening for enzymes and its purification, Characterization and use. **Agric boil chem.** (10) 2621-2631.

61. Hosono, K., Kajiwara, S., Yamazaki, Y., Maeda, H. (1990). Construction of a bioreactor for production of (R)-(-)-mandelate,a a typical speciality chemical. **J Biotechnol.** (14) 149-156.

62. Hummel, W., Schütte, H., Kula, M-R. (1988). R-(-) Mandelic acid dehydrogenase from *Lactobacillus carvatus*. **Appl Microbiol Biotechnol.** (28) 433-439.

63. Ward., D.E., Donnelly, C.J., Mullendore, M.E., van der Oost, J., de Vos, W.M., Crane, E.J., III (2001). The NADH oxidase from Pyrococcus furiosus: implications for the protection of anaerobic hyperthermophiles against oxidative stress. **Eur J Biochem.** (268) 5816-5823.

64. Hummel,W., Riebel, B. (2003). Isolation and biochemical characterization of a new NADH oxidase from Lactobacillus brevis. **Biotechnol Lett.** (25) 51-54.

65. Riebel, B.R., Gibbs, P.R., Wellborn, W.B., Bommarius, A.S. (2002). Cofactor regeneration of NADR from NADH: novel water-forming NADH oxidases. **Adv Synth Cat.** (344). 1156-1169.

66. Geueke, B., Riebel, B., Hummel, W. (2003). NADH oxidase from Lactobacillus brevis: a new catalyst for the regeneration of NAD. **Enzyme Microb Technol.** (32) 205-211.

67. Riebel, B.R., Gibbs, P.R., Wellborn, W.B., Bommarius, A.S. (2003) Cofactorregeneration of both NADR from NADH and NADPR from NADPH:NADH oxidase from Lactobacillus sanfranciscensis. **Adv Synth Catal.** (345) 707-712.

68. Filho, M.V., Stillger, T., Mu¨ller., M., Liese, A., Wandrey, C. (2003). Is logP a convenient criterion to guide the choice of solvents for biphasic enzymatic reactions? **Angew Chem Int Ed Engl** (44) 2993-2996.

69. Gro¨ger, H., Hummel, W., Buchholz, S., Drauz, K., van Nguyen, T., Rollmann. C., Hu¨sken H, Abokitse, K. (2003). Practical asymmetric enzymatic reduction through discovery of a dehydrogenasecompatible biphasic reaction media. **Org Lett.** (5) 173-176.

70. Niefind, K., Mu¨ller, J., Riebel, B., Hummel, W., Schomburg, D. (2003). The crystal structure of R-specific alcohol dehydrogenase from Lactobacillus brevis suggests the structural basis of its metal dependency. **J Mol Biol.** (327) 317-328.

Advances in Biochemistry and Biotechnology Volume 1 *Pages 183-194*
Edited by: **Dr. Chiranjib Chakraborty**
Published by: **DAYA PUBLISHING HOUSE**

8

Use of Immunostimulant in Aquaculture

Arup Tewary & Bidhan C. Patra*
**Aquaculture Research Unit, Department of Zoology, Vidyasagar University, Midnapore- 721 102, West Bengal*
E-mail : bcpatra@yahoo.com

Introduction

The goal of aquaculture industries are to optimize and to produce high quality fish as in all farming, the outbreak of diseases in fish farming can be a major concern. The high susceptibility of fish to stress and the rapid spread of diseases in water have forced aqua culturists to concentrate their efforts on maintaining their fish in good health in order to achieve sustainable economic performances. Diseases are the major limiting factor in the aquaculture industry particularly the shrimp industry. In India, the loss of shrimp production during 1995–1996 due to disease is estimated to be Rs. 600 crores and the losses continues around this level since then.

Several years ago, during the early development of the salmon industry, antibiotics are commonly used in the treatment of disease. However, the consumption of drugs has progressively been reduced owing in part to environmental and regulatory concerns and in part to increased resistance of pathogens. Furthermore the curative effect of oral drugs is minimized by the fact that diseased fish frequently do not feed properly, on the other hand intensively raised fish may be exposed to stressful

* Correspondent Author

situation which often result in a depressed immune status. Good management practices reduce stress and therefore help to maintain healthy animals. However, since not all stressing situations can be avoided, fish with enhanced defense mechanisms will be better prepared to combat the negative effect of stress.

The nutritional quality of the feed is a major factor in sustaining healthy fish; the use of immunomodulator or immunostimulant in fish culture or in aquaculture of other species for the prevention of disease is a promising new development. (Anderson 1992; Onarheim, 1992; Raa *et al.*, 1992). Generally immunostimulants are, antioxidant vitamins, carotenoids and other feed additives.

The combination of good management, vaccination and nutritional prophylaxis will insure higher survival rates and improve growth in intensive farming systems.

Nutritional prophylaxis
Vitamins, Carotenoids, Immunomodulators

⇩

Good Management ⇨ Immune response ⇦ Vaccination

⇩

Health Status

Increased Survival rate & Growth

This communication highlights the importance of various synthetic product and vitamin (basically vit–C, E and A) as an immunomodulator or immunostimulant and a key nutritional element in modern fish farming in promoting sustainable aquafarming.

Fish are the most primitive vertebrates and an important link between invertebrates and higher vertebrates. They posses the non-specific defense mechanisms of the invertebrates such as the phagocytic mechanisms developed by macrophages and granular leukocytes, but were also the first animals to develop both cellular and humoral immune responses mediated by lymphocytes.

The non-specific immune system is responsible for initiating a rapid and general response to invading microorganisms. On the other hand non-specific immune mechanism defined also as natural or innate immunity of host–defense mechanisms that do not require a specific recognition of the antigen as occurs in humoral and cell-mediated responses of the specific defense. The most important cells involved in this defense is the phagocytes, which are helped by several soluble factors such as complement and lysozyme. There are also minor cell mechanisms and a battery of soluble factors that intervenes at various levels.

Non-specific Immune Response

Natural barriers: skin and mucus

When Micro-organism Penetrate the Body

Cellular mechanism:

Macrophage and Neutrophil Activities

Phagocytosis

Phagocytosis is the process where by cell internalize, kill, and digest invading micro organisms. It can be divided into three main phases: attachment of the cell surface, Ingestion involving the formation of phagosome, and lastly, breakdown of the particle within the phagosome. *In vivo* and *In vitro* studies have shown that monocytes or macrophages and granulocytes (Nutrophils and in some cases eosinophils) are phagocytic and will ingest a wide range of inert and antigenic particle (Ainsworth, 1992; Secombs and Fletcher,1992), and soluble ligands (Dorin *et al.*, 1993; Dannevig *et al.*, 1994).

Killing

Phagocytes are able to kill pathogens using a variety of killing mechanisms that can be broadly categorized as oxygen dependent or oxygen independent. In many instances, where the pathogen is a microorganism, killing by fish phagocytes is an intracellular event and can occur relatively quickly. There are two types of killing mechanism, *i.e.*

Oxygen Dependent Mechanism

When phagocytes ingest particles, there is an increased oxygen uptake that is independent of mitochondria respiration (*i.e.*, is not prevented in the presents of metabolic inhibitors). For *e.g.*, rainbow trout head kidney phagocytes consume some 12nmol $O_2/min/10^7$ cells after stimulation with zymosan compared with an uptake of 1.3 nmol $O_2/min/10^7$ cells in resting cells. Soluble membrane stimulants such as Phorbol esters can also elicit this response. This so called respiratory burst is used in the generation of a number of oxygen and nitrogen free radicals known to be toxic for bacteria and protozoan parasites.

Oxygen Independent Mechanism

Virtually nothing is known about pahgocyte oxygen-independent killing mechanisms. It is known that fish phagocytes possess a large array of enzyme that would be potentially bactericidal following lysosomal fusion with phagosomes. They also possess lysozyme, which is known to be bactericidal for a number of fish pathogens (Grinde, 1989) and can be secreted to allow extra cellular killing.

Extra cellular killing

Natural killer cells

Eosinophils

Soluble factors

Complement

Lysozyme

C-reactive protein

Transferrin

Lactoferrin

Ceruloplasmin

Lectins natural agglutinins

Interferons

The Specific Immune Response

When an infectious agent penetrate the organism, non-specific defense mechanisms are stimulated. Their sole activation might be sufficient to stop the infection. If not, the disease will develop, leading to the induction of specific defense mechanisms. These will then lead to the cure at the disease and the setup of an immunological memory, blocking the development of a new infection caused by the same pathogen, *e.g.* vaccination.

The macrophages act as antigen presenting cells in the specific immune response. Their role is to process the antigen and to present the processed antigenic determinants in association with recognition molecule to the lymphocytes. Subsequently T lymphocytes are activated by interaction with the antigenic determinants and factors secreted by macrophages (interleukin). The activated T lymphocytes also called helper cells, stimulate the differentiation and proliferation of B lymphocytes by secretion of interleukins. One of the subset of activated T lymphocyte known as helper T cell (T_H) stimulate the proliferation and differentiation of B lymphocyte by secretion of interleukins.

Depending on the circumstances B lymphocyte will develop in to long lived B memory cell or short lived plasmocytes. These plasmocytes secrete huge amounts of specific antibodies (Immunoglobulin) of type M(I_gM) and will bind or kill invading micro-organism presenting the by corresponding determinants.

Modification of Fish Immune System

Many occasions arise in the course of fish culture that calls for enhancement of immune response. These includes strengthening of the normal immune response in order to enhance immune protection and reduce immunosuppressive conditions (Galeotti, 1998). Enhancement of immune system by means of specific as well as non- specific way plays an important role in fish health. Non-specific mechanisms are important in the defence of all multicellular animals against invading pathogenic microorganism. It is well established that mononuclear phagocytes or macrophages play a central role in the cellular part of the non- specific defence. The non-specific immune responses such as phagocytosis and the production of oxidative radicals are quickly activated by the immunostimulants and help to protect the host against pathogens (Anderson,1992).

Factors Influencing The Immune Response

Many factors can influence the immune response of fish. Among them are stressors and environmental factors of natural origin. Nutrients, micronutrients and substances of no nutritional values can also modulate the immune response. Depending on their type, the amount and the duration of exposure, their effects can be either negative or positive. Substances with immunomodulating properties can compensate the immunodepression caused by other factors, *e.g.* the immunodepression caused by a stressor can be compensated by an increased intake of vitamins before a predictable stress event. There are some substances (plant, animal, as well as synthetic product), which able to stimulate the immune response known as immunostimulant. It has been extensively studied in fish species both at whole animal and on cellular level. Fish depends more heavily then mammals do, on non-specific defense mechanisms (Anderson-1992), and therefore, immunostimulants play an important role in health management of aquaculture.

Use of Immunostimulants in Aquaculture Systems

The contribution of aquaculture to fish production is steadily increasing. The increase would have been much more but for the major constraint of losses in culture production due to disease of fishes particularly in shrimp.

In recent years, the application of vaccination in respect of finfish and immunostimulants in respect of shrimp/finfish for disease management in aquaculture is being increasingly recognized. Generally, immunostimulants enhance individual components of non-specific immune response but this does not always translate into increased survival. In addition, immunostimulants fed at too high dose or for too long duration can be immuno- suppressive (Gannam *et al.*, 1999). Immunostimulants enhance the humoral and cellular immune response in both specific and non specific way. These agents are widely used for impaired immune function and to stabilize the improved immune status (Simoens and Rosenthal, 1997).

The use of immunostimulants in fish culture or in aquaculture of other species for the prevention of disease is a promising new development (Anderson, 1992; Onarheim, 1992; Raa *et al.*, 1992).

Some Important Immunostimulants and their Role

Glucans

Glucans are major structural polysaccharides derived from yeast cell walls (Duffus *et al.*, 1982). Yeast cell walls are constructed almost entirely of two classes of polysaccharide, mannose polymer covalently linked to peptides for mannoproteins and glucose polymer (or glucans). The glucans and mannoproteins occur in roughly equal amounts in the wall. Glucan composed of several glucose unit, which are held together through β-1,3 and β-1,6 glucosidic bonds (Rosenberg, 1976). Among various types of glucans, peptide Glucan and yeast Glucan have been well studied and reported by many scientists (Adams *et al.*, 1997). Lyophilised whole yeast (*S. cerevisiae*) able to stimulate the cellular innate immune response. Yeast cell walls are constructed almost entirely of two classes of polysaccharides *i.e.* mannose polymer (mannoprotein) and glucose polymer (glucan). Little amount of N-acetylglucosamine and chitin are also present. Indian major carps *Labeo rohita* were fed with four different diets for four weeks supplemented with yeast. After one, two and four weeks Total Lymphocyte

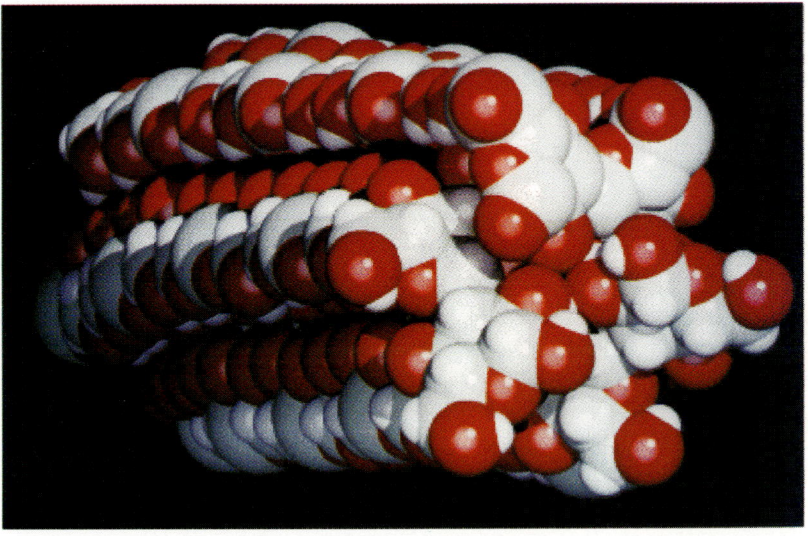

Fig. 8.1: Molecular Model of a Non-branched, Triple Helical (1-3)-β-D-glucan Polymer

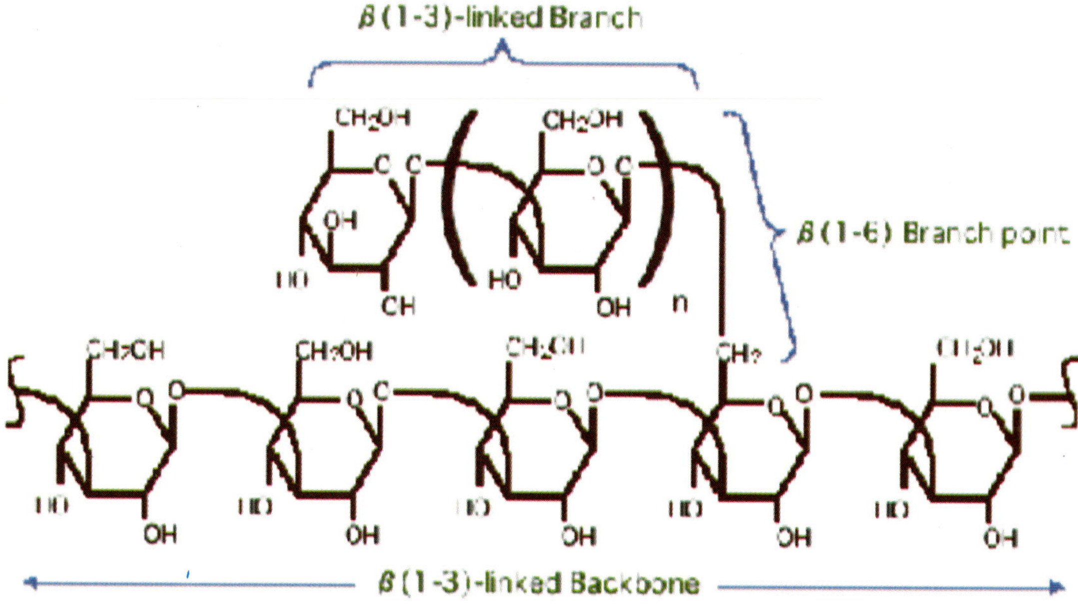

Fig. 8.2: Primary Structure of Yeast (1-3)-β-D-glucan

Count (TLC), phagocytic morphology and killing activity, respiratory burst activity were evaluated. The results show that yeast cell wall able to enhance the cellular innate immunity (Tewary *et al.*, 2003). Cook *et al.* (2003) used Ecoactiva as an commercial source of glucan in Snapper (*Pagnus auratus*) and shows the enhanced growth rate, macrophage and respiratory burst activity.

Chitin

The chitin polymer has been less studied for its role as immunostimulant but it plays a clear role in enhances defense and resistance to infection (Sakai, 1999; Esteban *et al.*, 2001). Chitin is a cell wall component of a large array of fungi and component of exoskeletons of insects and crustaceans.

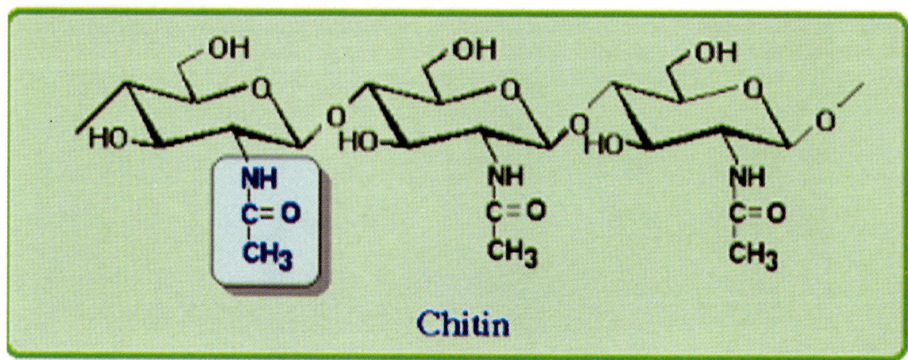

Fig. 8.3: Chemical Structure of Chitin

Immunostimulatory roles of chitin came into highlight only in 1985 when Nishimura *et al.*, reports that it increases the non-specific immune response in guineapig. Sakai *et al.* (1992) reported that, rainbow trout injected with chitin, showed stimulated macrophage activity and imported an increased resistance to *Vibrio anguillaruna* infection. They found that activated macrophages secret some factor like IL–1; which promotes inflammatory responses; IL–6, which also promote inflammatory response in many tissues especially endothelium; interferon–∝ (IFN–∝), which activate cellular genes, resulting in the production of protein that confer a temporary antiviral state, on the recipient cell.

Kawa Kanai *et al.* (1998) reported chitin's immunostimulatory roles against the *Pseudomonas piscicida*. They also reported that chitin can increase IL–1 secretion by activated macrophages.

Chitosan

Chitosan is a chitin derivative (Voet and Voet, 1995) having slightly more chemical complexity over chitin. Siwiki *et al.*, 1994 reported that, the oral administration of chitosan increases the phagocytosis in rainbow trout and gives elevated resistance to *A.salmonicida* infection. The almost similar result was reported by Anderson *et al.*, 1995 in same species but with intraperitonial injection.

Vitamin–C

With the exception of perhaps two or three species, vitamin-C biosynthesis does not occur in fish due to the lack of the last enzyme to the biosynthetic pathway: L–gulono lactone oxidase. Vitamin-C must therefore be supplied via the feed. Major signs of ascorbate deficiency include reduced growth, lordosis, internal and fin haemorrhage, distorted gill filaments, fin erosion, anorexia and increased mortality. Vitamin-C is an antioxidant, so it able to stimulate the immune response non-specifically as well as specifically. Vitamin-C able to enhance the rate of phaogcytosis (Blazer,1992), oxidative burst (Verlac *et. al.*, 1997), complement activity (alternative and classical) pathway (Verlhac *et. al.*, 1995), vitamin-C treated fished able to make antibody response against *V. salmonicida, V. anguillarum, A. salmonicida*, (Waagbo *et al.*, 1993, Lal *et al.*, 1989,). High vitamin C concentrations improved macrophage migration in guinea pigs (Ganguly *et at.*, 1976) and enhanced neutrophil mortality in higher vertebrates (Le Grusse and Watier 1993). Similarly, gilthead sea bream head kidney leukocyte migration increased after incubation in vitro with vitamin C (Mulero *et al.*, 1998) although the present results indicate that vitamin C dietary intake does not significantly affects this activity in these cells. It has been demonstrated that this activity is not dependent on this vitamin in *Salmon salar* (Thompson *et al.*, 1993).

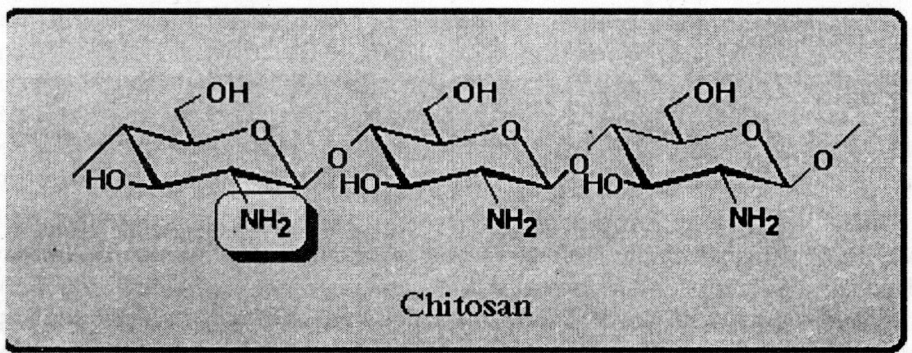

Fig. 8.4: Chemical Structure of Chitosan

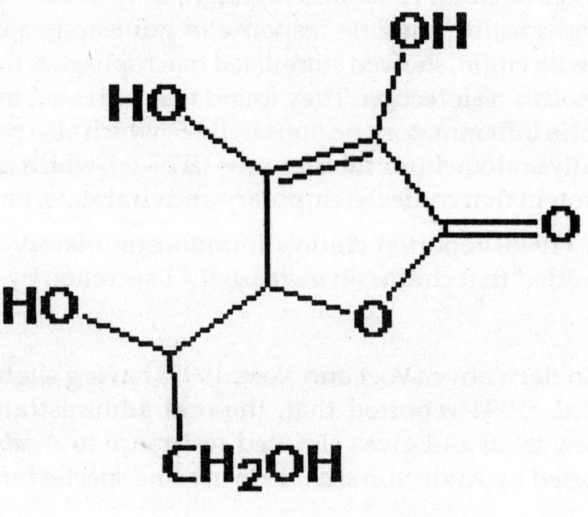

Fig. 8.5: Vitamin C-2D Structure–m.p. 190-192 °C., C_6H_8

Respiratory burst, phagocytosis and bactericidal activities were not affected in different fish species as consequence of a high dietary intake of vitamin C (Hardic *et al.*, 1991; Thomson *et al.*, 1993) or an invitro incubation of their leucocytes with this vitamin (Hardie *et at.*, 1993; Mulero *et al.*, 1998). The present results indicate that respiratory burst activity of gilthead sea bream head kidney leucocytes was, on the contrary, affected by the dietary intake of vitamin C. Perhaps this is due to the method applied in the present study because, at present, DHR is considered the most sensitive flow cytometric indicator for the detection of oxygen reactive species in activated leucocytes, the oxidation of non-flurescent DHR to the green fluorescent rhodamine 123 being highly specific and sensitive for respiratory burst activity (Vowells *et al.*, 1995).

Vitamin–E

Vitamin-E plays an important role in the fish immune response, as it is an important component of the cell membrane and has a specific role as an antioxidant, controlling preoxidation of unsaturated fatty acids. Since fish have high levels of unsaturated fatty acids to maintain cell membrane fluidity especially at low temperatures (Blazer, 1992), it is assumed that vitamin E plays an important role. Indeed, pathological sign, such as growth reduction, haematological alteration, anemia, epithelial pigmentation loss, ascites, exophthalmia and apathy appear in fish fed vitamin E deficient diets (Wilson *et al.*, 1984; Hamre *et al.*, 1997). In addition, a marked immuno-depression has been observed when vitamin E levels are depleted (Blazer and Wolke, 1984; Hardie *et al.*, 1990), a finding which confirms its instruction with the immune system.

It is well established that high dietary vitamin E supplementation enhances both humoral and cellular defenses in mammals (Tengerdy *et al.*, 1981; Panush and Delafuente, 1985; Tengerdy, 1989; Bendich, 1990; Moriguchi *et al.*, 1990). Taking into account that fish possess a well developed immune system as well as hydrolytic and transport mechanisms for vitamin E similar to those found in other vertebrates (Gallo–Jorres, 1980), the fish immune system might be expected to respond in the same way as in homeotherms. Contrary to the above hypothesis, Hardie *et al.* (1990) observed no

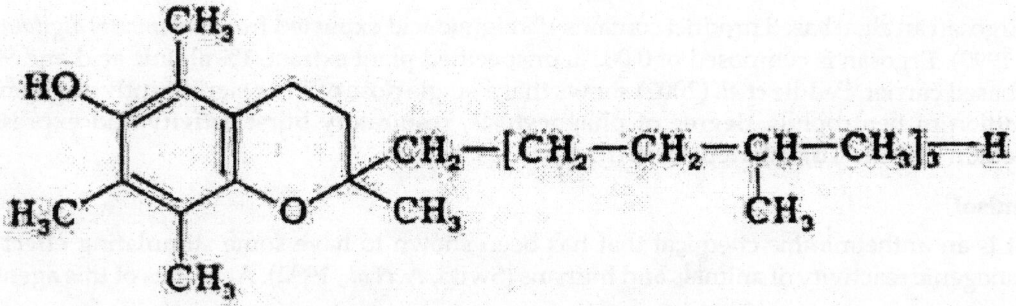

Fig. 8.6: Chemical Structure of Vitamin E

immunostimulation in rainbow trout fed a high dose of vitamin E. However, few studies has paid attention the effects of a high dietary vitamin E supplementation on the immune system of farmed fish (Monters *et al.*, 1999). In an attempt to shed more light on this question, the aim of this study was to provide more information related to the impact of vitamin E on the non-specific immune system of teleosts. For this, different dietary vitamin E doses, above the minimum requirement established for other species, were tested over various times in order to study their effect on both the humoral and cellular non-specific immune response of the gilthead sea bream (*Sparus aurata* L.) Vitamin E-depleted diets have been reported to reduce macro-phage phagocytosis and lymphocyte T and B function in rainbow trout immunised against *Yersinia ruckeri* (Blazer and Wolke, 1984), decrease complement activity in Atlantic salmon, *Salmo salar* (Hardie *et al.*, 1990), impair antibody response of rainbow trout (Verlhac *et al.*, 1993), reduce bacterial killing activity of isolated head-kidney macrophages of Atlantic salmon (Waagbo *et al.*, 1993), reduce peritoneal macrophage function in channel catfish (Wise *et al.*, 1993a) and reduce alternative complement activity in gilthead sea bream (*Sparus aurata*) (Montero *et al.*, 1998). Montero *et al.*, 2001 shows that low levels of vitamin E depleted alternative complement pathway activity and reduce stress resistance.

Vitamin–A

Vitamin–A has been widely related to the vertebrate immune system (Ross,1992). In mammals vitamin A displays potent anti-tumour activity and slightly enhanced phagocytosis activity (Goldferb and Herberman, 1981).

In Teleosts, as in other animals, vitamin A is incorporated directly from the diet or metabolised from carotenoids (Schiedt, 1985). This makes it an essential micro-nutrient for fish. Increased disease resistance of salmonids fed vitamin A has been reported (Malikova, 1965). However, despite these preliminary results and the great interest that vitamin A (and other vitamins) might have on the fish immune system, there are very few studies of the immunomodulatory effect of this vitamin in fish (Amar,2000). Furthermore, the available data are controversial and, to our knowledge, only one paper describes a clearly positive effect of retinal acetate on the fish immune system or, more specifically, on the natural cytotoxic activity of channel catfish (Evans, 1984). According to Cuesta *et al.* (2002) retinol acetate plays an important role in the gilthead seabream nonspecific cellular immune system due to its antioxidant properties. They also point to the importance of the way in which it is administered, by dietary uptake or intraperitoneal injection.

Ergosan

Ergosan an algal based product contains 1% alginic acid extracted from *Lamimaria digitata* (Lewis *et al.*, 1990). Ergosan is composed of 0.002% unspecified plant extract, 1% alginic acid and 98.998% algal based carrier. Peddie *et al.* (2002) shows that a single dose of 1 mg significantly augmented the proportion of neutrophils, degree of phagocytosis, respiratory burst activity and expression of interleukin –1B (1L–1B); interleukin –8 (IL–8).

Levamisol

It is an anthelminthic chemical that has been shown to have some stimulating effect on the immunogenic reactivity of animals and humans (Swiki, A. *et al.*, 1992). Activities of this agent are:

1. Enhancement of cell mediated cytotoxicity, lymphokine production and suppressor cell function,
2. Stimulation of phagocytic activity of macrophage and neutrophils.

Leaf Extract of Ocimum Sanctum

Effect of leaf extract *Ocimum sanctum* on (1) the specific and non-specific immune response and (2) disease resistance against *Aeromonas hydrophila* was investigated in *Oreochromis mossambicus* (Logamble *et al.*, 2000). It stimulated both antibody response and neutrophil activity. Dietary intake also enhanced the antibody response and disease resistance to Aeromonas. Possibility of using *O. sanctum* as immunostimulant in the maintenance of finfish health in intensive freshwater aquaculture is suggested

Conclusion

Non- specific immunity plays an important role in the defense mechanism of fish and it is the sole immunological mechanism by which invertebrates protects themselves from diseases. So, the use of immunostimulant such as glucan, chitin, chitosan and various antioxidant vitamins such as Vitamin-C, Vitamin- E, Vitamin- A able to enhance the different non-specific immune parameters such as macrophage, respiratory burst, TLC, Albumin Globulin ratio. Enhanced immune system develops stronger primary defense system in fish through which it can protect themselves primarily from various diseases. Many of the commonly used immunostimulants (Glucan, Chitin, etc.) are not economically viable for the farmers, so it requires to develop commercially available and economically viable immunostimulats, which could be frequently used in aquaculture industry. It is expected that during the coming years, immunostimulants will find more application to make aquaculture increasingly sustainable.

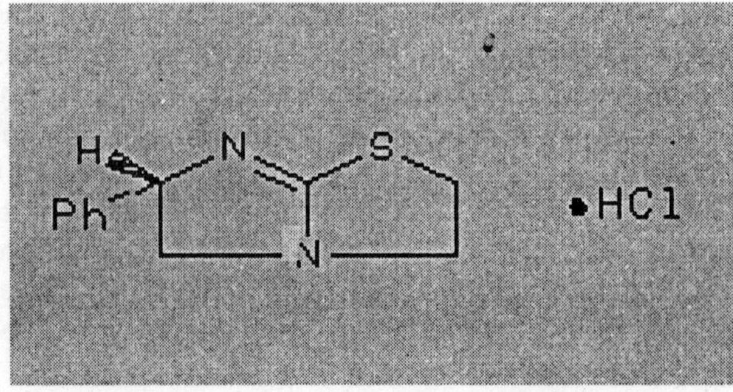

Fig. 8.7: Chemical Structure of Levamisol

Table 8.1: Exampl϶ of Induction of Immunoresponse in Crustaceans

Crustacean Animals	*Inducing substances*	*Response Induced*
Cray fish (*Parachaeraps bicarinatus*)	Formalised cells of *Pseudomonas*	Protection against Injection
Spiny lobster (*Panulirus argus)*	Bacteria	Bactericidin
Squilla mantis	Vibrio bacteria	Bactericidal and agglutination activity
Penaeus monodon	V. alginolyticus	Bacterial activity
	Bacterin Glucan	Protection against
	Vibrio bactarin	*V. vulnificus*
	Yeast β–1, 3 glucan	*Chanllenge*
		Vibriocidal activity
		Oxygen sp. induction

Acknowledgement

The authors are thankful to the Department of Zoology for extending all sorts of working facilities and the Research Scholars of the Aquaculture Research Unit, Department of Zology, Vidyasagar University, Midnapore, West Bengal for their valuable co-operation to prepare this manuscript.

REFERENCES

1. Adams, S.M., ed., 1990. " Biological Indicators of Stress in Fish " American Fisheries Society. Bethesda. Maryland
2. Anderson, D. P., 1992. Immunostimulants, adjuvents, and vaccine carriers in fish: applications to aquaculture, Aun. Fish Dis. 2: 281-307.
3. Anderson, D. P. and Siwicki, A. K., 1994. Duration of protection against *Aeromonas salmonicida* in book trout immunostimulated with glucan or chitosan by injection or immersion. The progressive fish culturist, 56: 258–261.
4. Blazer, V.S., 1992. Nutrition and disease resistance in fish. Ann. Rev. Fish Dis. 1992. 309-323
5. Cook,T.M., Hayball, J.P., Hutchinson, W., and Hayball,D.J., 2003. Fish Shellfish Immunol. 14, 333-345
6. A. Cuesta, J. Ortuno, A. Rodriguez, M.A.Estaben and J. Meseguer., 2002. Changes in some innate defence parameters of seabream (*Sparus aurata* L.) induced by retinol acetate. Fish Shellfish Immunol. 13, 279-291
7. Esteban, M.A., Cuesta, A., Ortuno, J., Meseguer, J., 2001. Immunomodulatory effects of dietary intake of chitin in gilthead seabream (*Sparus aurata* L.) innate immune response. Fish Shellfish Immunol. 11, 305-315.
8. Gannam, A. L., Schrock, R. M. 1999. Immunostimulants in fish diet. J. Appl. Aquacult.,9 (4): 53-70.
9. Montero, D., Marrero, M., Izquierdo, M. S., Robaina, L., Vergara, J. M. and Tort, L. (1999). Effect of vitamin E and C dietary supplimentation on some immune parameters of gilthead seabream (*Sparus aurata*) juveniles subjected to crowding stress. *Aquaculture* 171, 269-278.

10. Ross, A. C. (1992). Vitamin A status: Relationship to immunity and the antibody response. *Experimental Biology and Medicine* 200, 303–320.

11. Schiedt, K., Leuenberger, F. J., Vecchi, M. and Glinz, E. (1985). Absorption, retention and metabolic transformations of carotenoids in rainbow trout, salmon and chicken. *Pure and Applied Chemistry* 57, 685–692

12. Siwicki, A.K., Anderson, D.P., Rumsey, G.L., 1994. Dietary intake of immunostimulants by rainbow trout affairs non-specific immunity and protection against furunculosis. Vet. Immunol. Immunopathol. 41, 125-139.

13. Tewary, A., Bandyopadhyay, P. and Patra, B. C., 2003. Yeast, *Saccharomyces cerevisiae* stimulate the cellular innate immune response of IMC (*Labeo rohita*). In Abstract In. International Seminar on "International Symposium on Biochemicales Roles Of Cell Surface Molecules" during 16–18 January 2003 at Indian Institute of Chemical Biology, Kolkata. pp 44.

Advances in Biochemistry and Biotechnology Volume 1 ***Pages 195-213***

Edited by: **Dr. Chiranjib Chakraborty**

Published by: **DAYA PUBLISHING HOUSE**

9

Application of Environmental Biotechnological Tools in Water Resource Development

Biplab Sarkar, Debabrata Mukherjee, S. Adhikari & S. Ayyappan*

**Deputy Director General, (Fisheries), ICAR, Krishi Anusandhan Bhaban, PUSA, New Delhi*

ABSTRACT

Biotechnological tools are well applicable in water resource development. In the development of water resource wastewater treatment is an important aspect. Among the primary, secondary and tertiary treatment of wastewater, secondary treatment is the biological–based on the exploitation of catabolic activities of microbes. Physical process of wastewater treatment is unable to remove major portion of organic matter rather remain as colloidal or dissolved solids. Removal of such can be achieved through bacterial oxidation. The conventional system can be done by two ways- fluidized or suspended microbial culture and film-flow system. Algal treatment can precipitate toxic substances like heavy metals from wastewater, at the sane time it can act as a source of by-products in water system. Algae and phosphate metabolizing bacteria are considered to be a biological means of nitrogen and phosphate removal. Application of virus

* Correspondent Author

like Cyanophages can control excess algal growth in eutrophic water. Many aquatic phytoplanktons and benthics can act as natural bio-scavengers of effluents. Biomechanisms of metal chelation detoxification can be through adsorption and absorption. Cell immobilization is a very important tool in wastewater treatment involves the use of biocatalysts. GEMs (genetically engineered microbes) can be prepared biotechnologically which can play important role in bio-treatment of waste. In toxicity assessment bio-assay is an important criterion. Bioluminescent (Lux reporter gene) technology is sensitive and attractive means of monitoring contamination of ground water and surface soil. Biosensor is an analytical device, which has a great application in the realm of environmental monitoring and pollution detecting. Bioremediation and biorelamation are new technologies and need proper application and scale up. Biomining is another prime area, which has the potentiality to utilize microbial system to isolate metal ores from raw metals, with minimum toxic input to the environment. Above all, ex situ and in situ conservation of both economically important plants and animals for sustaining aquatic bio-diversity and water quality, are the others thrust area to be cover up.

Keywords: *Waste water treatment, Bioscavenger, Cell immobilization, Metal chelation, LRG, Biosensor, GEMs, Bioremidation, Biomining, Aquatic biodiversity, Conservation, Water quality management.*

Introduction

Biotechnology is recognized as a discipline where basic and applied sciences have emerged and are working hand in hand. Several basic scientific disciplines like physics, mathematics, chemistry, biochemistry, biophysics and engineering have an interface with biotechnology. According to U.S. National Foundation –biotechnology is the controlled use of biological agents, such as micro-organisms or cellular components, for beneficial use. Biotechnology in particular is very popular and it has several affiliated branches such as biogenetics, bioinformetics, microbiology, genetic engineering and so on. Biotechnology is one of the fastest growing field today. Primary objective of is to develop processes whose large scale application results in social benefits through health, agriculture, aquaculture, environment, forest and industry etc. In the environment sector use of biotechnology is magnanimous and well diversified. Various environmental problems are being solved through biological processes. Biotechnological methods have been devised for some of harmful environmental problems like:

1. Pollution control,
2. Depletion of natural resources for non-renewable energy,
3. Biodiversity conservation, and
4. Restoration of degraded lands.

As for example microbes are being used as bio-pesticides, biofertilisers, biosensor, etc.They are also being used in metallurgy for recovery of metals, cleaning of spilled oils and for a variety of other purpose. For the purpose of biomonitoring microbes are being used in industries. Biomass is being produced and it is used as renewable source of energy. For the reclamation of degraded land root nodules of both leguminous and nonleguminous plants are being used.

Due to rapid industrial development there is a constant threat on environment and also the depletion of natural resources. Pollution in every sphere *i.e.* air, water and land is a common phenomenon nowadays. To combat against this various environmental hazards biotechnological tools are being used. Different biological processes are being exploited to get rid of pollution.

Here, in this chapter the environmental biotechnological tools used in water resource development will be discussed.

Conventional Wastewater Treatment

In order to control pollution of water, industrial effluents with toxic substances should be treated to make harmless or less harmless. Most effluents are heterogeneous in nature. They contain both dissolved and suspended matters, be it organic or inorganic.

The waste characteristic of industrial effluents are:

1. Suspended solids or soluble organic matters, which can undergo progressive decomposition therefore result in oxygen depletion and production of noxious gases.
2. Heavy metals, cyanides, and other toxic substances (*e.g.* chlorinated compounds) which are deleterious to aquatic life.
3. Undesirable level of nitrogen (N) and phosphorus (P), which enhance eutrofication and stimulate undesirable algal growth.
4. No biodegradable chemical and volatile materials like H_2 S and SO_2. In the conventional treatment procedures, three steps are usually followed (Dart and Sretton, 1994).

Primary Treatment

This is essentially a physico-chemical process where sedimentation, chemical coagulation and precipitation are followed to remove the coarse and fine suspended solids. This process has an advantage of low capital investment and ease of adsorption and operation. By primary treatment B.O.D is brought down to 5,000 to 10,000 ppm, which is too high to disposal, keeping the standard fixed (100 ppm for land and 30 ppm for water.

Secondary Treatment

It is essentially a biological treatment process involves the removal of colloidal and dissolved organic substances or some other toxic substances. It follows the principles of biodegradation and bioabsorption. This treatment involves anaerobic digestion (biomethanation) and it may be used to bring down the B.O.D further from primary treatment. DGTD recommended secondary treatment as a national program in India.

It is mainly based on the exploitation of catabolic activities of microbes including bacteria, algae, fungi, protozoa, rotifers and nematodes. Adequate supply of microbes, ensuring their contact with effluents and oxygen supply is needed in this process–Foster (1980). Biotreatment of waste water is done in bioreactors (controlled environment). Bioreactors shorten the reaction time and manipulation of the process. Lagoon and stabilization ponds are natural system for this process. Treatment in natural system take much longer time and tough to perform in populated area.

Tertiary Treatment

This is an advance treatment also known as "polishing". In this treatment procedure further purification of water is done depending on the reuse of that water. By this type of treatment excess nutrients like N_2, P_4 and toxic elements like metals are removed.

Conventional treatment processes are comprised of two main types:

1. Fluidized or suspended microbial culture (includes activated sludge process), and
2. Film-flow system (includes trickling filter and rotating biological contactor RBC).

Activated Sludge Process

This aerobic process is basically utilize in secondary treatment plant. Here, organic matters are brought in contact with diverse types of microbes in presence of a mechanical aerator.

The essential features are:

1. Aeration tank,
2. Solid/liquid separation, and
3. Sludge recycle (Rao, 1991).

Different Steps

1. The effluents from aeration tank containing the sludge (flocculant microbial biomass) is separated in a settling tank. In this tank the separated sludge is without contact with organic matters and become activated.
   (a) Microbes are agglomerated in the form of *flocs* each having a diameter of ca. 0.1 mm or larger.
2. A part of this activated sludge is recycled in aerated tank and thus "seed" or "inoculum" is formed. Rest of the sludge is waste and deposited later.
3. High O_2 level is maintained by constant mechanical aeration either by bubbling or through the use of rotating paddles.

Indicator

Development of protozoa indicates the healthy condition.

Advantage

It is mainly used as fertilizer in crop lands or as landfills after drying. It may be "lagooned"(Lowries and Wells,1991).

Another big advantage of this process is its compactness for a given rate or removal of substrate. This process has removal potency of upto 95% of BOD_5, 98% of bacteria, particularly coliform and 95% suspended solids.

In paper industry, waste bleach water are treated currently using *aerated lagoon* (having bacteria) and activated sludge process. These methods however not able to purify high-molecular mass chlorinated materials.

Trickling Filter Process

This process is comparatively older than previous process and is used in small sewage works. First biological filter was made in 1890.

Formation of filter

1. This is an oxidation unit.
2. Organic matters are absorbed and oxidized when it passes through the filter medium containing a microbial slime layer.
3. The media are normally crushed stones or granite rocks, plastics and slags.
4. Conventional trickling filter contains a rock bed (1-3 m) with an opening for aeration.

Method of Filtering

1. The influent is sprinkled over a biological film.
2. As the liquid trickles through the biofilm, and as O_2 and organic matters diffuse in it, microbial metabolism renders it into CO_2 and NO_2.

Ultimately the slime gets thickened and it is sloughed and settles in settling tank(Fair *et al.*, 1968)

Bacteria that are contained in microbial film, *viz.* Flavobacterium, pseudomonas, alcaligenes, and algae like ulothrix, phormidium, chlorella and fungae like yeasts.

Maximum removal rate occurs when slime thickness is 70–100 um.

Milk processing, pharmaceuticals and paper mill wastes are treated in trickling filters.

Rotating Biological Contactor (RBC)

It is also a film-flow bioreactor.

Formation

Microbial slime film is built upon a partly submerged support medium. The medium is of a series of light weight disks of polythene/PVC mounted on a horizontal shaft in a tank through which waste water is allowed to flow. Rotation of the shaft is slowly done by a low speed motor. RBCs can be linked in parallel or series.

RBC was first constructed with circular steel disks and evaluated on experimental basis in 1925(Domain,1929).

The design and operation of plant for this purpose has been reviewed by Antonie(1976) and the factors controlling the performance by Poon *et al.* (1979).

Factors Effecting the Treatment Performance

1. Rotational speed
2. Wastewater retention time
3. Staging
4. Temperature
5. Disk submergence.

Application

This type of plant is widely used in the treatment of:

1. Municipal wastewater
2. Industrial like pulp, textile, meat and vegetables

Its efficiency has been noticed in the removal of phenols from the industrial effluents by the use of suitable microbes (Banerjee,1994).

Use of Biotechnological Tools in Water Pollution Monitoring

Bioassay

It is not just a tool, but a very important criterion to allow/disallow the discharged of a toxic substance (Chatterjee, 2003).

It has now been recognized that no water pollution studies are complete until biological aspects are included (James and Evison,1979; Peakall,1994).

Biomonitoring or screening of xenobiotics in the environment with living taste systems is one of such facet (Maltby and Caloy,1989).

The technique of bioassay largely encompasses three types of criteria, *viz.*, symptoms of visual damage, genotoxicity test and assessment of the responses at the metabolic levels (Wang and Williams, 1988).

Visual Rating

At the initial stages of bioassay (Doudoroff *et al.*, 1951), it was largely based on visual rating only. Hence, the concept of LD_{50} or LC_{50} emerged for fishes, *i.e.* the dose at which 50% of test organism is affected. Computer programmes for the determination of these values have also been developed (Trevors, 1986). For this purpose the growth rate, productivity, phenology are considered. Presence or absence of a particular species in a polluted environment (indicator) is also another good criterion (Austin and Deniseger,1985). No single indicator species can fulfilled all the purposes equally well hence multispecies testing concept has emerged (Cairns, 1985).

Genotoxicity Test

Damage at the cellular and sub cellular level is assessed through this process. Since cells are the ultimate sites of interaction between chemical and biological system, cellular changes due to the chemical effects and the ultimate changes in the metabolism, morphology of the organism, species or population level are assessed by this toxicity assay.

Cytotoxic tests depend on the estimation of chromosome damage including breakages, sister chromatid exchange or micronuclei counting (Wolf,1997; Sharma,1987; Grant,1993; Sanat Kumar and Chatterji,1993)

DNA probe is being considered as a parameter, particularly in the identification of pathogens in water samples.

Metabolic Rating

Biomonitoring involves both qualitative and quantitative assessment of the physiological and biochemical mechanisms of metabolic processes. They include changes in the contents of chlorophyll molecule and its fluorescence kinetics (Lichtenthaler and, Rinderle, 1988; Alam *et al.*, 1996), contents of soluble protein; nucleic acids; changes in the activities of key enzymes, their reaction kinetics and electrophoretic assessment of isoenzymes (De *et al.*, 1985; Mukherji, 1990) or the induction of some stress proteins (Grill *et al.*, 1985).

Usefulness of Bioassay

Plant Test Aystem

Due to the sensitivity, easy availability and culturing facilities in limited areas algal bioassay is fairly used among plant test system. This is considered as a good indicator of pollution.

1. Algae commonly used as test material are Chlorella, Navicula, Spirulina, Anabaena, Ulva, and Codium etc.
2. Cyanophyceae (blue-green algal group) is a good bio-indicator of pollution. Dominance of Cyanophyceae in the phytoplankton community of a water body indicates its eutrophic nature (James and Evision, 1979).

3. For the detection of faecal pollution in potable water, bacterial bioassay is a good method. It is basically done by popular "coliform" test. Bacterium Salmonella has been used in Ames Test to detect atmospheric mutagens with the help of different mutant strains(Ames *et al.*, 1975).
4. Bacterial bioluminescence is being used as a good indicator in the analysis of atmospheric gases and other compounds like SO_2.
5. Lichens (represents commensalisms between algae and fungi) dwindle in areas where the SO_2 concentration in the air is relatively high (Ferry *et al.*, 1973).
6. A number of forest and aquatic mosses such as Bryum, Stereophyllum, Frontinalis and others have been used to monitor metal pollution in rivers by measuring their shoot tip growth (Say *et al.*, 1981).
7. Aquatic ferns like Azolla(Sella *et al.*, 1988) and Salvania (Chatterji *et al.*, 1995) have shown promise in metal pollution monitoring.
8. Latest and most important biochemical parameter added to the list is the detection of pollution –induced small peptides within the cells called Phytochelatinsin plants and Metallothioneins in animals (Grill *et al.*, 1985; Steffens,1990).

Animal Test System

1. Among the faunal species, toxicity to fishes have been considered since 1950s (Doudorioff *et al.*, 1951) as a good criterion of xenobiotic bioassay to provide the basis for strategy formulation in the release of toxicants.
2. In the monitoring of phenolic compounds, pesticides, waste of tanneries and textile industries, toxicity assessment of fishes is a basic tool.
3. Toxicity assessment has been done on morphology, behavior, changes in different organs, muscles and metabolisms well (Chatterji 2003). AchE, a neurotransmitter is a good marker of pesticide pollution (Kumar *et al.*, 1995).
4. Common fish types in bioassay are Catla, Labeo, Channa, Clarias, and Tilapia.
5. In protozoa Ciliates are good bioindicators (Ricci, 1995).
6. Zooplanktons like *Daphnia monga* and other crustacean species are well known for evaluating toxicant effects in aquatic system.
7. Rotifers are a kind of helminthes among the zooplanktons and grow mainly among aquatic vegetation. They constitute another class of fauna, which have proved to be good indicators of saprobity and trophic levels of water (Sladecek, 1995).
8. Cairns (1985) has advocated multiple species toxicity testing– some species within a genus may behave as more responsive indicators than others.

Cell Biology in Environment Monitoring

Cell biology with other allied branches of biology coupled with instrumentation have broaden the scope of screening environmental mutagens and carcinogens.

Since 1940s, various test systems have been developed using microorganisms, plants, laboratory animals and human cell cultures to assess the toxic mutagenic potency of numerous chemicals and radiations (Sharma, 1984, Zimmerman and Taylor-Meyer, 1985)and even "living mutages" or virus and bacteria-induced mutation (Manna, 1982).

Ames Test

It involves the utilization of a series of the bacterium *Salmonella typhimurium* with the specific ability to detect certain types of chemical mutagens. It is a widely used preliminary short-term testing and screening of drugs, agricultural chemicals, cosmetics, food additives, inorganic metals and other pollutants (Tazima, 1982).

Principle

Based on the frequency of reversion of mutations at the histamine locus (His^- to His^+).

Cytogenic Assay

In the higher plants and animals where there is longer life-span this process is a criterion of assessment of mutagenecity or carcinogenecity.

Among some other plants Pea, maize, soybean are used in screening and monitoring of pollutants and other harmful elements in the environment.

For mutagenecity testing "stamen hair" assay and micronucleus test are ideal.

The stamen hair test is highly sensitive to gaseous chemicals like Feron-22, Benzene, Ozone, NO_2, SO_2 and they serve as indicator by changing their color (Grant, 1993).

This can be done by the following ways –

Chromosome Damage

This may be done by fragmentation of chromosome, bridge formation and disruption of the normal cell division.

Micronucleus Test

This is now routinely used for screening mutagenic compounds due to relative simplicity and rapidity (Schimid, 1982). It involves large-scale fragmentation of chromosome.

Sister-chromatid Exchange

This is an effective parameter of cytogenetic assay (Latt, 1973; Wolf, 1977). The principle is based on the damage caused to DNA molecule and misexchange of chromosome segments.

Membrane Damage

Since cell membrane is the first structure encountered by the toxic agents, structural modification of the membrane and change in its permeability is the measurement factors of the toxicity.

Loss of semipermiability of the membrane can be measured by the leakage of the enzyme LDH or through the uptake of trypan blue. Fragility of lysosome can be assayed by neutral dye. It can be used as biomarker of cell viability(McCarthy and Roch, 1983).

Molecular Biological Tools in Environment Monitoring

It can be done by the following ways:

DNA Probe

It is used widely in biotechnological studies. It can be done through the following different steps:

1. Extraction, separation and isolation of DNA fragments.
2. Denauturuation of double stranded DNA into single strand structure and thus preparation of DNA probe.

3. Labelling of the probe.
4. DNA hybridization.
5. PCR technique for DNA amplification.

Using both DNA probe and PCR technique, it is now possible to detect bacteria and viruses for water quality monitoring up to 1/100 ml approximately (Chatterji, 2003).Bej *et al.* (1990) could identify coliform bacteria (*E. coli*) and enteric pathogens like Shigella and Salmonella in water through the use of lac Z and lam B gene analysis.

Immunoassay

It is based on antigen- antibody reaction. Immunological techniques are widely used in the toxicity analysis of pesticides/herbicides and also in the identification of different pathogens of food and water.

Mainly two types of immunological techniques are widely applied:

Enzyme immunoassay (ELA)–here an enzyme is used as marker *e.g.* Glucooxidase Alkaline phosphatase etc.

Latex agglutination test–it is a simpler process. Here, polysterene latex particle are coated with antibodies.

Assay have been developed for cholera toxin, Salmonella, Listera and Yersinia species (Behrens and Kramer, 1990; Vaheri *et al.*, 1991).

Immunoassay technique can detect pesticides concentration much lower in water than other conventional techniques.

Lux Reporter Gene Technology

It is a completely new concept and is based on plasmid encoded certain gene sequences that produce assayable flurogenic signals whenever the genes for catabolism are expressed in luminescent bacteria like Vibrio or Photobacterium (Burlage *et al.*, 1994).

It is very much sensitive in monitoring ground water and surface soil.

Biosensor in Detecting Environmental Water Pollution

Biosensors are biophysical devices, based on sensing response of biological materials (sugars, proteins, hormones in human body) in combination with an electro-chemical transducer to convert the biological reaction into a digital electronic signal. The digital signal is proportional to concentration of the biological material to be tested.

It has a great specificity and accuracy in detection. The concept of biosensor is usually attributed to Clark and Lyons (1962). The enzyme electrode was christened by Updike and Hicks (1971).

Different parts of the biosensor are:

1. A biological sensing agent— it may be of different kinds like–enzymes, immune-agents, lectin, DNA, whole microbial cell etc.
2. Physical component—it is a transducing elements. The function of this is to convert biochemical signal into electrical or optical signal.
3. The interface—it is a polymeric film or membrane. It links the biological part to the transducer.

Sensors work on the principle of gas electrodes.

Types of Biosensor

1. Enzyme electrode
2. Immobilized cell biosensor
3. Optical biosensor
4. Ion-sensitive field-effect transistor
5. H_2O_2 biosensor
6. Microbial biosensor and
7. Gas phase biosensor.

Application of Biosensor

Biosensors have a wide use in water resource development and also in other environmental monitoring processes. Studies are going on to find out the different new applications of these modern technique in controlling environmental pollution.

1. To measure the organic pollution BOD_5 is a widely used test. Conventional test of BOD_5 takes five days while BOD sensor takes only 15 minutes. Here, Trichosporon is used as O_2 probe.
2. Pesticides-specific antibodies have been utilized to detect (Vanderlon *et al.*, 1988) the presence of herbicides like triazines and carbamates in ground water with the sensitivity of 0.13 to 0.7 mg/l within 16 minutes.
3. Heavy metals can be detected and estimated using enzymes with –SH group. Cu^{2+} is detected using enzyme tyrosinase.

Biotechnological Tools in Water Pollution Abatement

Abatement of pollution is well done by the different biotechnological methods particularly in aquatic system.

Reduction of CO_2 Emission

For the doubling of CO_2 level in the atmosphere global temperature can rise by around 1.5-5^0 C (Bolin *et al.*, 1986). It can be controlled biotechnologically by increasing the rate of photosynthesis.

Biotechnologically photosynthetic CO_2 fixation is improved. Enzyme RUBP-case is responsible for CO_2 fixation. It has two subunits (large and small) controlled by two genes, one is nuclear and another is cytoplasmic.

Genetically manipulation of these genes can control CO_2 absorption.

1. Microbial photosynthesis is an efficient method for CO_2 reduction, due to their high rate of photosynthesis.

   It has been estimated that the amount of CO_2 released from a power plant of 150 MW capacity would require an algal culture volume of 150000 m^3.

   (a) Using gene cloning technique it may perhaps be possible to develop strains with faster growth rate to yield larger biomass (Borowitzka and Borowitzka, 1988) of the concerned algae.

2. Biological calcification can reduce CO_2 emission in sea water.

$$H_2O + CO_2\,(aq) \text{——} H_2CO_3$$

$$H_2CO_3 + Ca_2 \text{———} CaCO_3 + H_2O + CO_2$$

This is achieved with the help of calcifying organisms like corals and a number of calcareous green and red algae (Dubnisky, 1991).

Algal Photosynthesis in Wastewater Treatment

Wastewater can be categorized in to different types according to their sources. It can be municipal, agricultural or industrial in origin. Depending on their sources water contents are also varied. Since physical process of wastewater treatment cannot remove major portion of organic matters, rather it needs mineralization of water *i.e.* biodegradation.

Biodegradation is basically done by bacterial oxidation.

Hermann and Gloyna, 1958 first showed that micro-algae can supply oxygen that is needed by the bacteria and protozoa in the process of biodegradation.

In the treatment of sewage algal treatment is done as secondary treatment where biooxidation is done.

In the biotreatment of waste water "green algae" like Chlorella, Euglena, Chlamydomonous, Ulothrix, Tribonema are good oxygen donor. Oxygen donation depends on the steady supply of CO_2 and also on the pH of the pond. 10-20 ppm of CO_2 is optimum for efficient O_2 donation. Moreover, through this algal treatment it is possible to precipitate some of the toxic substances like heavy metals from wastewater. This is due to the high negative charge on algal cell-surface against the strong positive charges of heavy metal cataions which get absorbed and settle out (Rai *et al.*, 1981).

After the purification of wastewater, the purified water is also the source of different by-products. Algal cell rich in protein could serve as food and feed or be used as manure.

Single Cell Protein (SCP)

The term SCP was coined at Massachusetts Institute of Technology by a group of scientists in 1966.But the term is quite misleading as it is "any microbial biomass from both uni-and multicellular bacteria, yeasts, filamentous fungi or algae which can be used as food or feed additives".

Biological Treatment for Eutrofication and Algal Bloom

Eutrofication is the stage at which water is enriched with undesirable organic or inorganic nutrients specially phosphorus and an amount of nitrogen. Undesirable excessive amount of these elements lead to the algal bloom. Algal bloom increases BOD of the water body by decreasing the amount of dissolved oxygen. Approximately 20 algal species are capable of forming bloom. Among them Cyanophyceae are predominant. Storage of phosphorus as polyphosphate in algal cell can favour algal growth when it is above 6 ppm.

Algae are considered to be a biological means of N_2 and P_3 removal, which in turn also helps checking eutrofication (Doran and Boyle, 1979).

Algae can be controlled chemically in eutrofic water. Algicides like copper sulphate, sodium arsenate can be used.

Removal of Phosphorus from Wastewater

Phosphate metabolizing bacteria can remove phosphate biologically. The energy required for this is made available by the release of phosphorus bound as polyphosphate in volutin granules in the bacterial cytoplasm (Comeau *et al.*, 1985).

Biotechnological Tools in Combating Metal Pollution

Heavy metals are a considered as a major field of research as they are increasingly polluting the environment *i.e.* air, water and soil. Some of the heavy metals as dangerous that they can even cause cancer. Cd has now been black-listed with Pb and Hg in this respect (Berry and Burank, 1972).

Toxicity of the metals is not only dependent on concentration and exposure, but is also known to increase with electropositivity of metals, *viz.* Hg > Cd > Zn.

Some of the bioprocesses can reduce or check the metal pollution.

Bio-methylation

it involves the transfer of methyl group from organic compounds to metals primarily due to the microbial activities in the soil and water. Through this process some metal ions can be detoxified, yet others turn into more toxic *e.g.* E. coli for Hg and Pseudomonas for Cd and Pb.

Bioscavenger

Many of the aquatic plants, including phytoplankton and benthics are gradually showing up their unique sorption potential of metals from the banthic media and thus act as a natural bioscavengers.

(a) Biosurveillance of toxic metals is still in the informative stage and is expected to turn out to be a real cost-effective (about 50-70 per cent) device in the long run.

Natural Scavengers (Aquatic Macrophytes)

Water Hyacinth (Eichornia crassipes)

It has the capacity to absorb Cd from the bathing medium at the rate of 9.1ug/l and to help build up metal concentration at the tip of the plant at 6.1ug/g of dry matter within 24 hours was first shown by Wolverton and McDonald, (1979).

Chigbo *et al.* (1982), on the other hand, have shown the remarkable capacity of the plant to absorb relatively high concentration of metals like Cd, As, Pb and Hg over a short period of time.

Duckweeds

Same as like water hyacinth, members of the duckweeds family (Lemnaceae) high promise of heavy metals sequestration.

Chatterji and his group noted that Spirodela polyrrhiza could effectively remove Pb and As from the aquatic medium.

These plants are small in size but have very strong in-built capacity of resistance against metals. Hyacinth and duckweeds are widely used in wastewater treatment containing metallic effluents.

Water Lettuce (Pistia stratiotes)

This weed is common in tropical environment. It is a good metal scavenger. Recently (Basu *et al.*, 1999) noted the capacity of arsenic removal by Pistia up to 82% in 6 days under lab condition.

Apart from all these all these plants Hydrilla, Vallisneria and Potamogeton have indicated their bioaccumulation capacity of toxic metals (Jana and Choudhury, 1984; Chatterji and Nag, 1991).

Planktonic or Benthic Microphytes and Microbes

Ferns

The role of two fern genera Salvinia and Azolla are important in this regard. Two species of Salvinia can uptake metals like Cr, Ni and Pb.

Azolla show more accumulation of metals like Cu, Cd and u in their roots than in shoots.

Microbes

Trollope and Evans (1976) noted concentration of metals in freshwater "Algal bloom". Green algal genus Chlorella has been noted to accumulate metals like Cu, Hg and U (Verma and Dutt, 1988). C. pyrenoidosa is used in recovering metals from sewage sludge (Wrong *et al.*, 1984).

Among the marine green algae Fucus, Ulva, and Codium have shown great role in minimizing the metal pollution load of rivers (Barnet and Ashcroft, 1985).

Selective accumulation of heavy metals is also done by some fungal genera like Aspergillus, Pennicillium, Rhizopus, Mucor and Neurospora etc (Nakajima and Sakaguchi (1982).

Bacterial cell-wall has a major role in the accumulation of metals. Beveridge and Murray (1980) have shown the site of metal deposition in the bacterial cell- wall. Actinmycetes is an important group of bacteria in this respect.

Metal Chelation and Detoxification

The uptake of xenobiotic metals could be either through adsorption and/or absorption and is possible through some physiological adaptiveness and homeostasis (Woolhouse, 1983). This phenomenon can be categorized in the following steps–1. Metal binding on the cell wall: Cell wall acts as the first barrier of the metals in the process of encountering with the cell. Heavy amount of cataions get bound with the cell walls at the cataion binding sites of cell wall.

Metal Binding

With cell wall is rather common in lower group of plants, particularly fungi, bacteria (Gadd and White, 1992) and algae (Jensen *et al.*, 1982). Higher plants too have been reported to "fix" such metals in the cell wall components and thus help in detoxification (Wierzbieka, 1987)

Compartmentalization

This is an another balanced device for plants to combat against metal toxicity. Plants can transport metals to various cellular compartments. This has been proved in the case of Zinc tolerance.

Cell Immobilization in Wastewater Treatment

Biotechnological process can use biocatalysts both in the form of specific enzyme or as the whole cell.

Either the naturally occurring biodegrading microbes or their genetically improved counterparts (GEMs) can be maintained in substantially "unchanged" form to have continued expression of biological activities and effectivity inside bioreactors through this novel technique (Linko and Linko, 1984).

Technique of Immobilization

It involves "fixation" of cells or enzymes on to a support system. This can be achieved by entrapment and/or attachment (Klein and Vorlop, 1985). Induced immobilization is induced through either physical or chemical means.

Advantage of Cell Immobilization

It offers great advantage over free cell suspension

1. Recoverable and reusable,
2. Avoid wash out of cells during operations,
3. Protection against extreme pH and temperature of the medium,
4. Possibility of easy separation from bathing solutions or reactors,
5. To have a continuous process operation in biological reactors.

Immobilized Microbial Removal of some Toxicants (after Fletcher, 1992):

Sl.No.	*Immobilized Microbes*	*Support*	*Substrate*
1.	Cryptococcus elinovii	Activated carbon	Phenol
2.	Pseudomonus putida	Sintered glass	Phenol
3.	Enterobactor, Alcaligenes spp.	Ca-alginate	4-chloro-2-phenol
4.	Anthrobactor spp.	Polyacrylamide	Triethyl lead
5.	Thiosphaera spp.	Agarose	Ammonia
6.	Rhizopus arrhizus (fungus)	Polyacrylamide	Cu-effluent

In wastewater treatment many pollutants like phenolics, cyanides, nitrates, ammonia, chlorinated compounds and heavy metals are treated by this technique.

Biodegradation as a Tool in Biotechnology

Some microbes are can transform organic compounds biologically. Biodegradation is a good tool in sewage treatment Degradation depends on the chemical properties of organic compounds. Such as:

1. Aliphatics are degraded easily than the aromatic compounds.
2. Water-soluble compounds are easier to degrade than isoluble forms.
3. Molecular orientation of aromatic compounds is a factoring biodegradation.
4. Halogen, nitrogen and sulfonate substitution inhibit biodegradability.

Some of the microbes which can degrade various chemicals are (Brown *et al.*, 1987)

Sl.No.	*Chemicals*	*Microbes*
1.	Hydrocarbon	Pseudomonas, Arthobactor, Mycobacterium.
2.	PCBs	Pseudomonas, Candida, Alcaligens.
3.	Polycyclic aromatics	Arthobactor, Nocardia, Pseudomonas.
4.	Napthalene	Pseudomonas, Nocardia
5.	Organophosphates	Pseudomonas
6.	Benzene	Mycobacterium, Alcaligens

Biological degradation can be of two types:

Aerobic

1. Here, substrate is used as the source of carbon and energy.
2. The extent of degradation is associated with the O_2 consumption.
3. Two enzymes are di- and mono-oxygenase.
4. Peroxidase is another class of enzyme. It has ability to degrade lignin.

Anaerobic

1. This process is slow and needs long retention time.
2. It is more advantageous than aerobic process as it is independent of O_2.
3. Materials like fats and cellulose can degrade by this process.

Three temperature ranges are used in anaerobic digestion-

1. Cold digestion at about 20° C
2. Mesophilic digestion at 20°-40°C
3. Thermophilic digestion at 40°-55° C

Sequential degradation is another technique where both aerobic and anaerobic processes are used sequentially. Since it is done in the case of wastewater treatment (pulp and paper industry) reduction of BOD is taken care of in each stages.

Biodegradation of polycyclic aromatic hydrocarbon like naphthalene phenanthrene etc. can be done by Pseudomonas spp. Incubated Pseudomonas at 30°C gives optimum result.

Biohydrometallurgy and Biomining

Biometallurgy is a new discipline of mineral science and is interdisciplinary in nature where microorganisms are used in the solubilization, prospecting and exploration of mineral deposits (Ehrilich, 1981, Brooks, 1983; Troma, 1987).

Metal recovery by microbes may involve two processes, *viz.* bioleaching (Brierley, 1978) and biosorption.

Bioleaching

Basically it is a process of microbial extraction or solubilization of minerals from ores by their metabolic activities.

For this leaching process the solution contains live bacteria particularly of chemolithotropic and heterotropic type. The solution also contains different nutrients like ferric ions, dil. H_2SO_4, organic acids, proteins polysaccharides and other metal-binding agents that are metabolic product of bacteria.

Most popularly used bacteria in bioleaching is Thiobacillus ferrooxidans. It can precipitate iron from the ore that can create problem in the processing of metals.

Metals like nickel, cobalt, and silver, gold, arsenic etc. can be extracted by this technique.

Biosorption

It involves microbial cell-surface adsorption of metals from mine wastewater. It offers alternate possibility for detoxification and recovery of valuable but toxic metals (Zajic and Chiu, 1972).

Biosorbent based granules have been developed for wastewater treatment and metal recovery (Brierley *et al.*, 1986).

Genetically Engineered Microbes in Treatment of Waste

Microbes have their own sets of genes, which can provide them protection against hostile environmental factors like different pollutants. These genes can reside on the linear DNA of the bacteria or on the circular extra chromosomal DNA *i.e.* plasmid.

It is now known that in bacteria, majority of the degradative genes for organic matter are primarily located in plasmids (Timmis and Puller, 1979), rather than chromosomal DNA and biodegradation of organics or hazardous wastes is essentially a plasmid-mediated process (Broda *et al.*, 1981; Plemberton, 1983).

Gene transfer is possible among microbes with wide range of gene distribution. This was revealed from the observation of "antibiotic resistant" plasmid. It was first noted in Shigella and E. coli and was found to be related to transferable plasmids (Watanabe, 1963). Plasmids can be transferred among various species of bacterium by three processes *viz.* conjugation, transformation and transduction. Pioneer work in the making of GEMs was done by A.M. Chakraborty (Chatterji, *et al.*, 1981) in the use of Pseudomonas putida as a biodegrader of petroleum hydrocarbon in spills (Super bug). Now by the use if recombinant DNA technology plasmid with specific gene can be inserted in to the host genome and can multiply thus confer desired resistance.

Potential GEMs with their substrate (Chatterji, 2003):

Sl.No.	GEMs	Substrate
1.	Pseudomonas putida	Mono- and dichloroaromatics
2.	P. oleovorans	Alkanes
3.	P. diminuta	Parathion
4.	P. cipacia	2, 4, 5-T
5.	Alkaligenes sp.	2, 4-D
6.	Acinetobactor	4-chlorobenzene

To avoid the risks involved in the release of GEMs, the following steps may be adhered to (Beringer and Bale, 1992):

1. Gain maximum information about the donor and the recipient organism.
2. Genetic exchange schedule should be highly specific.
3. Recipient organism should be locally adapted.
4. Proper marker character should be there in the altered organism to trace it in the environment.
5. Constant monitoring of its fate and interaction in the new ecological niches is absolutely necessary.

Biotechnological Tool in the Conservation of Biodiversity

Ex situ and *in situ* conservation of both economically important plants and animals for sustaining aquatic biodiversity and water quality, are the major areas that can cover by biotechnology. Rich biodiversity of marine and fresh water ecosystem is greatly supported by different biotechnological tools particularly in the conservation of endangered species. Germplasm conservation, gene bank are the developed modern technique by which both ex situ and in situ conservation can be done.

Biofertilizer

These are the organisms, which can bring about soil nutrient enrichment with their biological activities. These are divided in three types:

1. N_2 fixing bacteria
2. N_2 fixing cyanobacteria
3. Fungi or mycorrhiza.

N_2 Fixing Bacteria

It may be symbiotic or non-symbiotic type.

1. Symbiotic-Rhizobium, Franka (found in the roots of non-symbiotic plants), Xanthomonas, Mycobacterium.

   Non-symbiotic–Azotobactor, Beijerinckia and Bacillus polymyxa etc.
2. Cyanobacteria- it may be symbiotic or non-symbiotic. These have the potential to fix 7 to 8 lakh tones of atmospheric N_2 per year.

   Anabaena is the symbiotic bacteria.

   Aulosira ferlissima is the most active free-living N_2 fixing cyanobacteria of rice field.
3. Mycorrhiza is the symbiotic association between the fungus and roots of higher plants.

The detoriation of environmental health is a fact now whether it is slow or high in speed. The traditional and modern physio-chemical technologies are of high cost and not always environment friendly. So, now it is an essential to apply biotechnological tools to combat against different environment hazards. Although biotechnology has already shown promises in the environmental problems still enough scopes are there for further development.

REFERENCES

1. Dart, R.K. and Sretton, R.J.(1994) **Microbial aspects of pollution control**, Suravi publication 118-157
2. Foster, R. (1980) **Water Pollution Control** 79:28.
3. Rao, C.S.(1991) Environmental Pollution Control Design, Wiley Eastern Ltd., New Delhi
4. Lowries,P. and.Wells,S. (1941) **Microorganisms, Biotechnology and Disease**, Cambridge University Press, Cambridge.
5. Doman, J. (1929) Result of operation of experimental contact filter with partially submerged rotating plates. **Sewage Works Jour.** (3)555.
6. Banerji, G. (1994) PhD. Thesis, IIT, Kharagpur, India.
7. Chatterji, A.K. (2003) **Introduction to Environmental Biotechnology**, Prentice/Hall of India Private Ltd. 60-63
8. James, A. and Evison, L. (Eds) (1979) **Biological Indicator of Water qualit.** John Wiley and Sons, New York.

9. Maltby, L. and Calow, P. (1989) The application of bioassay in the resolution of environmental problems: past, present and future. **Hydrobidioga** (188/189) 65-76.

10. Wang, W. and Williams, J.M. (1988) Screening and biomonitoring of industrial effluents using phytotoxicity tests. **Environ, Toxicol. Chem.** (7) 654-662.

11. Doudroff, F.P., *et al.* (1951) Bioassay for the evaluation of acute toxicity of industrial wastes to fish. **Sewage and Industrial** (23)1380-97.

12. Trevors,J.T. (1986) A basic programme for estimation of LD_{50} values using IBM-PC. Bull. **Environ. Contam. Toxicol**. (37) 18-26.

13. Austin, A. and Deniseger, J. (1985) Periphyton community changes along a heavy metal gradient in a long narrow lake. **Envtl. And Exptl., Bot.**(25) 41-52.

14. Sharma, A.K. (1987) Cell biology and monitoring of environment. In: **Pespectives in environmental management**. T.N. Khoshoo (Ed),Oxford and IBH, New Delhi, 228-233.

15. Grant, W.F. (1993) Plants as detectors of atmospheric mutagens. In: **Handbook of hazardous materials,** Morton Corn (Ed.), Academic Press, New York.

16. Lichtenthaler, H.K. and Rinderle, U. (1988) The role of chlorophyll fluroscence in the detection of stress conditions in plants. **CRC critical review of analytical chemistry** (19) 529-585.

17. Alam *et al.* (1996) Screening of suitable biomarker for monitoring aquatic Cr (VI) pollution in Spirodela polyrrhiza. **Proc. acad. environ. bio**. (5) 177-181.

18. De *et al.* (1985) Studies of toxic effects of Hg (II) on Pistia stratiotes. **Water, air and soil pollution** (24) 351-360.

19. Mukherji, S. (1990) Changes in some physiological and biochemical characters in a few plants as bioindicators of atmospheric pollution and selection of protection species. In: **Recent res. In ecol.,env. and pollu.** (4) 153-170.

20. Grill, E., Winnacker, E.L. and Zonk, M.H. (1985) **Science** (230) 674-76.

21. Ferry, B. W., Baddely, M.S. and Hawksworth, D.L. (1973) **Air pollution and Lichen,** Althlome press, London.

22. Kumar, S., Thembre, M. and Satyadevan, S. (1995) Acetylecholine esterase inhibitory kinetics for monitoring pesticide toxicity to fish, In. **Pollution and biomonitoring,** B.C. (Ed.), Tata-McGraw-Hill, New Delhi, 426-446.

23. Ricci, N. (1995) Protozoa: Biomonitors of environmental monitoring and pollution control. In: **Pollution and biomonitoring,** B.C. Rana (Ed.) Tata-McGraw-Hill, New Delhi.

24. Sladecek, V. (1995) Rotifers as test system for pollution monitoring. In: **Pollution and biomonitoring,** B.C. Rana (Ed.) Tata McGraw-Hill, New Delhi.

25. Zimmerman, F.K. and Taylor-Meyer, R.E. (1985) **Mutagenecity testing in environmental pollution control**. John Wiley and Sons, New York.

26. Manna, G.K. (1982) **Curr. Sci**. (51) 1087

27. Tazima, Y. (1982) A brief sketch of environmental mutagen studies in Japan. In: **Environmental mutagens and carcinogens,** T. Sugimura et al. (Eds.), University of Tokyo Press, Tokyo, 91-100.

28. McCarthy, J.F. and Roch, M. (1983) **Biomarkers of Environmental Contamination,** CRC Press, Boca Raton, USA.

29. Bej, A.K., Steffen, R.J., Dicesare, J., Haff, L. and Alas, R.M. (1990) Detection of coliform bacteria in water by PCR gene probes. **Appl. Envtl. Microbiol.** (56) 307-314.

30. Behrens, D. and Kramer, P. (Eds.) (1990) **DECHEMA Biotechnology Conf.** Vol. 4(A), Frankfurt, Germany.

31. Vaheri, A., Tilton, R.C. and Barlows, A. (Eds.) (1991) **Rapid methods and automation in microbiology and immunology Springer–Verlag**, Barlin.

32. Burlage, R.S., Palumbo, A.V., Heitzen, A., and Sayler, G.S. (1994) **Appl. Biochem. Biotechnol.** (45\46) 731.

33. Clark, L.C. Lyons, C. (1962) Electrode system for continuous monitoring in cardiovascular surgery. **Annals N.Y. Acad. Sci.** (102) 29-45.

34. Rai, L.C., Gaur, J.P. and Kumar, H.D. (1981) Phycology and heavy metal pollution. **Biol. Rev.** (56) 99-151.

35. Doran, M.D. and Boyel, W.C. (1979) P_3 removal by activated algae. **Water Res.** (13) 805.

36. Comeau, Y., Rabinowitz, B., Hall, K.J. and Oldham, W.K. (1985) Phosphorus release and uptake in enhanced biological phosphorus removal from wastewater. Proc. 8th Sympo. On Waste water treatment, Montreal, 301-313.

37. Basu, A., Chatterji, A.K. and Mukherji, S. (1999) Arsenic reduction from aqueous environment by water lettuce, **Journal of Environmental Science and Health** (part A), USA-comminucated.

38. Jana, S. and Choudhry, M.A. (1984) **Water, Air and Soil Pollution** (21) 351-57.

39. Chatterji, A.K. and Nag, U. (1991) **Intl. J. Toxicol. Occup. and Environ. Hlth.** (1) 166.

40. Trollope, D.R. and Evans, B. (1976) **Environmental Pollution** (1) 109-16.

41. Verma, S.R. and Dutt, D. (1988) **Curr. Sci.**, 804-805.

42. Wong, M.H., Chu, L.M. and Chan, W.C. (1984) **Environmental Pollution** (34) 55-71.

43. Barnett, B.E. and Ashcroft, C.R. (1985) **Environmental Pollution** (B), (9) 193-213.

44. Nakajima, A. and Sakaguchi, T. (1986) **Appl. Microbiol. Biochem.** (24) 59.

45. Beveridge, T.Z. and Murry, R.G. (1980) **J. of Bacteriology**. (127) 1502.

46. Woolhouse, H.W. (1983) Toxicity and tolerance in renponses of plants to metals. **Encyclopedia of Plant Physiology**, New Series, (12C) 245-300.

47. Gadd, G.M. and White, C. (1992) **Jour. Chem. Tech. Biotech.** (55) 39-44.

48. Jenson T.E., Baxter, M., Rachlin, J.W. and Jam, V. (1982) **Environmental Pollution** (A), (27) 119-27.

49. Wierzbieka, M. (1987) **Can. Jour. Bot.** (65) 1851-60.

50. Linko, P. and Linko, Y.Y. (1984) Industrial application of immobilized cells. **CRC Crit. Rev. Biotech.** (1) 289.

51. Klein, J. and Vorlop, K.D. (1985) Immobilization technique for cells. **The Principles of Biotech**: Pergamon Press, Oxford, 191-201.

52. Ehrilich, H.L. (1981) **Geomicrobiology**, Marcel Dekker, New York.

53. Troma, A.E. (1987) Impact of biotechnology on metal extraction. **Mineral Processing and Extractive Metallurgy Rev.** (2) 289-330.

Advances in Biochemistry and Biotechnology Volume 1 *Pages 214-244*
Edited by: **Dr. Chiranjib Chakraborty**
Published by: **DAYA PUBLISHING HOUSE**

10

Induced Breeding Techniques for Aquaculture

M.A. Haniffa
Centre for Aquaculture Research and Extension, St. Xavier's College, Palayamkottai, Tirunelveli–627 002, India
E-mail: haniffacare@hotmail.com

ABSTRACT

India needs to frame important strategies for aquaculture through biotechnology, which will offer increased production and income generation. Biotechnology is the very basis of hi-tech aquaculture, which enables production of quality seeds. Seed requirements of hi-tech aquaculture are not only limited to the quality of the seed but also involve availability of large quantities of specific size at specific times. Genetic improvement in fish includes various disciplines such as selective breeding, hybridization, ploidy induction, nuclear engineering and gene engineering. Induced breeding is the basic and initial step and in recent years the break through in induced breeding has initiated the various disciplines of aquaculture biotechnology.

Hormone–Induced spawning has been attempted in various fish species since 1934. Surprisingly the same procedures with only minor modifications have been used to spawn an entire range of fishes from the ancient sturgeon and paddle fish to carps, catfishes, salmon, sea bass, red fish, snook, mullet, and murrels. Induced breeding experiments attempted in fishes using pituitary extract indicated that an increased blood gonadotropin hormones (GtH) which act on the ovaries and testes is a pre-requisite for ovulation. GtH act on the local ovarian mediators (steroids and prostaglandins) causing release of the eggs. Natural (pituitary extract, human chorionic gonadotropin (HCG), Leutinizing hormone releasing hormone analog LHRHa) and synthetic hormones (ovaprim, ovatide, Wova-FH) administered to various fish species produced good results. But the ability of the injected hormones depend on many variables viz; condition of fish, sexual maturity size of the fish and water quality parameters.

At Centre for Aquaculture Research and Extention (CARE) native fish species of different categories (threatened, vulnerable and endangered) have been collected from the wild and their maturity was assessed by external morphological characteristics and hand stripping. Natural and synthetic hormones where injected intramuscularly into the dorsolateral region of both males and females. Immediately after administering the hormones the breeding sets were released into cement tanks. The egg output, latency period, incubation period and percentage of hatching were estimated as a function of quality and dose of hormone. Breeding and spawning behavior were reported in murrels. No doubt induced breeding and larval rearing techniques under captivity are the viable aids in conservation and rehabilitation of the endangered species.

Introduction

The development of induced breeding techniques to induce reproduction in fish is one of the most significant advancements in Aquaculture. By induced breeding techniques, fish farmers profitably breed and culture the commercially important species. Fisheries is an important segment of our country. It helps not only in increasing food supply but also generating job opportunities and earning foreign exchange. As a result of concerted efforts by the Central and State Governments of India, fish production has been continuously been increasing reaching a record level of 5.96 million tonnes during 2000-2001. India ranks as the 4th largest producer of fish in the world and the second largest producer of freshwater fish after China.

The freshwater aquaculture resources of India are huge in terms of 2.36 million hectares of ponds and tanks and 1.07 million hectares of beels, jheels and derelict waters in addition to 0.12 million km of canals and 2.05 million hectares of reservoirs that could be put to different fish culture practices or even culture- based capture in case of large water bodies(1). To increase the production levels further and meeting the goals it will be imperative that newer technologies be applied for enhancing aquaculture production. Hence techniques such as sex control through hormonal and chromosomal manipulation(2,3,4) and breeding technologies(5) have been applied.

One of the most significant technological breakthrough in fish farming which has revolutionized freshwater aquaculture is the development of fish breeding and larval rearing techniques of Indian major carps. However while for Indian major caps the seed production technique has been extensively applied, that of catfishes has not yet found wider acceptance and large scale production facilities have not come up (6).

In fish the nervous and endocrine systems act together to coordinate reproduction. Environmental factors (photoperiod, temperature and rainfall) regulate reproduction by triggering internal mechanisms into action. *i.e.*, the resulting neural signals reach the hypothalamus and stimulate the pituitary through chemical messengers namely the releasing hormones (gonodotropin releasing hormone: GnRH). These releasing hormones stimulate the pituitary to release hormones (gonadotropins: GtH) which act on testes and ovaries through steroids, and prostaglandins for spermeation and ovulation (7) (Fig. 10.1). Those which spawn in wild and fail to respond in captivity, hormone induced spawning is the only reliable method to induce reproduction. Many variables *viz.*: condition of the

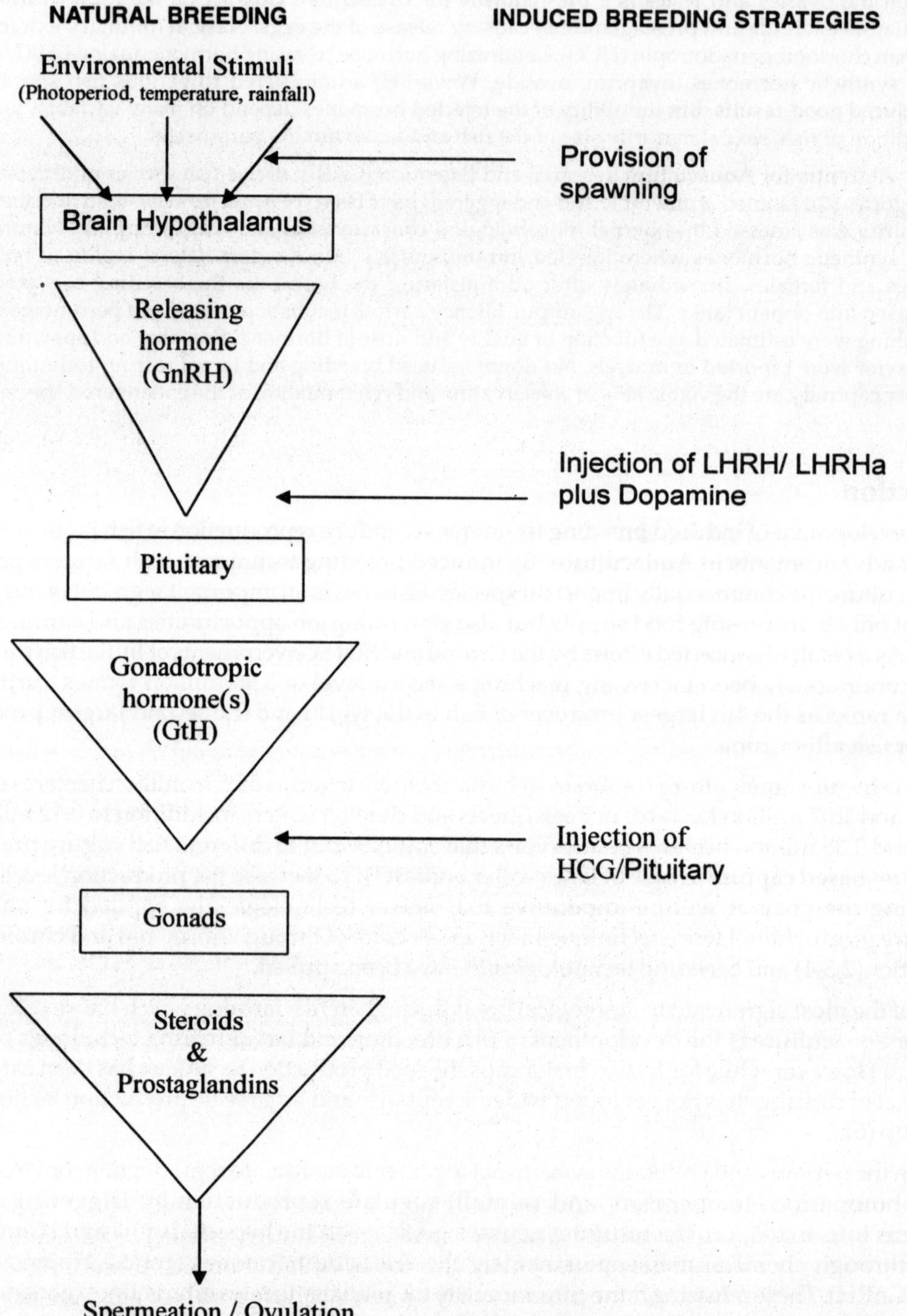

Fig. 10.1: Natural Breeding and Induced Breeding Strategies in Fish

fish, stage of sexual maturity, size of the fish, previous spawning history, water temperature and season of the year, influence the ability of the hormone to induce spawning (8).

Ovulating Substances

During the past two decades induced breeding by carp pituitary has been attempted on Indian major carps, Chinese carps murrels and catfishes. The ever increasing cost of donor pituitary and the cumbersome process forced experts to test alternative hormones such as ovaprim, ovatide, Wova FH, human chorionic gonadotropin (HCG) and luteinizing hormone releasing hormone (LHRH). Male and female steroids when administered over longer periods bring about regressive changes in the ovaries by virtue of their feedback effect on pituitary (9). Among the corticosteroids, mineralocorticosteriods (e.g deoxycorticosterone acetate-DOCA; 17- hydroxycorticosterone; 11-dehydrocorticosterone) play a better performance than glucocorticosteroids (*e.g.* prednisone; prednisolone; cortisone; hydrocortisone) in hypophysectomised catfish since they act directly on ovaries instead of pituitary. At present various ovulating substances such as clomiphene citrate, tamoxiphen, progesterone, methyl testosterone, oestradiol benzoate and testosterone propionate have been used to induce breeding in fish.

Maturity and Determination of Spawning

Seven stages of gonadal maturation in carps (10), 6 stages in murrels(11) and 6-7stages in catfishes (12) have been reported. The gonadosomatic index is more pronounced in females (more than 30) when compared to males. Sexually matured brood fish can be selected by morphological features. The dorsal surface of the pectoral fin is rough in mature male especially during the breeding season in the case of carps. In males vent does not protrude whereas in ripe females vent is bulged and reddish in colour. A more reliable method is drawing an egg sample from the posterior end of the ovary either by applying slight pressure or using a catheter and then observing the position of the nucleus under the microscope. Acentric or peripheral location of nuclei is a good indication of readiness of the fish for spawning (10).

Historical Resume

In early thirties, Brazilian fish-culturists first successfully implemented the technique of pituitary treatment. Houssay (13) of Argentina succeeded egg production in catfish *Cresterodon decemmaculatus.* Brazilian scientist Ihering(14), Russian scientist Gerbil'ski (15), and American scientists Hasler *et al.* (16) also attained success in induced breeding. Khan (17) of India failed in his attempts on spawning in Indian major carp *Cirrhinus mrigala* by the injection of mammalian pituitary gland. But Chaudhuri(18) succeeded in induced spawning in minor carp *Esomus donrius* by catla pituitary extract. In India the first success in induced breeding in carps (*C. mrigala, C. reba, Labeo rohita, L. bata* and *Puntius sarana*) was made at Central Inland Fisheries Research Institute, Cuttach by Chaudhuri and Alikunhi (19). Administration of pituitary extract is still widely practiced all over the world.

Induced Breeding Method

The most commonly adopted induced breeding method is hypophysation *i.e.*, injecting the mature female and male with pituitary extract of same species (homoplastic) or different species (heteroplastic). To gain access to the pituitary gland, the top of skull is chopped off exposing the brain. The brain is carefully removed and the pituitary may or may not have a stalk of attachment with the brain depending on the species, (Fig. 10.2). To avoid loss of water-soluble gonadotropins, care has to be exercised ensuring the gland is not damaged. Pituitary glands are collected from mature and gravid fishes and preserved in acetone or absolute alcohol in vials. If acetone drying is preferred, the glands

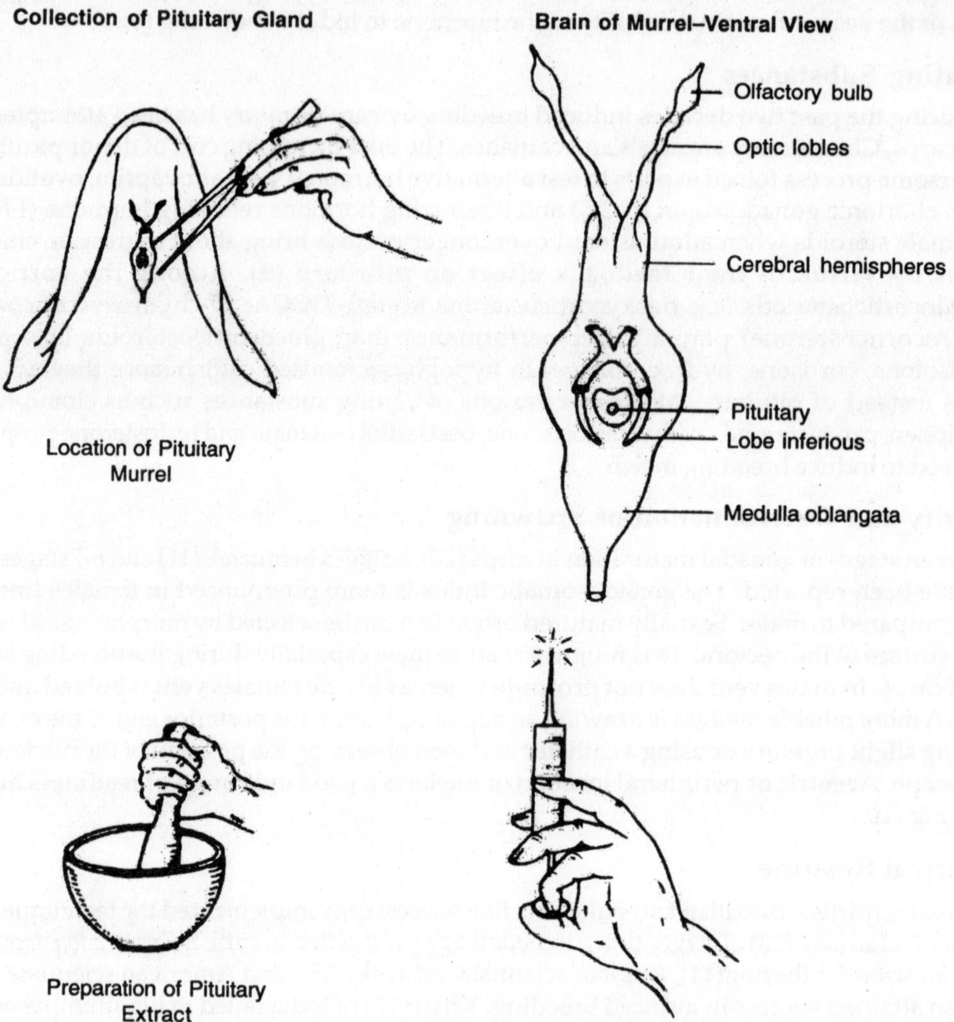

Collection of Pituitary Gland

Brain of Murrel–Ventral View

Olfactory bulb

Optic lobles

Cerebral hemispheres

Pituitary

Lobe inferious

Medulla oblangata

Location of Pituitary Murrel

Preparation of Pituitary Extract

Pituitary Suspnsion in a Syringe

Routes for Injecting Hormone

Fig. 10.2: Induced Breeding Technique in Murrel

must be placed in acetone in 10 times their volume and the latter should be replaced with new acetone after 12 hours. The acetone is again changed 8-12 hours later. After this the glands are air-dried on filter paper and stored in tightly sealed vials. The pituitary stored at room temperature may retain its potency upto a year but preservation using a refrigerator is known to retain its efficiency for more than three years (10). Since hormone content is the greatest in sexually mature fish, it is always preferred for collection of pituitary. Gonadotropin from pituitary extracts of phylogenetically remote fish (Heteroplastic) such as salmon, tilapia, catfish and even some marine fish has been found to be effective in carps (10), whereas major carps, chinese carps and goldfish are more likely to spawn successfully when injected with pituitary extracts from carp. In general the closer the donor species is related to the recipient fish, the greater the chance of successful induced spawning. At present common carp pituitary or salmon pituitary extracts are available in market.

With regard to Chinese and Indian major carps, a system of two injections the first one the stimulating (priming) dose (2-4 mg/kg weight) and the second one the resolving dose (8-12 mg/Kg) is practiced. The prime dose is usually given at 1600–1800 h and the resolving dose after 6 h. Injected brood fish (Chinese and Indian major carps) are released into the breeding tanks in paired sets (one female and two males whose combined weight equals to female weight). Injections may be administered either by intramuscular (<2 ml) or intraperitoneal (>2ml) routes. Intramuscular (IM) injection is given between the base of the dorsal fish and the lateral line at the base of pectoral fin or between the anal fin and lateral line (Jhingran and Pullin, 1985), whereas intraperitoneal (IP) may be at the base of either the pectoral fin or pelvic fin (Fig. 10.2).

Precautions and Care

It is better to choose live or freshly killed fish for collection of pituitary. Otherwise fish freshly preserved in ice or cold storage is preferable. Pituitary gland must be collected from sexually mature unspent fish. A relatively lower stage of maturity and cooler water may require greater dosages of injections(10). Care must be taken to avoid puncturing the heart while inserting the needle when intraperitoneal route is selected. After the withdrawal of the needle a gentle massage may help even distribution of the hormone into musculature. Backflow of hormone could be prevented by adding glycerine in the ratio of 7:3 (water: glycerine). Depending upon the size of the brooder and volume of fluid, the size of the syringe and diameter of the needle could be selected. Bruising, abrasion and scale loss must be avoided while handling since bacterial injection may ultimately lead to death of the fish. To avoid bacterial injection, the brooder could be given a dip in acriflavin (0.01%) or potassium permanganate before injecting the hormone.

Anesthetics such as Traicaine methane sulfonate (MS 222), chlorobutanol and quinaldine are available in the market. Since chlorobutanol is carcinogenic, it must be avoided. CO_2 bubbled through the water causes sedation and leaves no chemical residue but it is impractical. A cheaper anesthetic is 2-phenoxyethanol, the only liquid dissolves in water, which is used (200- 500 ppm) as an antibacterial and antifungal agent. Other anesthetics are metamidate hydrochloride (Marinil; Hypnomidate; 5-20 ppm) and benzocaine (100 g/l ethanol), which are also used as a topical human anesthetic (*e.g.* in throat lozenges). A role of cotton filled cloth soaked in 0.04 M solution of MS 222 (Sandoz) and quinaldine (50 –100 mg/l water) inserted into the mouth of the fish is recommended while weighing, biopsy and injection. Metomidate hydrochloride is very useful since it causes sedation within seconds after the solution passes over the gills. Since anesthetics leave residues in flesh, dead brood fish should not be eaten.

The stress of capture using nets and handling by workers strip away the protective mucous coat leading into infection. Careless handling may result in not only reduced feeding or infection but the quality of eggs in induced spawning (7). The right hormone at the right time may fail unless otherwise handling stress is avoided. A moist towel is always recommended during handling to minimize scale and mucous loss. Eye of the fish could be covered with a wet cloth whenever possible. Fish could be kept rather inverted than upright while handling. Noise could be minimized since fish have very acute sense of hearing. Rinsing eggs in milk powder solution (200ml powder/30 L water) may be used to remove the adhesive gelatinous covering of the fertilized eggs (20).

Artificial Insemination/Stripping

If spawning has not taken place after the 2nd injection, artificial insemination/stripping has to be practiced. The female fish is held head towards up, tail down and belly facing the vessel (Fig. 10.3) Applying gentle pressure on the abdomen the eggs are collected in the vessel. Similarly the milt is also collected in the same vessel and they are mixed using a feather (Fig. 10.3). After fertilization, the eggs are washed in distilled water to remove excess milt and released into hatching tanks. According to Harvey and Carolsfeld (7) artificial fertilization has unique advantages such as requirement of no spawning enclosures, easy handling of fertilized eggs and efficient usage of milt when it is scarce. Fertilization could be more than 90% in the case of artificial insemination. The most effective way is mixing the milt with eggs before any water is added and later natural spawning water is added to wet them thoroughly and washed with larger volumes of water after several minutes. When milt is collected in vessel the sperm cells are activated (motile) and motility of spermatozoa losts within a minute. Anyhow fertilization requires only a few seconds to contact the eggs. According to Harvey and Carolsfeld(7), the dry method is the best for fertilization.

Carps Breeding

The Indian major carps and the Chinese carps breed naturally only in flowing waters of their natural habitat, the rivers in association with monsoon floods. Until the introduction of the techniques of induced breeding, the only source of their seed for aquaculture was the rivers (10). The carp culture industry is rapidly expanding in India. But shortage of carp seeds is one of the major constraints in the development of this industry. Successful hormone induced breeding of common carp (*Cyprinus carpio*), Chinese carps (*Ctenopharyngodon idellus, Hypopthalamichthys molitrix* and *Aristichthys nobilis*), Indian carps (*Labeo spp, Cirrhinus mrigala* and *Catla catla*) and Thaicarp (*Puntius gonionotus*) have been widely reported for more than a decade.

The techniques of induced breeding of Indian major carps by administration of pituitary extract have been adopted by Alikunhi *et al.* (21); Chaudhuri (22,23); Chondar (24); Bimachar and Tripathi (25); Bardach *et. al.* (26) and Jhingran (27). The present trend is toward standardization and reduction of cost. Hence with the objective of finding out an efficient substitute for pituitary glands, more and more studies are aimed at application of synthetic hormones in induced breeding. Woynarovich and Horvath (28) worked out hypophysation techniques for carps. Ngamvongchon *et al.* (29) administered LHRH analogue (5+15μg/Kg) to bighead carp, noticed cent percent ovulation after 4-8 hours and suggested that LHRHa has to be combined with fresh silver carp pituitary for better results. LHRHa was injected intraperitonially twice (20 mg+40-50 μg and 0.5-1.5 mg carp pituitary) to grass carp (*Ctenopharyngodon idellus*), and bighead carp (*Aristichthys nobilis*) by Kumarasiri and Senevivatne (30). Both males and females showed multiple spawnings (3 times between April and October) and fecundity indicated that both species spawned completely in each cycle. Leelapatra (31) reported spawning success by administration of ovaprim to Rohu (*Labeo rohita*), Thai carp (*Putinus gonionotus*), mrigal (*Cirrhinus mrigala*), Silver carp (*Hypophthalamichthys molitrix*), and grass carp.

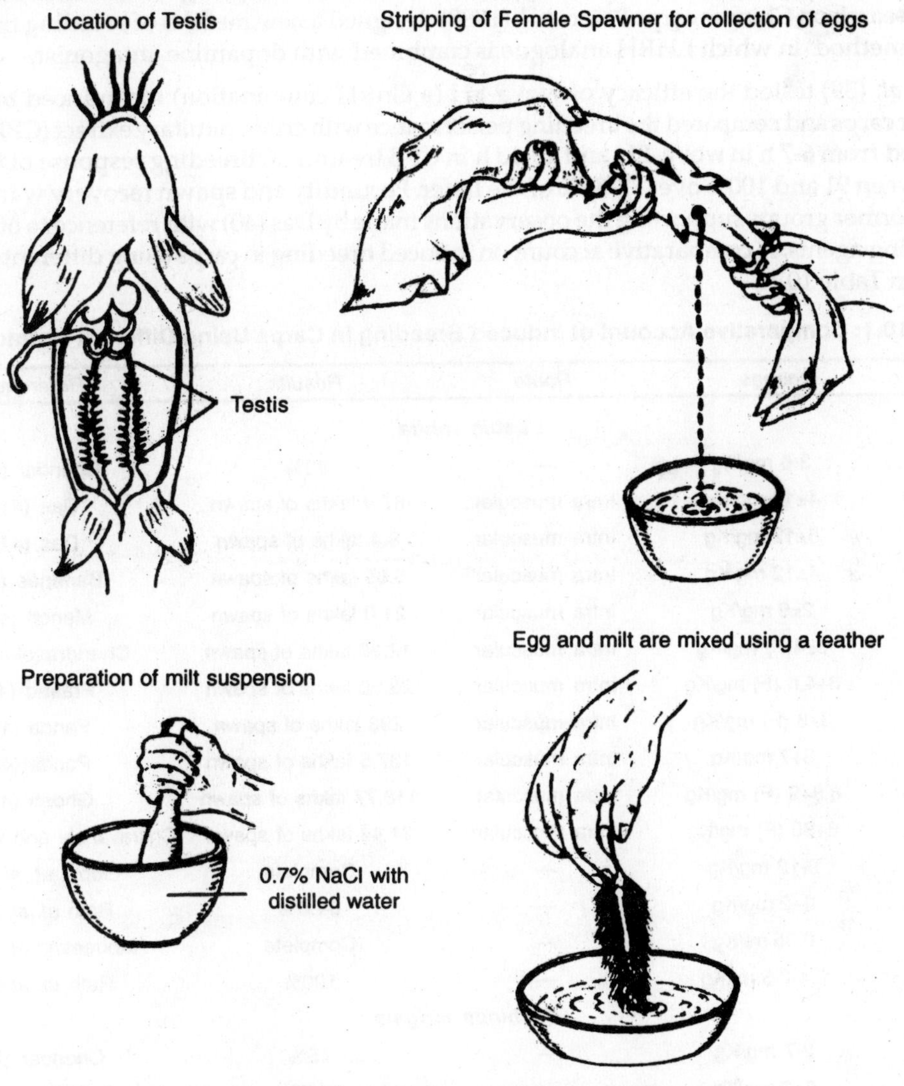

Fig. 10.3: Artificial Fertilization in Catfish

Nandeesha *et al.* (32) tested the induced breeding efficacy in *L. rohita*, *C. mrigala* and *Catla catla*. The breeding trials were conducted by pairing each female with 2 males in a cloth hapa measuring 2.5 × 1.2 × 1 m fixed in ponds provided with moderate circulation water. Based on their results, they recommended an optimum dosage of 0.5 ml ovaprim/Kg to Indian major carps. Based on Harvey and Hoar (33), Jhingran and Pullin(10) reported the protocols for Indian major carps. The prime doses of 2 mg/Kg, 1-2 mg/Kg and 1-1.13 mg/Kg followed by resolving doses of 8 mg/Kg. 6-8 mg/Kg and 6 mg/Kg were suggested for rohu, catla and mrigala respectively. Privious investigations have shown the potentiated action of analogues when they are combined with dopamine antagonists like pimozide or domperidone (34-37). Sahu and Biswal (38) obtained a high rate of spawning success in all the three Indian major carps by combining 20 mg/Kg of LHRH with 10 mg/Kg of Pimozide. Based on the extensive research on Chinese carps Peter *et al.* (1988) designed a new method of breeding fishes called the ' Limbe method' in which LHRH analogue is combined with dopamine antagonist.

Rath *et al.* (39) tested the efficacy of wova-FH (a GnRH combination) for induced breeding of Indian major carps and compared the breeding performance with crude pituitary extract (CPE). Latency period varied from 6-7 h in wova-FH and 13-13 h in CPE treatment. Breeding response of the former ranged between 91 and 100% over 77-84% of the latter. Fecundity and spawn recovery ware found to be better in former groups supporting the observations made by Das (40) with reference to other GnRH based inducing agents. A comparative account on induced breeding in carps using different hormones is reported in Table 10.1.

Table 10.1: Comparative Account of Induced Breeding in Carps Using Different Hormones

Hormone	*Dosage*	*Route*	*Results*	*Reference*
		Labio rohita		
Pituitary	3-6 mg/Kg	—	90%	Chondar (24)
Pituitary	4±14 mg/Kg	Intra muscular	87.4 lakhs of spawn	Das (41)
Pituitary	6±12 mg/Kg	Intra muscular	8.4 lakhs of spawn	Das (41)
Pituitary	4±12 mg/Kg	Intra muscular	5.65 lakhs of spawn	Banerjee (41)
Pituitary	2±9 mg/Kg	Intra muscular	21.0 lakhs of spawn	Menon (41)
Pituitary	3±4(F) mg/Kg	Intra muscular	12.30 lakhs of spawn	Chandrasekar (41)
Pituitary	3+4.6 (F) mg/Kg	Intra muscular	29.50 lakhs of spawn	Prasad (41)
Pituitary	3+8 (F) mg/Kg	Intra muscular	298 lakhs of spawn	Panda (41)
Pituitary	3+7 mg/Kg	Intra muscular	137.5 lakhs of spawn	Panda (41)
Pituitary	8.8+9 (F) mg/Kg	Intra muscular	118.77 lakhs of spawn	Ghosh (41)
Pituitary	6+30 (F) mg/Kg	Intra muscular	21.44 lakhs of spawn	Chitravanshi and Verma (41)
Pituitary	9-12 mg/Kg	—	Complete	Gupta *et. al.* (42)
Pituitary	6+2 mg/Kg	—	84.6%	Rath *et. al.* (39)
Ovaprim	0.35 ml/Kg	—	Complete	Nandeesha *et. al,* (32)
Wova-FH	0.4-0.5 ml/Kg	—	100%	Rath *et. al.* (39)
		Cirrhinus mrigala		
Pituitary	2-7 mg/Kg	—	75%	Chondar (24)
Pituitary	6+2 mg/Kg	—	80.9%	Rath *et. al.* (39)
LHRH	5 to 100 mg/Kg	—	1,20,000 ova/Kg	Kaul and Rishi, (43)
LHRH a+ PIM	10+10 μg/mg	—	Partial	Kaul and Rishi (43)
Ovaprim	0.35 ml/Kg	—	Complete	Nandeesha *et. al,* (32)
Wova-FH	0.4-0.5 ml/Kg	—	91.6%	Rath *et. al.* (39)
		Catla catla		
Pituitary	6+2 mg/Kg	—	77.7%	Rath *et. al.* (39)
LHRH a+ PIM	20+10 μg/mg	—	Complete	Sahu and Biswal, (38)
Ovaprim	0.4 ml/Kg	—	Complete	Nandeesha *et. al,* (32)
Wova-FH	0.4-0.5 ml/Kg	—	95.4%	Rath *et. al.* (39)

Contd...

Table 10.1–Contd..

Hormone	Dosage	Route	Results	Reference
		Hypopthalmicthys molitrix		
LHRH a+ PIM	100+10 µg/mg	—	Complete	Lin *et. al.* (34)
Ovaprim	0.7 ml/Kg	—	Complete	Nandeesha *et. al,* (32)
Ovaprim	0.5 ml/Kg	—	Complete	Peter *et. al.* (37)
		Ctenopharyngodon idella		
Ovaprim	0.7 ml/Kg	—	Complete	Nandeesha *et. al,* (32)
		Mylopharongodon piceus		
LHRH a+ PIM	100+10 µg/mg	—	Complete	Peter *et. al.* (44)
		Aristichthys nobilis		
LHRH	5+15 mg/Kg	Intra muscular	100%	Ngamvongchon *et. al.* (29)
Ovaprim	0.6 ml/Kg	—	Complete	Nandeesha *et. al,* (32)
		Labio dussmieri		
Ovaprim	0.15+0.5 ml/Kg	—	100%	Madhusoodana Kurup, (45)
Pituitary	4+8 mg/Kg	—	73.5-97%	Madhusoodana Kurup, (45)
Pituitary + HCG	4mg/Kg (P) 800-1000 IU/Kg (HCG)	—	71-74.2	Madhusoodana Kurup, (45)
LHRH	5 mg/Kg + 2.5 mg/Kg)	—	83.4	Madhusoodana Kurup, (45)
		Puntius japonicus		
Ovaprim	0.4 ml/Kg	—	Complete	Azad and Shimray (46)

* Chakrabarti (41)

Murrels Breeding

In recent years, the wild stocks of cultivable and commercially important fish species from natural water bodies are declining due to introduction of alien species, over exploitation, loss of habitat, disease, pollution, siltation, poisoning and other anthropogenic factors. Artificial fecundation and ranching of the depleted water bodies with the fingerlings are the envisaged strategies for the conservation and rehabilitation of the endangered species (47,48). Murrels breed naturally during southwest and northeast monsoons in flooded rivers and ponds in India (49,50). According to Qasim and Qayyum (51), the moderate climate and abundant rainfall in peninsular India provide favourable conditions for breeding of murrels throughout the year, in contrast to Haniffa and Sridhar (52) who have reported that murrels breed naturally only during monsoon. Since monsoon failure often limits murrel seed production, Banerjee (53) induced the spotted murrel *Channa punctatus* to spawn by administrating carp pituitary glands. Parameswaran and Murugesan (54) successfully induced the giant murrel *C. marulius*, stripped murrel *C. striatus* and spotted murrel *C. punctatus* to spawn at Bhadra Reservoir by hypophysation using carp pituitary glands.

Haniffa *et al.* (55) selected *C. striatus* brooders by sexual dimorphism, injected ovaprim (0.5 ml/kg) and introduced three sets of brood fish each consisting one female and two males into breeding ponds of CARE Aquafarm (each 16 m × 7.5 m × 1.5m) holding clear water of 0.8 m of depth. Aquatic macrophytes like *Hydrilla verticillata* and *Eichorrnia crassipes* were introduced into each breeding pond. After 24h of injection, eggs were noticed floating among the weeds. The diameter of eggs ranged between 1.2 mm and 1.4 mm. The different stages of development noticed in the eggs collected from the same brood fish indicated that the eggs were released in batches during the prolonged spawning act. Haniffa *et al.* (49) injected natural hormones (pituitary extract and HCG) and synthetic hormones (LHRHa and ovaprim) to *C. striatus* and released them into breeding ponds of CARE aquafarm. Water (dissolved oxygen: 6.1-6.8 ppm; pH 7.8-8.1; salinity 1.01-1.04%; temperature 27-290C) was pumped upto a level of 1m depth from a nearby well within the campus. When compared to LHRHa and ovaprim, the latency period was longer in pituitary- (24h) and HCG injected (26 h) fish. In the pituitary injected *C. striatus* the percentage of fertilization was the lowest (60-68%) but the duration of hatching was the longest (34-43 h) followed by HCG (36-38hrs). LHRHa- (34-36h) and ovaprim injected (21-23hrs) individuals. In terms of fertilization and hatching, ovaprim yielded better results (95-98%). Ova showed the maximum diameter (1.34–1.45mm) in *C. striatus* injected with ovaprim, followed by HCG (1.22–1.30 mm) and pituitary (1.21–1.27mm). The lowest ova diameter (1.07–1.09 mm) was observed in those injected with LHRHa. A comparative account on induced breeding in murrels using different hormones is reported in Table 10.2.

Table 10.2: Comparative Account of Induced Breeding in Murrels Using Different Hormones

Species	*Temperature (ºC)*	*Pituitary Extract*		*Latency Period (hr)*	*Fertilization Rate (%)*	*Reference*
		Prime Dose (mg/Kg)	*Resolving Dose (mg/Kg)*			
Channa marulius	—	20	60–100	—	23.3–34.7	Banerjee (53)
Channa punctatus	26–29	5–10	10–160	9–14	27.9–100	Parameshwaran & Murugesan (65)
Channa punctatus	—	50	—	28±0.76	85.0±4.52	Marimuthu (69)
Channa punctatus	—	100	—	25.8±0.58	95.0±3.60	Marimuthu (69)
Channa punctatus	—	150	—	26.8±2.08	88.6±2.51	Marimuthu (69)
Channa striatus	—	2–20	5–380	13–17	0–100	Parameshwaran & Murugesan (54)
Channa striatus	23–33	20	250	13–17	27–100	Parameshwaran & Murugesan (65)
Channa striatus	—	6	—	32	71	Francis *et. al.* (70)
Channa striatus	28–30	10	100	22.5±0.1	70.3±3.0	Haniffa *et. al.* (49)
Channa striatus	—	10	50	24.3±0.3	60.0±5.0	Haniffa *et. al.* (49)
Channa striatus	—	10	150	23.0±0.2	68.6±2.0	Haniffa *et. al.* (49)
Channa striatus	27–29	10	50	23–24	60	Haniffa *et al.* (49)
		HCG (/Kg.bw)				
Channa striatus	—	1000 IU		33	78	Francis *et al.* (70)
Channa striatus	28–30	5000 IU		23.0±0.7	79.0±5.0	Haniffa *et. al.* (49)
Channa punctatus	—	1000 IU		—	—	Haniffa and Sridhar (52)
Channa punctatus	28–30	2000 IU		28–34	75.5	Haniffa and Sridhar (52)

Contd...

Table 10.2–Contd...

Species	Temperature (ºC)	Pituitary Extract Prime Dose (mg/Kg)	Pituitary Extract Resolving Dose (mg/Kg)	Latency Period (hr)	Fertilization Rate (%)	Reference
Channa punctatus	—	3000 IU		28–34	78.0	Haniffa and Sridhar (52)
Channa punctatus	—	1000 IU		32.3±2.51	73.0±2.64	Marimuthu (69)
Channa punctatus	—	2000 IU		28.1±1.25	85.6±3.05	Marimuthu (69)
Channa punctatus	—	3000 IU		28.8±2.75	78.6±3.51	Marimuthu (69)
		Ovaprim (ml/Kg.bw)				
Channa striatus	—	0.5		24	—	Haniffa *et. al.* (55)
Channa striatus	—	0.4		21	93	Francis *et. al.* (70)
Channa striatus	28–30	0.5		24.0±0.9	98.0±3.0	Haniffa *et. al.* (49)
Channa striatus	—	0.3		—	—	Haniffa *et. al.* (49)
Channa striatus	—	0.7		230±0.9	95.3±2.5	Haniffa *et. al.* (49)
Channa punctatus	28–30	0.3		28—34	73.5	Haniffa and Sridhar (52)
Channa punctatus	—	0.1		—	—	Haniffa and Sridhar (52)
Channa punctatus	—	0.5		28–34	75.0	Haniffa and Sridhar (52)
Channa punctatus	—	0.2		29.16±1.0	83.3±2.08	Marimuthu (69)
Channa punctatus	—	0.4		24.50±0.5	97.6±1.52	Marimuthu (69)
Channa punctatus	—	0.6		26.0±2.5	78.3±6.6	Marimuthu (69)
		LHRHa+Pimozide				
Channa striatus	27–29	40 µg+5 mg		20.0±1.0	75.5±2.0	Haniffa *et. al.* (49)
Channa striatus	27–29	50 µg+5 mg		18.0±0.9	84.0±2.0	Francis *et. al.* (70)
Channa striatus	27–29	60 µg+5 mg		19.3±0.8	80.3±2.0	Haniffa *et. al.* (49)

Marimuthu *et al.* (56) made an attempt to study the spawning behaviour and parental care in induced bred *C. striatus* and *C. punctatus*. Brood fishes of *C. striatus* (600g-800g) and *C. punctatus* (50-90 g) were injected (Fig. 10.4a) with a single dose of pituitary extract (150 mg/Kg) and HCG (2000 IU/KG). *C. striatus* were released into breeding ponds (16 m × 7 m × 1.5 m) and *C. punctatus* were introduced into breeding tanks (3m × 1m × 1m). Spawning behaviour was observed at every one hour interval after the hormone injection until egg laying. The active male chased the female and frequently excited movement of the paired breeders commenced from 10 -12 hr after the hormone injection. In all the spawning attempts, the male was more actively involved in the courtship and spawning. It was seen hitting the female snout and vent region more frequently. The spawning activity continued till the releasing of gametes. The more active and aggressive male paired with the female and the spawning activity continued till the releasing of gametes (57) (24 –30 hrs). At the culminating courtship, the male bent its body close to the female and breeders joined together (Fig. 10.4b) and the male released its milt and the female its eggs (Fig. 10.4c) and after thatexternal fertilization occurred. The male parent was found with eggs and hatchlings while the female parent was seen in the vicinity of the egg mass in the breeding tank (Figs. 10.4d and 10.4e). Das *et al.* (40) reported that aggressive behaviour was generally exhibited by female *C. striatus*. On the contrary Marimuthu *et al.* (56) noticed aggressive behaviour by the male taking care of the young ones supporting the previous observations of Hundingford (58) and

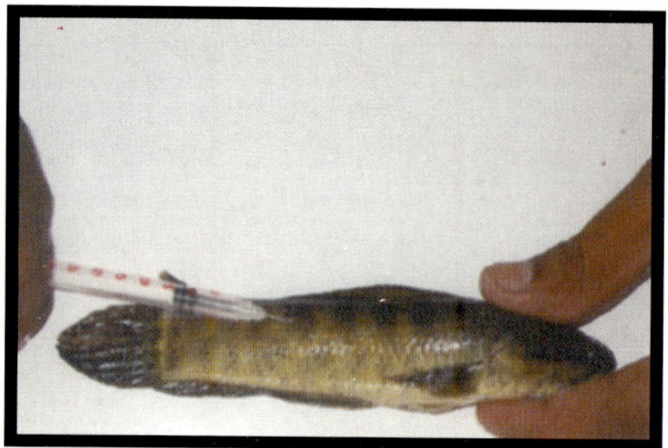

Fig. 10.4a: Intramuscular Injection in *C. punctatus*

Fig. 10.4b: Courtship Behaviour

Fig. 10.4c: Egg Mass

Fig. 10.4d: Egg Mass Guarded by Female *C. punctatus*

Fig. 10.4e: Hatchings Guarded by Male *C. punctatus*

Fig. 10.4: Spawning Behaviour and Parental Care in *C. punctatus*

opposing Parameswaran and Murugesan (54) who reported no parental care for the eggs and larvae of *C. striatus* and *C. punctatus* when the brood fish were induced bred.

Marimuthu *et al.* (59) reported low cost murrel seed production technique for fish farmers. The study was the first of its kind carried out in an earthen pond of 48 m2 using a rectangular drag net (2m × 2m × 1m) as a breeding hapa. The base of the drag net was fully submerged at the pond bottom using bricks. Two sets of brood fishes (four males and two females) weighing 600g–950g injected with ovatide intramuscularly (0.4 ml/KG) were introduced into the breeding hapa. Spawning was noticed 10 h after injection. The eggs hatched 24-26 h after spawning and it was possible to collect 750-1500 fry after one month by this technique.

Haniffa and Sridhar (52) selected *C. punctatus* breeders by external morphological characteristics and hand stripping (60) and injected HCG (1000 IU-3000IU/Kg) and ovaprim (0.1 ml–0.7 ml/Kg) intramuscularly in a single dose (55). Control fish was given corresponding volume of physiological saline solution. Egg output was determined by hand stripping for a separate set of three females for each treatment. The number of eggs released was calculated following the gravimetric method (61,62). Haniffa and Sridhar (52) reported a latency period 28-34 h for *C. punctatus* whereas Banerjee (53) reported 6-25 h. According to Haniffa and Sridhar (52) increase in dosage of the hormone produced elevated values in egg production (30-3276 due to ovaprim; 102 –1253 due to HCG). Irrespective of nature of hormone fertilization was 70% and above. The highest percentage of fertilization (78%) was noticed due to 3000 IU HCG injection. Haniffa and Sridhar (52) reported a fertilization rate of 60-70%, 65-79%, 75-80% and 95-98% in *C. punctatus* injected with pituitary extract, HCG, LHRHa and ovaprim respectively.

To study mass breeding Haniffa *et. al.* (63) administered a single intramuscular injection of ovaprim (0.5 ml/Kg) to 10 males and 6 females of *C. striatus* and introduced them into the breeding pond (50′ × 25′ × 3′) and allowed themto mix with 80 uninjected brooders. Spawning was noticed five times due to stimulation by uninjected brooders. Murugesan and Kumaraiah (64) too noticed mass breeding by injected and uninjected *C. striatus* due to the combined effect of stimulus created by the rush of water and the spawning activities of the injected fish. Haniffa *et al.* (63) suggested that pheromonal secretion might have increased the fish reproductive instinct. The mass induced breeding technique attempted by Haniffa *et al.* (63) in earthen ponds is simple and advantageous as it does not require expensive components like plastic pools, aquaria and hapa. Moreover the breeding environment helps the eggs to hatch out and young ones to grow under parental care for some time.

A good number of reports are available on embryonic and larval development of *C. punctatus* (53, 65, 66). Haniffa *et al.* (67) conducted induced breeding experiments in *C. punctatus* by administrating a single intramuscular injection of ovaprim (0.3 ml/kg). Spawning was observed 23-24 h after the injection at ambient temperature and the fertilized eggs were buoyant, non-adhesive and straw yellow in colour with diameter ranging between 0.9 mm and 1.10 mm. Embryonic development of *C. punctatus* is given in Table 10.3 (Fig. 10.5). The hatchlings were transparent and measured 2.7 mm-2.9 mm, with a large oval head, a well-defined yolk sac and short tail and the larvae attained free normal movement when 36 h old. The yolk got fully absorbed within 3 days and by this time mouth formation was complete and the larvae started exogenous feeding. Larval and post larval development of *C.punctatus* are reported in Table 10.4 (Fig. 10.6). The post-larvae were observed to start aerial respiration on 17th day. After 20 days the length of fry ranged between 7.5mm and 7.8mm and resembled the adult in its external features. Length of the fingerlings ranged from 8mm–21mm after 35 days and appeared just like an adult in all respects except sexual maturity (Table 10.4 and Fig. 10.6). Although a true metamorphosis is not generally described in fishes, the terms hatchling, post-larvae and fry are used to indicate different stages of development (68).

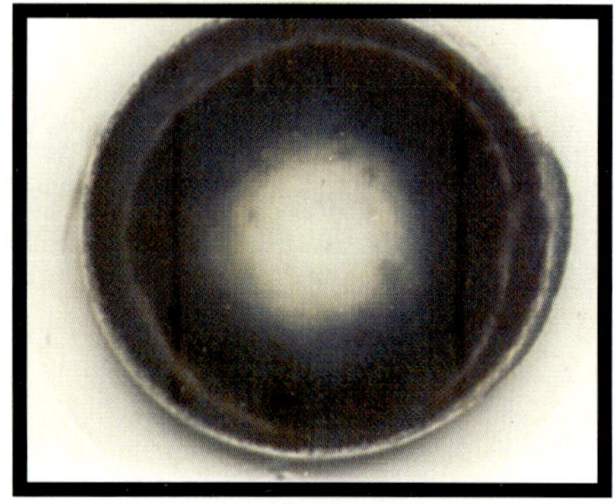

Fig. 10.5a: Fertilized Egg

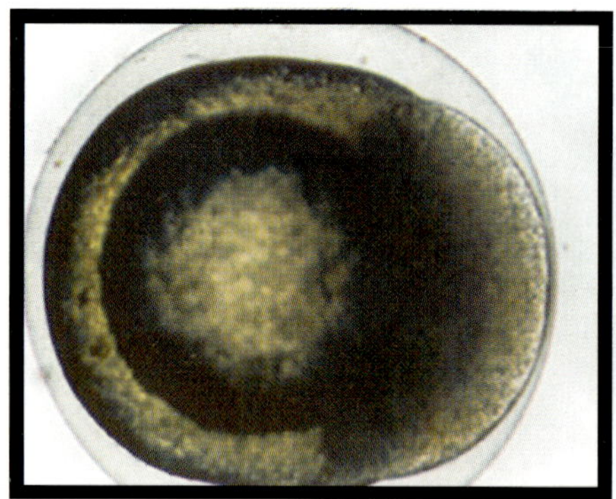

Fig. 10.5b: Morula Stage

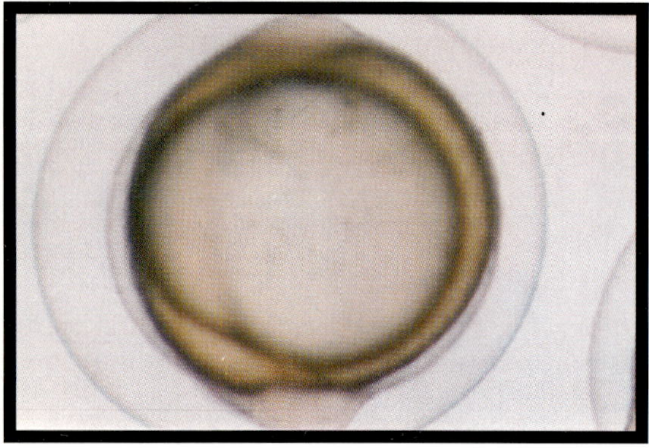

Fig. 10.5c: 8 Hours Old

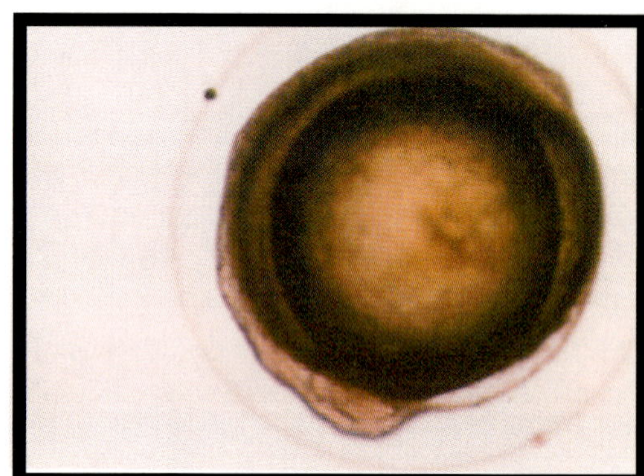

Fig. 10.5d: 10 Hours Old

Fig. 10.5: Embryonic Development of *C. punctatus*

Catfish Breeding

Catfishes belonging to the families Ictaluridae, Claridae, Pangasidae and Siluridae are widely distributed in different parts of the world. But the promise of catfish culture in India has yet to be fulfilled. Though catfishes have consumer preference and advantages like survival in oxygen-depleted waters, tolerance to crowding and accepting artificial feeds, tender flesh and delicious taste, their culture systems are yet to be established in many countries of Asia including India and the existing catfish culture industry needs considerable research and development to solve its problems.

Practical methods for inducing ovulation and spawning in species of *Clarias macrocephalus, C. batracus, C. gariepinus, C. fungus, Pangasius sutchi* and *P. pangasius* in Southwest Asia and India have until recently relied on hypophysation and HCG (7). Mollah and Tan (71) administered a single IM injection of HCG 2000 IU/Kg to Thai catfish *C. macrocephalus.* The fish spawned 10 hrs later with average hatching of 73% after incubation at 300 C. Saidin *et al.* (72) injected pituitary, HCG and LHRH intramuscularly to *P. sutchi* and *C. macrocephalus. C. macrocephalus* was strip-spawned whereas *P. sutchi* spawned spontaneously. HCG at 3000–4000 IU/Kg was more effective than LHRH analaque at 10-30 μg/Kg. Ngamvongchon *et al.* (29) selected *C. macrocephalus* brooders and injected LHRHa 1M in a single dose of 20 μg/Kg. Ovulation and strip spawning followed within 18 hours with a success rate of 70%. When HCG was administered at 4000 IU/Kg the result was cent percent ovulation.

Table 10.3: Embryonic Development of *C. punctatus*

Time after Fertilization (hr)	*Developmental Features*
0.15	Blastodisc formation
0.30	2 celled stage
0.40	4 celled stage
0.45	8 celled stage
0.50	16 celled stage
2.00	Morula stage
4.00	Germinal ring formed
5.00	Embryonic shield formed; more than half of the yolk invaded, anterior and posterior differentiation evident
9.00	Yolk invasion complete
10.00	Cephalic region broader with distinct forebrain
14.00	6 Somites and notochord formed
16.00	9 Somites present and cephalic region broadened
18.00	13 –15 Somites present and embryonic finfold formed
20.00	22 Somites present, lens fully formed in the eye; heart formed and blood circulation commenced
22.00	Commencement of heart beat

Table 10.4: Larval and Post-larval Development of *C. punctatus*

Larval Age (days)	*Length (mm)*	*Characters*
Just Hatched	2.7-2.9	Brown colour. The fin fold originates and heavy ovoid yolk sac.
1 day	4.3-4.5	Eye pigmented; heart beat 155-160/minutes.
2 days	4.7-4.9	Pectoral fin paddle shaped; Mouth formed with well developed lower jaw; feeding exogenous
Post larva 3 days	5.2-5.4	Yolk fully absorbed. Abdomen appears to be heart shaped when viewed from the ventral side.
6 days	5.5-5.8	Yellow pigments on the dorsal and lateral sides appear as bands
10 days	6.0-6.4	The colour pattern is now in the form of alternating lemon yellow and dark bands; rudiments of two rays seen in the caudal fin region.
15 days	6.5-6.9	Dorsal contour of the head convex; eight caudal rays distinguished.
20 days	7.5-7.8	14 rays in the caudal and 19 in the pectoral at this stage. Aerial respiration noticed; ventral fin not appeared.
35 days (fingerling)	18.3-21.0	Appears like adult.

Fig. 10.6a: 2 Days Old Larva

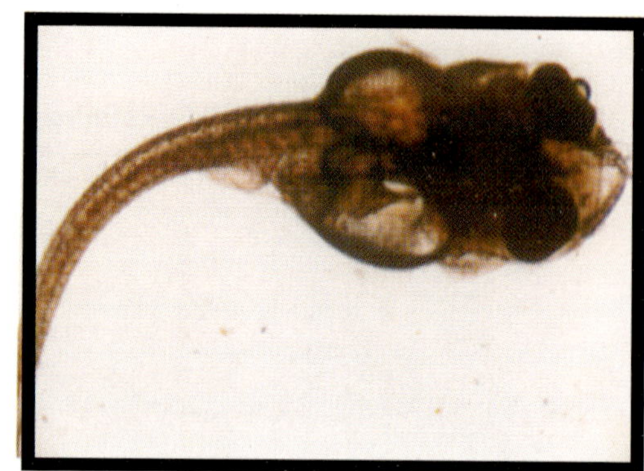

Fig. 10.6b: 3 Days Old Larva

Fig. 10.6c: Fry

Fig. 10.6d: Fingerlings

Fig. 10.6: Larval Development of *C. punctatus*

According to Viveen *et al.* (73) *C. gariepinus* broodfish can be artificially reproduced several times in a year. Ramamohana Rao (74) reported breeding and seed production in *C. batracus.* Among the catfishes which fetch high market value *C. batracus* is of importance. The demand for this fish is owing to its taste and medicinal values. However fish farmers are unable to take up *C. batracus* culture due to want of stocking materiel and still depend on wild collection which is unpredictable due to monsoon failure. Through studies conducted at CIFA, it has been found that a single dose to female fish with carp pituitary extract (alcohol preserved glands) at 30/mg/Kg is enough to bring about spawning in *C. batracus* females whilemales do not need administration of hormone. The optimum doses for HCG and LHRHa + Domporidone are 4000 IU/Kg and 0.04 mg + 300 mg/Kg respectively.

The stinging catfish *Heteropneustes fossilis* is widely distributed throughout India and is cultured along with other species in community ponds in southern India (75). Due to its high market value, low fat, high protein and iron content (76), it is recommended for intensive capture. In southern India, *H. fossilis* naturally spawns during the southwest monsoon (July–August) and spawning can be induced by manipulation of water levels (77). Ovulation can be induced during non-spawning season not only by environmental manipulation but also by hormonal stimulation.

Haniffa and Sridhar (52) selected *H. fossilis* brooders by external morphological characteristics and hand stripping. HCG (1000IU –3000IU/1 Kg) and ovaprim (0.3 ml–0.7 ml/Kg) were injected intramuscularly into the dorso lateral region of both males (80–105 g) and females (90–115 g) in a single dose. Control fish were given corresponding volumes of physiological saline solution. Eggs were collected from breeding tanks and the percentage of fertilization was estimated by examination of a sample of at least 150 eggs from each tank. Egg output was determined by hand stripping for a separate set of 3 females for each dosage. *H. fossilis* spawned after a latency period of 18-24 hrs irrespective of the nature of hormones. Increase in dosage of hormones produced elevated values in egg output and the enhancement was statistically significant ($P<0.05$). Salami *et al.* (78) reported a similar increase in egg output in *C. gariepinus*. The latency periods reported in the literature are 22-25 hrs for *H. fossilis* (79) and 16-20 hrs for *C. gariepinus* (66). The minimum HCG dosage required to induce ovulation in *C. macrocephalus* was 2000 IU/Kg (71). But *C. batracus* did not spawn at 2000 IU/Kg (80), suggesting that the number of ovulated eggs is related to hormonal dosage. Hence, Zonneveld *et al.* (81) recommended a dose of 3000 IU HCG for inducing ovulation in *C. batracus*. The dosage doubtless varies in different species, depending on how closely the endogenous gonadotropin is related to HCG. When compared to other hormones HCG injected fish showed a better performance with regard to fertilization, hatching and survival of hatchlings (52) supporting the previous observtions of Salami *et al.* (78) who reported a better performance due to HCG administration than carp pituitary extracts in *C. gariepinus*.

Vijayakumar *et al.* (82) designed a modified round plastic trough (25 l) as breeding chamber for *H. fossilis*. The breeding chamber contained a removable earthen tile to allow the fish to hide and is covered with a net to prevent them from jumping (Fig. 10.7). Each female laid 8000-1200 eggs (Fig. 10.8). The hatchery device consisting of a cylindrical glass tube was filled with well water (Fig. 10.9) and was placed in center of the breeding chamber. The breeding chamber had an outlet guarded by a fine net. The collected eggs were introduced into the hatching unit through the large opening at the top. The opening was plugged and water was pumped through the inlet. Using this breeding chamber on an average 30,000 eggs could be incubated at a time. Marimuthu *et al.* (83) reported the efficacy of ovatide to determine the optimum dose for a better performance of *H. fossilis*. The number of eggs laid increased from 1508 to 4358 with the increasing dosage of the hormone and they recommended 0.4 ml/Kg as optimum dosage. Jesu Arockiaraj *et al.* (84) attempted induced breeding in *H. fossilis* using non-piscine pituitary extract (toad and frog). Dosages ranging between 4-6 mg/Kg administrated in installments produced 6,000 to 10,000 eggs with fertilization ranging between 43% and 73% and hatching between 73% and 90%.

Ompok malabaricus is an endangered catfish species having high commercial value and during the last two decades, its wild population has declined due to various anthropogenic stresses. Hence Sridhar *et al.* (85) attempted induced spawning in *O. bimaculatus* by a single intramuscular injection of ovaprim (0.5 ml/Kg). Spawning was observed 5-6 h after injection. An average of 4,000 eggs were spawned by each female. The eggs were transparent and were found attached to the sand bed of the breeding tank. Egg diameter was 1.22±0.03 mm and the fertilization rate was 75%. Hatching occurred 24 h after spawning and the hatchlings were light yellow in colour with 55–60% survival. Haniffa *et al.* (86) reported fecundity ranging between 3700 and 4000 with 75% fertilization for *O. malabaricus* injected with o.5 ml ovaprim/Kg. When compared to other catfishes like *H. fossilis* (87) (12h) and *C. gariepinus* (78) (13h) the latency period was very less (5 –6 h) in the case of *O. malabaricus*. The survival at hatching varied from 55 to 60%.

Fig. 10.7: Breeding Chambers for *H. fossilis*

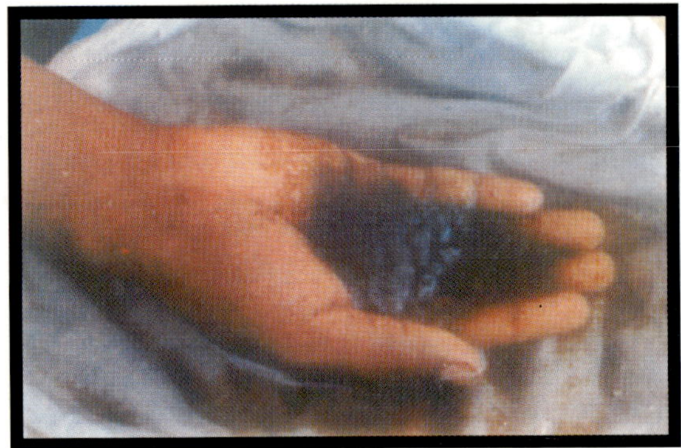

Fig. 10.8: *H. fossilis* Eggs Collected from the Breeding Chamber

Fig. 10.9: Glass Jar Hatchery in Operation

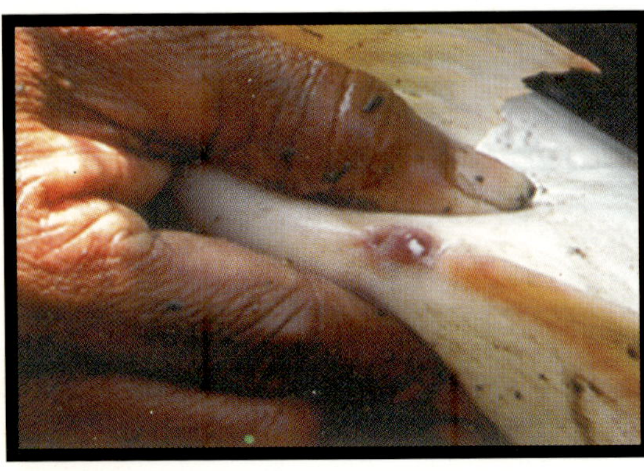

Fig. 10.10: Male *Pangasius pangasius*

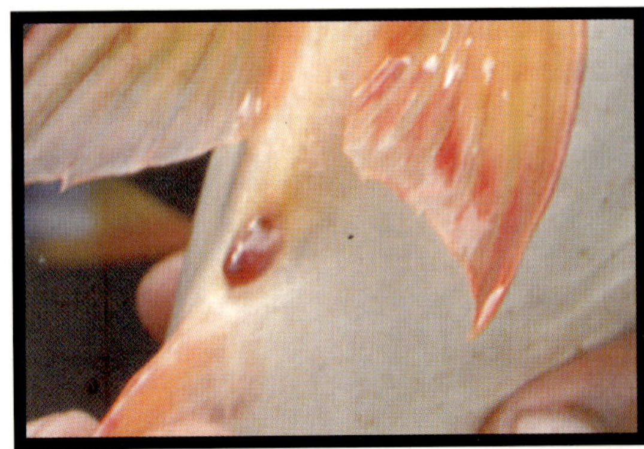

Fig. 10.11: Female *P. pangasius*

Haniffa *et al.* (86) reported that breeding of *Mystus montanus* takes place during monsoons but could be induced to spawn irrespective of monsoons. Mature fish weighing 15-16 g can be employed for captive breeding. *M. montanus* laid eggs 10-12 h after injection of HCG at 3000 IU/Kg. The fertilized eggs were golden yellow in colour, non-floating and sticky in nature. Hatching occurred 24-25 h after spawning. The fertilization rate, hatching rate and survival at hatching varied from 60-80 %, 76-84% and 69-87% respectively. Like other catfishes *M. oculatus* also do not breed in captivity easily, a major stumbling block in culture of this species. Spawning was successful with 61-63% fertilization in 2-6 mg/Kg carp pituitary injected individuals. The incubation period ranged between 23-28 h and hatching was between 88-92% (88).

Mystus gulio breeds naturally during northeast and southwest monsoons in rivers and ponds (with clay bottom). Review of literature shows that only a few reports are available on breeding, aspects of *M. gulio* and attempts have not been made to propagate it artificially in confined waters. Hence Haniffa *et al.* (89) selected brood fish of 15-19 cm length and 50-60 g weight by sexual dimorphism (gonadopodian structure in male and round and reddish vent in female). The incubation period (23-28) was more in ovaprim (0.3ml/Kg) injected individuals. Among the natural (carp pituitary extract and HCG) and synthetic (ovaprim, ovatide and wova-FH) hormones tested, ovaprim was found to be unique in showing better fertilization ability, hatching and survival. Hence a dosage of 0.3 ml ovaprim/Kg was recommended for better results (90).

Haniffa (91) reported spawning in *Mystus seenghala* 6-12 h after injection of HCG, ovaprim, ovatide and ova-FH at ambient temperature. Fertilized eggs were buoyant, sticky and golden yellow in colour with diameter ranging between 1.16 mm and 1.26 mm. The incubation period ranged between 22h–29h in all the treatments. The percentage of fertilization was the highest (80-95%) in ovaprim (0.5 ml/Kg) injected individuals.

Pangasid catfishes *Pangasius pangasius* and *P. sutchi* are recently adopted for culture in Bengal due to their remarkable growth rate (1kg/90 days) and the demand for their seeds is increasing day by day (20). They attain sexual maturity at four years (Fig. 10 and 11) when they normally reach a size of 7 kg. Chattopadhyay *et al.* (2002) injected a prime dose of 1.5mg crude pituitary extract/kg into mature females. After 5-6 hours the resolving dose of 6 mg CPE/Kg for female and 2 mg CPE/Kg for male were administrated and released into cement tanks/hapa. *P. sutchi* responded positively in all the trials and ovulated 5–6h after the second injection and the fertilization rate ranged between 95-100%. Rahaman *et al.* (92,93) too succeeded in their attempts on induced breeding in *P. sutchi*. According to Saidin and Othman (1986) the hatching period of *P. sutchi* ranged between 24–26h with a survival of hatchlings from 30–45%. According to them milt from one male was sufficient to fertilize the eggs of three to four females. A comparative account on induced breeding in catfishes using different hormones is reported in Table 10.5.

Conclusion

Aquaculture is a relatively recent, although fast growing industry, unlike the live stock industry, which is based on a few species there are around 25,000 species of fish. Nutrition, health and seed production are the three key biological areas that must be explored in the development of any viable aquaculture system.

Aquaculturists especially those involved in the rearing and propagation of commercially important fishes often fail in their culture practices, since they depend on seed collection from the wild. Hence techniques for production of adequate quantities of high quality seeds by captive breeding are necessary for large scale expansion of aquaculture. The lack of such techniques has often been important.

Table 10.5: Comparative Account of Induced Breeding in Catfishes Using Different Hormones

Species	*Temp (°C)*	*Hormone Used*	*Hormone Dosage (/body weight) (BW)*	*Latency Period (h)*	*Fertilization (%)*	*Reference*
Clarias gariepinus	28±1	CPE	4 mg/kg	—	92	Clemens and Sneed (94)
Clarias gariepinus	29±1	CPE	4 mg/kg	—	75–92	Hogendoorn (95)
C. gariepinus	25	LHRH + Pionozide	50μg/kg	12–13	—	De Leeuw *et. al.* (96)
C. gariepinus	29	Frog pituitary	4–6 mg/kg	—	47–82	Salami *et. al.* (78)
C. gariepinus	29	HCG	2000 IU/kg	—	90	Salami *et. al.* (78)
C. gariepinus	29	Catfish pituitary (*C. gariepinus*)	4–6 mg/kg	—	40–75	Salami *et. al.* (78)
Clarias gariepinus	30 ± 2	PE	5 mg/kg	20	61.9	Obi and Ewemade (97)
C. batracus	26–29	CPE	80–130 mg/kg	16–20	—	Thakur (98)
C. batracus	27–29	CPE	30 mg/kg	14	—	Rao *et. al.* (99)
C. batracus	—	CPE	50–200 mg/kg	16–20	—	Munshi and Hughes (66)
C. batracus	27–30	17 α hydroxy progestron	—	—	80	Zairin *et. al.* (80)
C. batracus	23–25	HCG	4000 IU/kg	12–16	—	Zairin *et. al.* (80)
Clarias batrachus	28 ± 1	Salmon gonadotropin (SG–G100)	10+15 μg/g	20–22	87.76 ± 1.02	Haider and Rao (100)
Clarias batrachus	28 ± 2	Ovopel	1.5 g/kg	—	—	Das (101)
H. fossilis	—	CPE	80–100 mg/kg	—	—	Ramaswamy and Sunderaraj (102)
H. fossilis	25–29	CPE	150 mg/kg	6–10	—	Thakur and Das, (103)
H. fossilis	30	SGnRH–A	1 ml/kg	14–18	100	Alok *et al.* (76)
H. fossilis	30	D–Lys6 SGnRH–A	25 μ/kg	14–18	60–70	Alok *et al.* (104)
H. fossilis	28–29	SgnRHa	0.1– 25 μg/kg	—	100	Alok *et. al.* (87)
H. fossilis	26–29	Ovaprim	0.5 ml/kg	—	80	Vijayakumar *et. al.* (82)
H. fossilis	28±2	Ovatide	0.5 ml/kg	10–14	65–93	Marimuthu *et. al.*, (83)
H. fossilis	28 ± 2	SGnRH–A	1 μg/kg	12	88.81	Nayak *et al.* (105)
H. fossilis	28±2	Non-piscine pituitary (Toad and Frog)	4–6 mg/kg	—	42.7–72.8	Jesu Arockiaraj *et. al.* (84)
Clarias macrocephalus	30	HCG	450–500 IU/kg	15	50–62	Carreon *et al.* (106)
C. macrocephalus	26–31	CPE	70 mg/kg	13	—	Mollah and Tan (71)

Contd...

Table 10.5–Contd...

Species	Temp (°C)	Hormone Used	Hormone Dosage (/body weight) (BW)	Latency Period (h)	Fertilization (%)	Reference
C. macrouphalus	30	Frog Pituitary	4–6 mg/kg	—	70–85	Mollan and Tan (71)
C. macrocephalus	26–31	HCG	4000 IU/kg	16	—	Nagmvongchon *et. al.* (29)
Clarias macrocephalus	27–30	LHRH–A + Pimozide	0.1 µg + 1 µg/g	16	36–39	Fermin (77)
Clarias anguillaris	20–23	HCG	3000 mg/kg	11–13	86	Inyang and Hettiarachehi, (107)
C. fuscus	28±2	Frog pituitary	4–6 mg/kg	—	70–85	Young *et. al.* (108)
Clarias gairiepinus	28–30	PE	3 mg/kg	14	79.6	Salami *et al.* (78)
C. isheriensis	28±2	Toad pituitary	4 mg/kg	—	70–85	Fagbenro *et al.* (109)
Mystus punctatus	28.5 ± 1.8	PE	3 mg/kg	8	80	Ramanathan *et al.* (110)
Mystus aor	29–30	PE	12–20 mg/kg	28–38	50	Gupta *et al.* (111)
M. tengara	29±1	Toad pituitary	4 mg/Kg	—	70–85	Fagbenro *et al.* (109)
M. montanus	28±2	HCG	3000 IU/kg	24	67–89	Haniffa *et. al.* (86)
M. oculatus	28±2	CPE	2-6 mg/kg	26	88–92	Haniffa (88)
M. gulio	29±2	Ovaprim	0.3 ml/kg	29	85–88	Haniffa (90)
M. seenghala	30±1	Ovaprim	0.5 ml/kg	22–29	80–95	Haniffa (91)
Mystus montanus			1000 IU/kg	11–12	84	Haniffa *et al.* (86)
Silurus asotus	—	PE	5 mg/kg	—	72–96	Choi *et al.* (112)
H. longifilis	29	HCG	250 IU/kg	—	75	Legender (62)
H. longifilis	30	Carp pituitary	6–8 mg/kg	—	75–92	Nwadukwe *et. al.* (113)
Heterobranchus longifilis	26–29	PE	4–8 mg/kg	7–11	73.6	Nwadukwe *et al.* (113)
Pangasius sutchi	30±2	Pituitary extract	8 mg/kg	24	92–95	Chattopadhyay *et. al.* (20)
Ompok malabaricus	28±2	Ovaprim	0.5 ml/kg	5–6	75	Haniffa *et al.* (86)
Ompok malabaricus	28 ± 2	HCG	3000 IU/kg	6–8	80–90	Haniffa *et al.* (86)
O. bimaculatus	29–30	Ovaprim	0.5 ml/kg	—	75	Sridhar *et. al.* (85)
Ompok bimaculatus	26–28	Ovaprim	0.5 ml/kg	5–6	55–60	Sridhar *et al.* (85)
Ictalurus punctatus	24–30	PE	4.4 mg/kg	—	72–96	Kim *et al.* (114)

HCG : Human Chorionic Gonadrotropin Hormone
LHRH–A : Leutinizing Hormone Releasing Hormone Analogue
CPE : Carp Pituitary
SGnRH–A : Salmon Gonadotropic Releasing Hormone Analogue
GNRH-A : Gonatotropin Releasing Hormone Analogue
SG–G 100 : Salmon Gonadotropin
Doc : Deoxy corticosteron
PE : Pituitary Extract

In the fifties the scenario of fish seed production changed with the development of induced breeding by administering carp pituitary extract (115). Virtually this was the turning point for captive breeding of Indian major carps and large scale seed production started taking off and during the last two decades, the technology has been wildly adopted and has become an organized industry (116). The scope generated from the captive fish breeding programme assures availability of desired seeds and bears immense importance of genetic upgradation. The induced breeding technique has proved its efficiency not only in the commercial venture of fish seed trade and aquaculture but also very much essential and effective in long term genetic evaluation of the progenies for traits of commercial importance (117). The key to further success lies in growth and survival during captive rearing especially at early stage. For wider acceptance of the new species, technology of seed production and culture should be commercially viable.

Though India is blessed with rich biodiversity, many fish species show declining trends due to various anthropogenic stresses and uncontrolled introduction of exotics. Several fish species are declining in their natural abundance, some even at the verge of extinction. The situation demands prioritization of such species for development of artificial propagation technology and captive breeding is the first milestone for artificial propagation of any species. In the case of conventionally bred organisms, restocking waterbodies with fish from poorly managed hatcheries may be intended to augment wild populations that have been depleted through over fishing or habitat destruction (118). The success of induced breeding open the avenues for ranging the hatchery produced stocks of threatened especially endangered fish species in depleted natural water bodies. No doubt induced breeding and larval rearing techniques under captivity are the viable aids in conservation and rehabilitation of the endangered species.

Acknowledgement

I am indebted to Management of St. Xavier's College for providing facilities to establish Centre for Aquaculture Research and Extension (CARE). I wish to acknowledge the financial assistance received from Department of Science and Technology (SSSP/RD/001/96), Indian Council of Agricultural Research (F.No. 27(28)/98/NATP/MM-III-18 and F.No.4-12/99- ASR-I) and Department of Biotechnology (BT/PR/1892/SPD/16/137/2000) for mass seed production techniques of murrels catfishes. Thanks are due to my research scholars especially M/S. M.Nagarajan, A.Jesu Arokia Raj and P.S. Allen Benziger for preparation of the manuscript. I am grateful to Dr.T.J. Pandian (ICAR National Professor, School of Biological Sciences Madurai Kamaraj University) for his valuable suggestion and encouragement.

REFERENCES

1. Ayyappan, S., Sudhir Raizada and Reddy, A. K. (2001) Captive Breeding and Culture of New Species of Aquaculture. In: *Captive Breeding for Aquaculture and Fish Germplasm Conservation.* eds. Ponniah, A. G., Kuldeep K. Lal, Basheer. V. S. NBFGR-NATP Publication No:3. Key paper No.1. pp 1-20.
2. Pandian, T. J. and Koteeswaran, R. (1998) Ploidy induction and sex control in fish. **Hydrobiologia,** (384) 167–243.
3. Haniffa, M.A., Sridhar, S. and Nagarajan, M. (2004) Hormonal manipulation of sex in stinging catfish *Heteropneustes fossilis*. **Curr. Sci,** 86 (7). 1012-1017.
4. Haniffa, M.A., Sridhar, S. and Nagarajan, M. (2004) Induction of triploidy and tetraploidy in stinging catfish *Heteropneustes fossilis* using heat shock. **Aquaculture Research**.

5. Garg, S. K. (2001) Environment Manipulation Techniques for Development of Captive Breeding Technologies of New Species Aquaculture. In: *Captive Breeding for Aquaculture and Fish Germplasm Conservation.* eds. Ponniah, A. G., Kuldeep K. Lal, Basheer. V. S. NBFGR-NATP Publication No:3. Invited paper No. 2. pp 1-20.

6. Godwin Christopher, J. and Ponniah, A. G. (2001) Relative Research Effort on Breeding and Larval Rearing of Indian Fishes. In: *Captive Breeding for Aquaculture and Fish Germplasm Conservation.* eds. Ponniah, A. G., Kuldeep K. Lal, Basheer. V. S. NBFGR-NATP Publication No:3. Contributed paper No.1. pp 1-6.

7. Harvey, B. and carolsfeld, J. (1993) Induced breeding in tropical fish culture Ottawa, Ont., IDRC, pp144.

8. Rottmann, R.W., Shireman, J. V. and Chapman, F. A. (1991) Hormonal control of reproduction in fish for induced spawning. SRAC publication No. 424 Institute of food and agricultural services, University of Florida.

9. Sundararaj, B. I. and Goswami, S. V. (1966) Effects of mammalian hypophysial hormones, placental gonadotropins, gonadal hormones, and adrenal corticosteroids on ovulation and spawning in hypophysectomized catfish, *Heteropneustes fossilis* (Bloch). **J. exp. Zool,** (161): 287-296.

10. Jhingran, V. G. and Pullin R. S. V. (1985) A hatchery manual for the common Chinese and Indian major carps. **ICLARM studies and reviews,** 11, pp191. Asian Development Bank, Manila and Philippines.

11. Merlinrose, T. (1998) Reproductive and developmental biology of an obligatory air-breathing fish *Channa striatus.* Manonmanian Sundaranar University, Thirunelveli, India.

12. Jesu Arockia Raj, A., Haniffa, M. A., Seetharaman, S. and Singh, S. P. (2003) Cyclic changes in gonadal maturation and histological observations of threatened freshwater catfish *Mystus montanus* (Jerdon) **Acta Zooloica.Taiwanica.**

13. Houssay, B. A. (1930) Accion sexual de la hipofisis en los perces y reptiles. **Rev. Soc. argent. Biol,** (6) 686–688.

14. Ihering R. Von. (1934) A method for inducing fish to spawn. **Progr. Fish. Cult,** (24) 15-16.

15. Gerbil'skii, N. L. (1938) Expedition for the study of the physiology of spawning. **Ryb. Khoz,** 18(10) 33-36 (In Russian).

16. Hasler, A. D., Meyar, R. K. and field, H. M. (1939) Spawning induced prematurely in trout with the aid of pituitary glands of carp. **Endocrinology,** 25(6) 978-983.

17. Khan, H. (1938) Ovulation in fish (Effect of administration of anterior lobe of pituitary gland). **Curr. Sci,** 7(5) 233-234.

18. Chaudhuri, H. (1956) Effect of pituitary injections and other factors on reproduction of pond fishes. **Ala. Pol. Inst. Bull.** (M. S. Dissertation).

19. Chaudhuri, H. and Alikunhi, K. H. (1959) Observations on the spawning in Indian carps by hormone injection. **Curr. Sci,** 26(12) 381-382.

20. Chattopadhyay, N. R., Mazumder, B. and Mazumdar, B. (2002) Induced spawning of *Pangasius sutchi* with pituitary extract. **Aquaculture Asia,** (1)10-13.

21. Alikunhi, K. H. and Sukumaran, K. K. (1964) Preliminary observation on Chinese carps in India. **Proc. Indian**. Acad. Sci, 60(3) 171 –88.

22. Chaudhuri, H. (1960) Experiments on induced spawning of Indian carps with pituitary injection. **Indian. J. Fish,** 7(1) 20-49.

23. Chaudhuri, H. (1967) Breeding and selection of cultivated warm water fishes in Asia and far East. **FAO. Fish. Rep,** 4(44) 33-66.

24. Chondar, S.L. (1970) Hyphophysation of major carps. Satish Book Entertprise, Agra-3, India. pp,24-99.

25. Bhimchar, B. S. and Tripathi, S. D. (1972) Hypophysation of fish with particular reference to India. Jawaharlal Nehru Krishi Vishwa Vidyalaya Jabalpur. **Rev. Ser. Dir. Res. Serv.** (1) pp 80.

26. Bardach, J. E., Ryther, J. H. and Mc Larney, W. O. (1972) Aquaculture, the farming and husbandry of freshwater and marine organisms. Newyork. Wiley- Interscience. pp168.

27. Jhingran, V. G. (1973) Fish and fisheries of India. Hindustan Pub. Corp. New Delhi, India.

28. Woynarorich, E. and Horvath, L. (1980) The artificial propagation of warm-water finfishes- A manual of extension. Food and Agriculture Organization of the United Nations, Rome, Italy. **Fisheries Technical Paper,** 201, pp183.

29. Ngamvongchon, S., Pawaputanon, O., Leelapatra, W. and Jhonson, W.E. (1988) Effectiveness of an LHRH analogue for the induced spawning of carp and catfish in northeast Thailand. Aquaculture, (74) 35-40.

30. Kumarasiri, W. S. A., Seneviratne, P. (1988) Induced multiple spawnings of Chinese carps in Sri Lanka. **Aquaculture,** (74) 57-62.

31. Leelapatra, W. (1988) Carp culture in Thailand with particular emphasis on induced spawning. In proceedings of the Aquaculture International Congress and Exposition, Vancouver, BC, 6-9 September, 1988, pp.331-337.

32. Nandeesha, M. C., Das, S. K., Nathaniel. D. E. and Varghese, T. J. (1990) Project report on breeding of carps with ovaprim in India. Special Publication no. 4, Asian Fisheries Society, Indian Branch, Mangalore, India.

33. Harvey, B. J. and Hoar W. S. (1979) The theory and practice of induced breeding in fish. IDRC- TX 21e. pp 48.

34. Lin, H. R., Kraak, G. V. D., Liang, J. Y., Peng, C., Lu, I. Z., Zhou, X. J., Chang, M. L. and Peter, R. E. (1986) The effects of LHRH analogue and drugs which block effects of dopamine of gonadotropin secretion and ovulation in fish cultured in China. In: *Aquaculture of cyprinids*, eds R. Billard, R. and Marcel, J. (INRA, Paris, France), pp. 139 -150.

35. Lin, H. R., Van Der Kraak, G., Liang, J. Y., Peng, C., Li, G. Y., Lu, L. Z., Zhou, X. J., Chang, M. L. and Peter, R. E. (1986) The effects of dopamine on gonadotropin secretion and ovulation in fish cultured in China, In: *Aquaculture of cyprinids*, eds R. Billard, R. and Marcel, J. (INRA, Paris, France), pp. 139-150.

36. Lin, H. R., Van Der Kraak, G., Zhou, X. J., Liang, J. Y., Peter, R. E., Rivier, J. E. and Vale, W. W. (1988) Effects of (D-Arg[6], Try[7], Leu[8], Pro[9] NEt)- leutinizing hormone–releasing hormone (LHRH-A), in combination with pimozide or domperidone on gonadotropin release and ovulation in the Chinese loach and common carp. **Gen. Comp. Endocrinol,** (69):31-40.

37. Peter, R. E., Lin, H. R., Van der Kraak, G. (1988) Induced ovulation and spawning of cultured freshwater fish in China: Advanced in application of GnRH analogues and dopamine antagonists. **Aquaculture,** (74)1-10.

38. Sahu, A. K. and Biswal, N. C. (1988) Use of pimozide in combination with LHRH-A on the induced spawning of Indian major carps, *Catla catla, Labeo rohita* and *Cirrhinus mrigala*. Proceedings of the national symposium on current status of general and comparative Endocrinology. 25-27 Nov., Delhi, pp. 113- 114.

39. Rath, S. C., Kumar, K. and Sarkar, S. K. (2001) Observation on induced breeding of Indian major carps using wova-FH and carp pituitary extract In: *Captive Breeding for Aquaculture and Fish Germplasm Conservation.* eds. Ponniah, A. G., Kuldeep K. Lal, Basheer. V. S. NBFGR-NATP Publication No:3. contributed paper No.17. pp. 1-3.

40. Das, S., Bhaumic, S. and Raut, S. K. (1998) Aggressive behaviour of *Channa punctatus.* **J. Bombay Nat. Hist. Soc,** 95 (3) 517-518.

41. Chakrabarti, N. M. (1998) Biology, Culture and Production of Indian major carps, Narendra Publishing House, Delhi.

42. Gupta, S. D, Khan, H. A., Tantla, M. S. and Das, R. C. (1990) Effects of Winstrol and progesterone on in vivo oocyte maturation and spawning in *Labeo rohita.* P. 24-28. In: *Carp seed production technology,* eds P. Keshavanath and K. V. Radhakrishnan Proceedings of the workshop on carp seed production technology, 2-4 September.

43. Kaul, M. and Rishi, K. K. (1986) Induced spawning of the Indian major carp *Cirrhinus mrigala* with LHRH analogue or pimozide. In *Aquaculture of Cyprinids* eds Billard, R., Marcel, J.(Institut National de la Recherche Agronomique, Paris, France), pp 167.

44. Peter, R. E., Chang, J. P., Nahorniak, C. S., Omeljaniuk, R. J., Sokolowska, M., Shih, S. R. and Billard, R. (1986). Interaction of catecholamines and sGnRH in regulation of gonadotropin secretion in teleost fish. **Recent Prog. Horm. Res,** (42) 513-548.

45. Madhusoodana Kurup, B. (2001) Early life history of endangered carp *Labeo dussumieri* (Val.) In: *Captive breeding for Aquaculture and Fish Germplasm Conservation.* eds. Ponniah, A.G., Kuldeep K. Lal and Basheer, V. S. NBFGR-NATP Publication No. 3 contributed paper 29, pp 1-4.

46. Azad, I. S. and Shimray, D. K. (1991) First success in induced breeding of Indian and exotic carps in Manipur using Ovaprim–c. **Fishing Chimes,** Oct–1991, 28-29.

47. Panday, A. K., Patiyal, R. S., Upadhyay, J. C., Tyagi, M. and Mahanta, P.C. (1998) Induced spawning of the endangered golden mahseer, Tor putitora, with ovaprim at the state fish farm near Dhradun. **Indian J. Fish,** 45(4) 457-459.

48. Haniffa, M. A. and Marimuthu, K. (2004) Induced Breeding in threatened, vulnerable and endangered fish species-A review. In: Fisheries Biotechnology. eds. Lakra, W.S., Abidi, S. A. H., Mukherjee, S. C. and Ayyappan, S. Narendra Publishing House, Delhi. pp 209-216.

49. Haniffa, M.A., Merlin, T. and Shaik Mohamed, J. (2000) Induced spawning of the striped murrel *Channa striatus* using pituitary extracts, human chorionic gonadotropin, luteinizing hormone releasing hormone analogue and ovaprim. **Acta Icht. Piscat,** 30(1) 53-60.

50. Marimuthu and Haniffa, (2004) Seed production and culture of snakehead. **INFOFISH International,** (2)16-18.

51. Qasim, S. Z. and Qayyum, A. (1961) Spawning frequencies and breeding seasons of some freshwater fishes with special reference to those occurring in the plains of northern India. **Indian J. Fish,** 8 (1) 24-43.

52. Haniffa, M.A., Sridhar S. (2002) Induced spawning of the spotted murrel *Channa punctatus* and the catfish *Heteropneustes fossilis* using ovaprim and human chorionic gonadotropin (HCG). **Vertrinarski Archiv,** 72(1) 51-56.

53. Banerjee, S. R. (1974). Hypophysation and life history of *Channa punctatus* (Bloch). **J. Inland Fish. Soc. India,** (6) 62-73.

54. Parameswaran, S. and Murugesan, V. K. (1976) Observations on the hypophysation of murrels (*Ophiocephalidae*). **Hydrobiologia,** 50(1) 81-87.

55. Haniffa, M.A., Shaik Mohamed, J. and Merlin Rose, T. (1996) Induction of ovulation in *Channa striatus* (Bloch) by SGNRH. **Fishing Chimes,** 16(5) 23-24.

56. Marimuthu, K., Haniffa, M.A., Muruganandam, M. and Jesu Arockia raj, A. (2001) Spawning and parental behaviour of the induced bred murrels. *Indian J. Fish,* 48 (4) 409-411.

57. Haniffa, M. A., Marimuthu, K., Nagarajan, M., Jesu Arockiaraj, A. and Kumar, D. (2004) Breeding behaviour and parental care of the induced bred spotted murrel *Channa punctatus* under captivity. **Curr. Sci,** 86(10) 1375-1376.

58. Huntingford, F. A. (1976) The relationship between antipredator behaviour and aggression among conspecifics in the three spined stickleback, *Gasterosteus acuteatus.* **Anim. Behav,** (24) 245-260.

59. Marimuthu, K., M.A. Haniffa, M. Muruganandam and A.Jesu Arockia Raj (2001) Low cost murrel seed production technique for fish farmers. **ICLARM NAGA,** (24) 21-22.

60. Billard, R., Reinaud, P., Hollebecq, M. G. and Breton, B. (1984) Advancement and synchronization of spawning in *Salmo gaireneri* and *trutta* following following administration of LHRHa-combined or not with pimozide. **Aquaculture,** (43) 57-66.

61. Lagler, K. F. (1982) Freshwater Fishery Biology. W.M. C. Brown Company Publishers. Iowa, pp.108-109.

62. Legender, M. (1986) Seasonal changes in sexual maturity and fecundity and HCG induced breeding of the *Heteropneustes longifilis* val. (Flarridae) reared in Ebrie lagoon (Ivory coast). **Aquaculture,** (55) 201-213.

63. Haniffa, M.A., Marimuthu K., and M. Muruganandam, (2001) Mass breeding of the striped murrel by the synthetic hormone GnRHa (ovaprim). **Fishing Chimes,** 21 (10 and11) 53-54.

64. Murugesan, V. K. and Kumaraiah, P. (1984) A simple technique for mass breeding of murrels. **J. Inland Fish. Soc. India,** (16) 15-18.

65. Parameswaran, S. and Murugesan, V. K. (1976) Breeding season and seed resources of murrels in stamps in Karnataka State.J. **Inland Fish. Soc. India,** (8) 60-67.

66. Munshi, D. T. S. and Hughes, G.M. (1992) Air breathing fishes of India. Oxford and IBH Publishing Co. Pvt. Ltd. New Delhi, pp.181–208.

67. Haniffa, M. A., Nagarajan, M., Marimuthu, K. and Jesu Arockia Raj, A. (2003) Embryonic and larval development of spotted murrel *Channa punctatus*. **Indian J. Fisheries,** 50 (3) 355-362.

68. Boglione, C. B., Bertolini, M., Russiello, S. and Cataudella (1992) Embryonic and larval development of the thick-lipped mullet (*Chelon lobrosus*) under controlled reproduction conditions. **Aquaculture,** (101) 349-359.

69. Marimuthu, (2002) Induced spawning in snakehead Channa sps using natural and synthetic hormone., Ph.D thesis, Manonmanium Sundaranar University, Thirunelveli, India.

70. Francis, T., Ramanathan, N., Athithan, S and Cheryl, H.F. (2000) Induced breeding of murrel, Channa striatus using various inducing agents. **Fishing Chimes,** 19 (10 and 11) 119-121.

71. Mollah, M. F. A. and Tan, E. S. P. (1983) HCG- induced of the catfish, *Clarius macrocephalus* (Gunther). **Aquaculture,** (35) 239-247.

72. Saidin, T. and Othman, A. F. (1986) Induced spawning of *Pangasius sutchi* (fowler) using an analog of luteinising releasing hormone and homoplastic pituitary extract, Proceedings of the first Asian Fisheries Forum. pp 687-688.

73. Viveen, W. J. A. R., Richter, C. J. J., Van Oordt, P. G. W. J., Janssen, J. A. L. and Huisman, E. A. (1997) Practical manual for the culture of the African catfish (*Clarias gariepinus*). Directorate General international Cooperation of the Ministry of Foreign Affairs, The Hague, the Netherlands. Department of fish Culture and Fisheries of the Agricultural University of Wageningen, the Netherlands.

74. Ramamohana Rao, S.D., Tripathi and Sahu, A.K. (1994) Breeding and Seed production of the Asian catfish *Clarias batrachus* (*Linnaeus*). Mnaual Series.3. Central Institute of Freshwater aquaculture, Bhubaneswar, India.

75. Sridhar, S. and Haniffa, M.A. (1998) Interspecific hybridization between the catfishes *Heteropneustes fossilis* (Bloch) and *Heteropneustes microps* (Gunter) (Siluridae: Heterponeustidae). **Curr. Sci,** (76) 871-873.

76. Alok, D., Krishnan, T., Talwar, G. D. and Garg, L. C. (1993) Induced spawning of catfish, *Heteropneustes fossilis* (Bloch), using D. Lys salmon gonadotropin- releasing hormone analog. **Aquaculture,** (115), 159-167.

77. Fermin, J. D. T. (1992) Induction of oocyte maturation and ovulation in the freshwater Asian catfish. *Clarius macrocephalus* by LHRHa and pimozide. **J.Appl. Ichtyol,** (8) 90-98.

78. Salami, A. A., Oyedapo, A., Frghero., Edibile, C. and Fagbemiro, S. O. (1994) Induced spawning of the African catfish *Clarias gariepinus* using non-piscine pituitary extracts. **J. World Aqua. Soc,** 25(1) 265-268.

79. Kohil, M.P.S., Goswami, U.C. (1987) Spawning behaviour of a freshwater air breathing Indian catfish *Heteropneustes fossilis* (Bloch), **Matsya,** (12) 180-183.

80. Zairin, M. Jr., Furukawa, K. and Aida, K. (1992) Induction of ovulation by HCG in the tropical walking catfish *Clarias batrachus* under 23–25^0 C. **Bull. Japan Soc. Sci. Fish,** (59) 1681–1695.

81. Zonneveld, N., Viveen, W. J. A. R. and Mudana, W. (1988) Induced spawning and egg incubation of the Asian catfish *Clarias batrachus.* **Aquaculture,** (74) 41-47.

82. Vijayakumar, C., Sridhar, S. and Haniffa, M. A. (1998) Low cost breeding and hatching techniques for the catfish (*Heteropneustes fossilis*) for small-scale farmers. **Naga. ICLARM,** 24 (4) 15-18.

83. Marimuthu, K., Muruganandam, M., Jesu Arockia Raj, A. and Haniffa, M. A. (2001) Induced spawning of the Indian catfish *Heteropneustes fossilis* using synthetic hormone, ovatide. **Fishing Chimes,** 19 (10and11) 105-106.

84. Jesu Arockia Raj, A., Haniffa, M. A., Raja Raja Perumalsamy, P., Marimuthu, K. and Muruganandam, M. (2001) Induced spawning of the freshwater catfish "singhi" *Heteropneustes fossilis* using non-piscine pituitary extracts. **Aquacult,** 2 (1) 5-8.

85. Sridhar, S., Vijayakumar, C. and Haniffa, M.A. (1998) Induced spawning and establishment of a captive population for an endangered fish, *Ompok bimaculatus* in India. **Current Science**, 75(10) 1066-1067.

86. Haniffa, M. A., Jesu Arockiaraj, A. and Arul Mozhi Varma, T. (2001) Optimum rearing conditions for successful artificial propagation of catfish. In: *Captive Breeding for Aquaculture and Fish Germplasm Conservation.* eds. Ponniah, A. G., Kuldeep K. Lal, Basheer. V. S. NBFGR-NATP Publication No:3. Invited paper No. 4. pp 1-11.

87. Alok, D., Krishnan, T., Talwar, G. D. and Garg, L. C. (1998) Multiple induced spawning of the Indian catfish *Heteropneustes fossilis* (Bloch) within a prolonged spawning season. **World Aqua. Soc,** 29(2) 252-254.

88. Haniffa, M. A. (2002) Germplasm inventory, evaluation and gene banking of freshwater fishes. Annual Progress Report, NBFGR, Lucknow.

89. Haniffa, M.A., Nagarajan, M., Marimuthu, K., Raja Raja Perumalsamy, P.and Jesu Arockia Raj, A. (2002) Induced Spawning and Larval Rearing of Indian Catfish *Mystus gulio* (Hamilton) Under Captive. **Aquaculture in the Tropics,** (Communicated).

90. Haniffa, M. A. (2003) Germplasm inventory, evaluation and gene banking of freshwater fishes. Annual Progress Report, NBFGR, Lucknow.

91. Haniffa, M. A. (2004) Germplasm inventory, evaluation and gene banking of freshwater fishes. Annual Progress Report. NBFGR, Lucknow.

92. Rahaman, M. K., Akhter, J. N., Mazid, M. A. and Halder, C. G. (1992) Culture feasibility of exotic catfish *Pangasius sutchi* (Fowler) in Freshwater Ponds of Bangladesh. **J. Inland Fish. Soc. India,** 25(2) 26-30.

93. Rahaman, M. K., Akhter, J. N., Mazid, M. A. and Halder, C. G. (1993) First record of Induced Breeding of Thai Panyas, *Pangasius sutchi* (Fowler) in Bangladesh **J. Inland Fish. Soc. India,** 25 (2) 26-30.

94. Clemens, H. B. and Sneed, E. K. (1971) Bioassay and use of pituitary materials to spawn warm-water fishes. **U.S. Fish and Wildlife Serv. Res. Rep,** (61) pp 30.

95. Hogendoorn, H. (1980) Controlled propagation of the African catfish *Clarias lazera* (C and V) III. Feeding and growth of fry. **Aquaculture,** (21) 233–244.

96. De Leeuw, R., Goos, H. J. T., Richter, C. J. J. and Eding, E. H. (1985) Pimozide–LHRHa induced breeding in the African cat fish, *Clarias gariepinus* (Burchell). **Aquaculture,** 44 (4) 295-302.

97. Obi, A. and Ewemade, C. (1999) Homoplastic hypophysation of African catfish *Clarias gariepinus* using male, female and a combination of male and female hypophysis. **J. Aqua. Trop,**14 (4) 261 – 267.

98. Thakur, N. K. (1978). Investigations on the biology of an airbreathing catfish *Clarias batrachus* (Linn.) Ph. D. thesis. Univ. of Bihar, Muzaffar pp 176.

99. Rao, G. R. M., Janakiram, K. and Muduli, H. K. (1989) Nat. Seminar Forty years of freshwater Aquaculture in India, Abstr. No. 111/3, 7-9.

100. Haider, S. and Rao, N. V. (1994) Induced spawning and maturity of Indian catfish *Clarias batrachus* (L) using low doses of steroid hormone and SGnRH. **Aqua. Fish. Manag,** (25) 401-408.

101. Das, S. K. (2000) Major carps and magur breeding success in Assam with Hungarian ovopel. **Fishing Chimes,** 20(6)10-11.

102. Ramaswamy, L. S. and Sundaraj, B. I. (1997) Induced spawning in the catfish *Clarias batrachus.* **Nature Wisseuchafton,** (44) 384.

103. Thakur, N. K. and Das, P. (1974) Length weight relationship of *Heteropneustes fossilis.* **J. Inland Fish. Soc. india,** (6) 95–96.

104. Alok, D., Krishnan, T., Salunke, D., Gupta, H., Talwar, G. P. and Garg, L. C. (1994) Precocious ovarian recrudescence and induced off season spawning of *Heteropneustes fossilis* by D-Lys[6] SGnRH–A. **J. Fish. Biol,** (45) 909–915.

105. Nayak, P. V., Mishra, J., Singh, B. N. and Ayyappan, S. (2000) Low doses of steroid hormones and SGnRH induced oocyte maturation in catfish *Heteropneustes fossilis.* **Indian J. Fish,** 47 (4) 321- 327.

106. Carreon, J. A., Estocapio, F. A. and Enderez, E. M. (1976) Recommended procedure for induced breeding and fingerling production of *clarias macrocephalus.* **Aquaculture,** (8) 269 –281.

107. Inyang, N. M. and Hettiarachehi, M. (1994) Efficiency of human chionic gonadotropin and crude extract of fish and frog in oocyte maturation and ovulations in African catfish *Clarias gariepinus,* Burchell. **Aquacult. Fish. Manag,** (25) 245–258.

108. Young, M. J. A., Fast, A. and Olin, P.G. (1989) Induced maturation and spawning of the Chinese catfish *Clarias fuscus.* **J. World Aqua. Society,** (20)7-11.

109. Fagbenro, O. A., Salami, A. A. and Sydenham, D. H. J. (1992) Induced ovulation and spawning in the catfish *Clarias isheriensis* (Claridae) using pituitary extracts from non-piscine sources. **Journal of Appl. Aqua,** (1) 15-20.

110. Ramanathan, N., Natarajan, P. and Sukumaran, N. (1985) Studies on the induced spawning and larval rearing of a freshwater catfish *Mystus punctatus* (Jerdon). **Indian Acad. Sci. (Anim. Sci.)** 94 (4) 389-398.

111. Gupta, S. D., Reddy, P. V. G. K., Rath, S. C., Dasgupta, S., Sahoo, S. K. and Sar. U. K. (1993). Induced breeding and larval rearing of the Indian catffish *Mystus aor* (Ham). **J. Inland Fish. Soc. India,** 52 (2) 51-53.

112. Choi, G. C., Kim, D. S., Jo, J. and Kim, J. M. (1992) Induced breeding and indoor culture of catfish *Silurus asohes.* **Aquaculture,** 5(2)117-126.

113. Nwadukwe, F., Ayinla, O. A. and Kalio, N. J. A. (1993) Effects of various dose of acetone dried powered carp pituitary extracts and season on hatchery propagation of *Heterobranchus longifilis.* **J. Aqua. Trop,** (8) 33-40.

114. Kim, D. S., Choi, G. C. and Kim, I. B. (1990). Induced spawning of channel catfish *Ictalurus punctatus.* **Aqauculture,** 3 (1) 25–30.

115. Chaudhuri, H. and Alikunhi, K. H. (1957) Observations on the spawning in Indian carps by hormone injection. **Curr. Sci.** 26:381-382.

116. Mahanta, P. C. (2001) Historical Perspective of Domestication of Fish in India and Some Prerequisite for Developing a Germplasm Repository. In: *Captive breeding for Aquaculture and Fish Germplasm Conservation.* eds. Ponniah, A.G., Kuldeep K. Lal and Basheer, V. S. NBFGR-NATP Publication No. 3 Contributed paper No.3. pp 1-4.

117. Sarangi, N., Tripathy, S. K., Ahlawat, S. P. S. and Mandal, A. B. (2001) Captive Breeding-A Boon for Mass Seed production and Genetic Improvement of Indian Major Carps- An Endeavour in Bay Islands. In: **Captive breeding for Aquaculture and Fish Germplasm Conservation**. eds. Ponniah, A.G., Kuldeep K. Lal and Basheer, V. S. NBFGR-NATP Publication No. 3 Contributed paper No.65. pp 1-4.

118. Penman, J. D. (1999) Biotechnology and Aquatic Genetic Resources: Genes and Genetically Modified Organisms. In: **Towards Policies for Conservation and Sustainable Use of Aquatic Genetic Resources**. *e*ds. by Roger S. V. Pullin, Devin M. Bartley, Jan Kooiman. Proceedings of an international conference held on 14-18 April 1998 in Bellagio, Como, North Itally. ICLARM, contribution No.1537. pp 17-22.

Advances in Biochemistry and Biotechnology Volume 1 ***Pages 245-280***
***Edited by:* Dr. Chiranjib Chakraborty**
***Published by:* DAYA PUBLISHING HOUSE**

11

Pharmacogenomics Based Emerging Drug Discovery Strategies

S.A. Bapat & G.C. Mishra*
National Centre for Cell Science, NCCS Complex, Ganeshkhind, Pune - 411 007, Maharashtra,
E-mail: gcmishra@nccs.res.in

ABSTRACT

High-throughput DNA sequencing has forced a re-definition of the paradigm for identification and validation of targets for drug development. The review highlights the emerging methodologies currently being applied in new drug design based on genomics and proteomics. The genomic target winnowing process is executed through three distinct phases *viz.* target screening, identification and validation, that finally leads to development of potential drug candidates. A plethora of novel approaches are being evolved and upgraded at each phase to address the specific queries that are associated with the emergence of a new generation of biopharmaceuticals. Some of the specific approaches at the target screening phase include database panning, transcriptome and proteome analysis, sequencing to locate polymorphisms, and monitoring of gene expression that can provide clues about the genomic response to disease

* Correspondent Author

and treatment. Further, the target identification phase involves elucidation of functional genetic networks and protein pathways to identify specific gene expression and protein–protein interactions. Finally, target validation through the down regulation of gene expression, protein inhibition and use of cellular assays and model organisms, help in arriving at molecular leads that could be optimized and developed as potential candidates with application in disease diagnosis and patient prognosis.

Keywords: Drug discovery, Genomics, Proteomics, target screening–identification-validation, biopharmaceuticals

Abbreviations

High-throughput screening (HTS); National Center for Biotechnology Information (NCBI); Haplotype Pattern Mining–HPM; Expressed sequence tags (ESTs); G-protein coupled receptor (GPCR); Age-related macular degeneration (AMD); Serial analysis of gene expression (SAGE); (RT-PCR); open reading frame (ORF); matrix-assisted laser desorption/ionization (MALDI); electrospray ionization (ESI); Protein Interaction Maps (PIMs); structure-to-activity relationships (SARs); Rolling Cirle Amplification (RCA); phenylisothiocyanate (PITC); polyvinylidene diflouride (PVDF); metabolic control analysis (MCA); Virtual screening (VLS); PISA (protein *in situ* array); green fluorescent protein (GFP); Thymidylate synthase (TS).

Introduction

The deciphering of the base sequence of the human genome has unveiled several new perspectives towards the understanding of human diseases and development of newer therapies [1]. The data published so far has startled biomedical researchers with the realization that there are only about 30,000 genes of which approximately 26,000 code for specifically identified proteins. RNA splicing generates multiple transcripts from a single gene, which, in turn, can lead to a greater number of expressed protein products than genes encoded in the genome [2]. Single nucleotide polymorphisms (SNPs) recur at an average spacing of 1 kb. Human populations are 99.99% identical, and most sequences are shared between people from different continents. The completion of the sequencing activity has thus, generated newer basic challenges in finding genes, locating their coding regions and predicting their functions. This will result in an increased understanding of human biology as well as in the design of new molecular structures in the fields of diagnostics, therapeutics and pharmacogenomics.

Emerging trends thus, indicate a shift in the scientific paradigm: in the pre-pharmacogenomics era, functional assignments were experiment-based, to prove certain hypotheses. The new approaches are based on high-throughput methods to predict genes, three-dimensional protein structures and describe protein networks on a proteome-wide scale. This information may then be used for the design of new hypotheses that are validated using classical physiological, biochemical and pharmacological tools. It is largely believed that the pharmacogenomic research over the last few years has the potential to expand the range of drug treatment options, as the tools of the genome program can be used to evaluate changes in the expression of many or all genes of an organism at the same time in order to investigate genetic pathways for normal development and disease. The ultimate impact and value of the human genome project will be recognized by the improvements seen in clinical outcomes which is the underlying feature of the emerging field of pharmacogenomics.

Pharmacogenomics is often understood on two different levels. In a research connotation, it is primarily defined as the identification of loci that are involved in determining the responsiveness and distinguishing responders and non-responders to a given drug. [3]. This is what is now identified as personalized medicine, taking individual variations into account up front and using them to decide not only what medicine to administer but also what tests to run, what lifestyle to recommend, and ultimately how to structure appropriate health and life insurance packages. Thus, pharmacogenomics predicts genetic risk factors and takes into consideration the hereto very often overlooked "protective genes" that also explains why certain high-risk patients do not display the symptoms of that particular disease.

Although environment, diet, age, lifestyle, and state of health all can influence a person's response to drugs, understanding an individual's genetic makeup is thought to be the key to creating new generation drugs with greater efficacy and safety. This is the other, more tangible interpretation of pharmacogenomics that underlines its application to identify potential novel targets for drug design. Traditionally, drug discovery programmes have been based on the high-throughput screening (HTS) of likely targets with the aim of identifying small-molecule antagonists or agonists at appropriate targets. Obviously, this approach requires a prior knowledge of the target. However, many of the 30,000 genes within the human genome that code for novel proteins could also be important targets for drug development. Without prior knowledge of the function of these gene products, classical pharmacological approaches are not feasible. Pharmacogenomic approaches involve the identification of which novel gene products might potentially be important in view of therapeutical applications [4]. Such approaches, in concurrence to the technical advancements in other fields such as pharmacodynamics, pharmacokinetics, pharmaceutics and toxicology will further enable earlier and improved selection of developing candidates, ending potentially in more effective medicines.

Biopharmaceutical research is thus on a steep trajectory towards drug discovery. A tremendous information overload has ensued due to the data harvested through bioinformatics, genomics and proteomics. A more focused approach hence becomes mandatory in order to identify specific molecular targets encoded by genes of known sequence and with a novel physiological function that can be validated. This is achieved by fragmenting the genomic target-panning process into three major tasks and critically evaluating the potential leads from each stage".

1. *Gene Target Screening*, which involves "earmarking" molecules that may be associated with a disease process,
2. *Molecular Target Identification*, the process of identifying molecules that clearly play a role in a disease process, and
3. *Drug Target validation*, the process of determining which among these molecules leads to a phenotypic change when modulated, suggesting it may have value as a therapeutic target.

Further, any validated drug target, in order to become commercially viable, must be amenable to drug development, and, most importantly, must address an unmet medical need of a large market size. The following review presents an update on the three main phases of target evaluation based on pharmacogenomics as well as the tools developed for analyzing the results of each phase.

Gene Target Screening

The role of screening for drug discovery has changed significantly in recent years as a result of the move to much higher throughput technologies, including miniaturization, greater diversity of lead sources, and increasingly sophisticated assays, as well as more extensive genomic information for identification and validation of targets. The output of the screening process is the identification of possible gene(s)/molecule(s) associated with a certain disease process, which can be used as targets in drug development. The following appoaches are being applied:

1. Bioinformatics
2. Mining of Differential Gene Expression Data and Gene Sequence Homology data
3. Mining of Differential Protein Expression Data and Protein Sequence Homology Data

Bioinformatics and Target Screening

One feature of the emerging pharmacogenomics based technologies, is the generation of a massive amount of data. Handling this data, analyzing it, and making it available to other researchers in the same/related fields is a herculean task that has been made achievable only through bioinformatics which is thereby defined as an integration of mathematical, statistical and computer methods to analyze biological, biochemical and biophysical data. Bioinformatics in fact, will continue to remain a key feature of the genomics and proteomics revolution (Table 11.1). Data can be queried from genomic DNA sequences, spatial and temporal patterns of mRNA expression, protein structure, immunological reactivity and clinical outcomes. For this, data processing using pattern recognition algorithms for areas such as nucleic acid or protein sequence assembly, sequence alignment for similarity comparisons or phylogeny reconstruction, structure or activity-based modeling, motif recognition in linear sequences or higher-order structure, and common patterns of gene expression is continuously being enhanced. [5].

Table 11.1: Bioinformatics in Pharmacogenomics

1.	Development of data mining tools,
2.	Identification of genes of interest,
3.	Establishing functions for the gene of interest- functional genomics and establishment of model systems,
4.	Diagnostic implications of new genes as early markers (cancer, obesity, diabetes, cardiovascular disease) in early intervention and patient stratification,
5.	Identify patient population that would respond to a given treatment because of specific genetic markers- *e.g.* Alzheimer's, Hypertension, cancer etc.,
6.	Medical informatics-clinician's ability to use genomics and bioinformatics to benefit patients,
7.	Genes to drugs-ability to take the new gene and convert it into a therapeutic target–for ex. antisense and gene therapy
8.	Identification of new drug targets for therapeutic intervention.

Thus, current drug discovery programs integrate bioinformatics, genomics and protein biosciences in establishing what is often referred to as the "gene-to-screen" process. Adoption of the "gene-to-screen" paradigm results in a dramatic increase in the efficiency of the process of converting a novel gene coding for a putative enzymatic or receptor function into a robust and pharmacologically relevant high throughput screen [6]. The "gene-to-screen" process begins with identifying the best targets from the genome. Repositories of genomic and proteomic information can be considered master "inventory lists" of genes, their maps, functions and products, which may be tapped for yielding specific information regarding particular disorders.

For embarking on the mining, identification and validation activities, one must have access to bioinformatic sources/databases of sequences. A large number of databases for this purpose lie in the public domain and are available for screening. Specific algorithms are also being developed on a case-basis for mining and analyses of the data collected. A case in point is the use of Haplotype Pattern Mining–HPM [7], for gene localization. In HPM, association is analyzed by searching all trait-associated haplotype patterns using data mining algorithms. Haplotype-trait associations can be measured by linear models, and marker-wise patterns of association are used for predicting the disease gene location. HPM is being applied extensively [8] and is found to be useful for fine mapping of mutations in DNA sequences.

Mining of Gene Sequences (Sequence Homology Data) and Gene Expression Data

Traditionally, gene targets for blockbuster drugs have been switches in mammalian physiology, which can be pharmaceutically modulated to produce a therapeutic effect [9]. A majority of diseases at a molecular level, involve quantitative and qualitative changes in gene expression. A high throughput analysis of gene sequences and differential gene expression is an essential focus from where drug evaluation can be initiated. In recent years, a variety of techniques have been developed to mine gene sequence data that can be categorized as:

1. Random cDNA sequencing,
2. mRNA display and
3. differential hybridization

Random cDNA Sequencing

These approaches involve sequencing and analyses of cDNA for identification of specific genes and is achieved through either of two techniques: expressed sequence tag and serial analysis of gene expression.

Expressed Aequence Tags–ESTs

ESTs, provide a quick and inexpensive route for discovering new genes, for obtaining data on gene expression and regulation, and for constructing genome maps. ESTs are sequences (usually 200 to 500 nucleotides long), generated by sequencing either one or both ends of an expressed gene. The database of human expressed sequence tags (dbEST) is a potential source for the identification of tissue specific genes. This database contains sequences from cDNA libraries from different tissues cell types and tumors. Computer based analysis can further be used to identify a cluster of sequence homologous ESTs, containing ESTs derived from tissue specific cDNA libraries. EST databases are probably the most abundant source of new coding sequences available today. By employing a database search tool, it is possible to conduct sequence homology searches against the database that has led to a revolution in the way new genes are identified.

There are several examples that stand witness to the widespread applications of EST databases. ESTs from solid tumors of six different organs were analyzed for differential expression [10]. Test genes were validated for expression specificity by RT–PCR, providing a proof of concept for gene discovery by Digital Differential Display. The study reported two hundred known genes and 500 novel sequences that were differentially expressed in select tumor-derived libraries. In yet another study, a new gene named as GDEP, has been recently identified as being differentially expressed in prostate carcinogenesis using such an approach [11].

EST database mining has been used for the identification of at least 14 new members of the G-protein coupled receptor (GPCR) superfamily [12]. The approach used proved to be especially useful for the detection of EST sequences that do not encode conserved parts of a protein, making it an ideal tool for the identification of members of divergent protein families or of protein parts without conserved domain structures in the EST database. In another report, a novel full-length human GPCR identified from the GPCR Resource has been identified, illustrating the utility of the method [13].

Age-related macular degeneration (AMD) is a multifactorial disorder affecting the visual system with a high prevalence among the elderly population but with no effective therapy available at present. An EST mining approach identified 39 transcripts with retina-specific expression; one of which was mapped to chromosome 18 and further characterized to display a complex pattern of differential splicing in the human retina. The various isoforms encode hypothetical polypeptides with no homologies to known proteins or protein motifs [14]. In another example, a data mining effort of a human failed heart cDNA library identified a novel kinase (MLK7) that is proposed to be a new member of the mixed lineage kinase family that modulates cardiac SAPK/JNK pathway and may play a role in cardiac hypertrophy and progression to heart failure [15].

New association rule mining methods for discovering nucleotide sequence patterns that appear in more sequences than expected within protein function classes are being developed. These involve derivation of associations between nucleotide patterns and protein function classes when applied to a database of human 3′ UTR sequences. Among previously identified patterns, the AU-Rich Element is found to occur within the 3′ UTR of cytokines, providing statistical validation of an association often reported in the literature [16]. The method has also identified some GC-rich patterns, found to occur within the 3′ UTR of homeodomain transcription factors and nuclear proteins and is predicted to lead to many types of regulatory element discovery.

The above examples are a mere representation of the staggering number of new transcripts/genes identified using ESTs. Undoubtedly, it is widely recognized that the generation of ESTs constitutes an efficient strategy that has also enabled the mapping of many genes to chromosomal sites, besides its primary role in gene discovery. However, some limitations are associated with the EST approach. Firstly, it is very difficult to isolate mRNA from some tissues and cell types. Another drawback is that, gene regulatory sequences may be found within an intron. Since ESTs are small segments of cDNA generated from a mRNA in which the introns have been removed, much valuable information may be lost by focusing only on cDNA sequencing. A constant upgradation in the computational skills being used (often on a individual case basis), is also required as the raw data deposited in the EST databases are to a large extent unorganised, unannotated, redundant and of relatively low quality [17]. Despite these limitations, ESTs continue to be invaluable in characterizing the human genome, as well as the genomes of other organisms.

Serial Analysis of Gene Expression–SAGE

SAGE™, is a commercially protected (U.S. Patent No. 5695937) sequence-based technology for gene identification and quantitation that provide a quantitative, alternative to the methods for deriving expression signatures from small amounts of biological material. In SAGE, simultaneous detection and measurement of the expression levels of virtually all genes expressed in a cell at a given time can be achieved. Three principles underlie the SAGE methodology:

1. A short sequence tag (10-14bp) contains sufficient information to uniquely identify a transcript provided that that the tag is obtained from a unique position within each transcript;
2. Sequence tags can be linked together to from long serial molecules that can be cloned and sequenced; and
3. Quantitation of the number of times a particular tag is observed provides the expression level of the corresponding transcript.

Since the first report of SAGE in 1995 [18], the technique has been instrumental in a several important discoveries. Some of the applications of SAGE in cancer biology, apoptosis, cell cycle regulation, expression of pleuripotency by embryonic stem cells, etc.are outlined in Table 11.2. One of the first reports of SAGE was to characterize the transcriptome expressed by the yeast genome [19]. In this study, the analysis of 60,633 transcripts revealed 4,665 genes with expression levels ranging from 0.3 to over 200 transcripts per cell. Such a wealth of information from one gene expression study was unprecedented and demonstrated the ability of SAGE to not only identify large numbers of transcripts at once but also to quantify their presence.

Table 11.2: SAGE Applications

1. Insights into p53-mediated apoptosis [20]
2. Identification and classification of p53-regulated genes [21]
3. Analysis of human transcriptomes [22]
4. Serial microanalysis of renal transcriptomes [23]
5. Genes Expressed in Human Tumor Endothelium [24]
6. Ferredoxin reductase affects 5-FU apoptosis [25]
7. Analysis of colorectal metastasis (PRL-3) [26]
8. Gene expression in primary human CD15(+) myeloid progenitor cells [27]
9. gene expression profiles in non-small cell lung cancer [28]
10. Mesothelin identified as a new marker for pancreatic adenocarcinoma [29]
11. Upregulation of Cystatin C, a cysteine protease inhibitor in neurons and glia in a rat model for mesial temporal lobe epilepsy [30]
12. Altered hippocampal gene expression prior to the onset of spontaneous seizures in the rat post-status epilepticus model [31]
13. Novel estrogen and tamoxifen induced genes in breast cancer(Seth P et al., , 2002)
14. Gene expression profiles in human cardiac cells subjected to hypoxia [32]
15. Investigation of the influence of the von Hippel-Lindau (VHL) gene on global gene expression profiles [33]
16. Gene transcript profiling in aging research [34]
17. Identification of gene transcripts with profiles unique to pluripotent mouse R1 embryonic stem cells [35]
18. Psoriasin expression in mammary epithelial cells in vitro and in vivo [36]
19. Differential gene expression in premalignant human epidermis [37]
20. cataloguing and measuring the relative levels of expression of the genes expressed in the human peripheral retina, macula, and retinal pigment epithelium (RPE) [38]

The original SAGE method contains several serious intrinsic problems, such as procedural difficulty in library construction, extremely short cDNA tags, etc. [39]. This has prompted the development of several case-specific value additions for instance, the development of GLGI–generation of longer cDNA fragments [40] in which, a cDNA fragment extending from the SAGE tag toward the 3′ end of the corresponding sequence can be generated. Application of the GLGI technique addresses two critical issues in SAGE: firstly, a longer fragment corresponding to a SAGE tag, which has no match in databases, can be generated for further studies; secondly, the specific fragment corresponding to a SAGE tag can be identified from multiple sequences that match the same SAGE tag. The technology is being continuously upgraded [41] to extend its current applications to large-scale identification of novel genes by converting novel SAGE tags into 3′ cDNAs.

The most attractive feature of SAGE is its ability to evaluate the expression pattern of thousands of genes in a quantitative manner without prior sequence information. This feature has been exploited in cancer biology for the definition of transcriptomes, the analysis of differences between the gene expression patterns of cancer cells and their normal counterparts, and the identification of downstream targets of oncogenes and tumor suppressor genes [42]. Such comprehensive analyses of gene expression not only will further the understanding of growth regulatory pathways and the processes of tumorigenesis but also may identify new diagnostic and prognostic markers as well as potential targets for therapeutic intervention.

mRNA Display

Although some changes in mRNA abundance may not have direct protein correlates, analysis of total mRNA present in a cell population has become a focal point of genomic research as it provides information about the dynamics of total genome expression in response to environmental changes and points to candidate genes responsible for the cascade of events that result in a disease state. A variety of methods usually based on the RT-PCR amplification of random subsets of genes from two or more RNA samples are available: differential display, RNA arbitrary primed PCR, gene expression fingerprinting and molecular indexing.

Differential Display

Since its invention [43], differential display has quickly overtaken most of the other approaches, to become the method of choice for cloning differentially expressed genes. The principle of mRNA differential display is that the mRNA content of two or more RNA extracts is compared following reverse transcription of the mRNAs with oligo(dT) primers anchored to the beginning of the poly(A) tail and subsequent amplification in the presence of a subsequent 10-mer arbitrary in sequence and a radiolabeled nucleotide. The amplified cDNA subpopulations of the 3′ termini of mRNAs as defined by the specific pair of primers are then fractionated on a sequencing gel and visualized by autoradiography. A comparison of the patterns enables the identification of PCR products that are differentially represented in the various extracts. The resulting PCR products can be further purified, re-amplified and either cloned or used as probes to screen cDNA libraries or carry out *in situ* hybridization. The use of numerous combinations of different primers in independent RT-PCR reactions ensures that each mRNA present in the extract is converted into a defined PCR product.

The differential display method is thus unique in its potential to visualize all the expressed genes in a eukaryotic cell in a systematic and sequence-dependent manner by using multiple primer combinations. It is often the method of choice for identifying differentially expressed genes because it:

1. Requires relatively little starting material
2. Can be used to screen a large proportion of the mRNA population and
3. Does not require the sequencing of large numbers of cDNA clones.

Unlike other techniques whose success is biased by relative mRNA abundances, differential display detects both abundant and rare mRNAs, as long as the arbitrary primer sequences match the target mRNA. Furthermore, the technique requires no previous knowledge of the genes within the study. The data presented in Table 11.3 highlights some of the recent applications of differential display technology for the elucidation of both drug efficacy and toxicity as well as novel candidate genes for pharmacogenetic analyses to assess individual variability to drug response.

Table 11.3: Applications of Differential Display Techniques in 2002

1. Expression of hGC-1 during normal myeloid-specific lineage differentiation [44]
2. Differential expression of E-selection in CNS and non-CNS microvascular endothelial cells[45]
3. Identification of a novel human liver cancer associated gene [46]
4. Specific proliferative dysregulation related gene expression during experimentally-induced atherogenesis [47]
5. Differential gene upregulation by 17-b estradiol in endothelial cells [48]
6. New intra-renal graft genes associated with tolerance or rejection [49]
7. Zfhep–suitable marker for angiotensin type 2 AT receptor activation [50]
8. Differential display of vincristine-resistance-related genes in gastric cancer [51]
9. Molecular Cloning and Functional Analysis of a Novel Cadmium-responsive Proto-oncogene [52]
10. Gene expression study in Down's syndrome [53]
11. Identification of the osteopontin gene as a direct target of TP53 [54]
12. Role of cyclophilin D in the ROS-mediated apoptotic effects of 4-HPR [55]
13. Importance of the interferon β as a negative regulator of hair growth [56]
14. prtb functions as a serum-responsive and adhesion-inducible gene [57]
15. 'VIT1', a novel gene associated with vitiligo [58]
16. Sjogren's syndrome (Tapinos NI *et al.*)Up-regulation of KEPI expression by morphine [59]
17. Co-relation of Urb upregulation in mice exhibiting mild late-onset obesity [60]

Despite its obvious advantages, numerous technical difficulties are associated with the use of the differential display method. These include the generation of irreproducible fragment patterns, the inability to resolve fragments adequately, and inaccurate or inefficient isolation of interesting bands. Several value additions are now commercially available including expanded coverage of expressed sequences, while simultaneously streamlining downstream characterization analysis.

RNA Arbitrary Primed PCR (RAP-PCR)

The products obtained from differential display gels derive almost exclusively from noncoding regions of genes, whose sequence analysis provides no clues regarding protein function. Thus, a different approach for fingerprinting of RNA populations using an arbitrarily selected primer at low stringency for first and second strand cDNA synthesis was developed [61]. During first-strand synthesis, a single 18-base arbitrary primer anneals and extends from sites contained within the mRNA. Second-strand synthesis is through a single round of low-stringency PCR with DNA polymerase. PCR amplification at high stringency proceeds by virtue of having incorporated the arbitrary primer in both ends of the cDNA. A template-dependent competition determines which potential PCR products will predominate. Differences in the electrophoretic patterns of PCR products amplified from two RNA samples arise from differentially expressed genes, as verified by Northern blot analysis.

The method has been applied to the detection of differences between RNA populations in a wide variety of situations ranging from heavy metal tolerance in *Acidithiobacillus ferrooxidans* [62], identification of differentially expressed genes in rheumatoid arthritis [63], isolation of genes of cholesterol metabolism-related proteins from rat liver [64], differential gene expression in primary, metastatic and recurrent lesions of endometrial cancer [65] and in attenuated and virulent strains of mycobacteria [66], etc.

The efficiency of this method is reflected from the fact that it can be applied when limiting amounts of mRNA are available (as in studies involving parasites, fetal tissues or tumor biopsies–[67]. Further, arbitrarily primed RNA fingerprinting can be applied for targeting coding regions of genes, as opposed to differential display techniques, which target 3′ noncoding cDNA. In order to be to permit a systematic survey of differential gene expression, RNA fingerprinting requires case-specific standardization and identification of a number of highly efficient and selective arbitrary primers [68]. With these, RAP-PCR fingerprinting promises to be a powerful tool for the isolation of differentially expressed genes in studies of neoplasia, differentiation, or development.

Molecular Indexing

Molecular indexing provides an ordered, non-cloning approach for defining, amplifying and isolating specific DNA fragments from complex genomic digests. Type IIS and Interrupted Palindrome (IP) restriction endonucleases cleave DNA at defined distances from their recognition sites, generating fragments with specific non-identical cohesive ends. (*e.g.* the Type IIS restriction enzyme *Fokl* recognizes the sequence GGATG and cuts 9 bases downstream, leaving a 4-base 9-cohesive end). These cohesive ends can be modified by ligation of specific indexers, to which various functional groups can be added to allow cohesive- end identification and fragment manipulation. Indexers consist of two annealed synthetic oligonucleotides: an indexing strand (typically 24 bases long) bearing a common primer-binding sequence with a variable 4-nucleotide 5′-cohesive end; and a complementary common primer strand (usually 20 bases). Selection of the appropriate indexer for each end of a targeted restriction fragment thus incorporates two common-primer-binding sequences onto the ends of that fragment, allowing specific amplification of the target fragment.

The integration of DNA indexing technology and cycle sequencing, provides an efficient system for informative differential display. Fundamental to the effective isolation of differentially expressed genes is the fact that indexing generates display patterns that are highly consistent and reproducible. Some of the recent applications of molecular indexing include the compilation of a comprehensive collection of mouse zinc finger motifs [69], screening of gene expression profiles in Alzeimer's disease [71], etc.

Development of indexing strategies has involved extensive troubleshooting and system diagnosis. Indexing reaction conditions have to be adapted with each model system. Due to the use of small numbers of common priming sequences for large numbers of amplifications of different target fragments, extensive measures are necessary to prevent PCR contamination. Long-term storage and purification procedures need to be developed to prevent low-level degradation of cohesive end sequences in digested DNA stocks. Automation would further be a significant value addition. Despites these limitations, the virtue of the indexing procedure remains in that it enables the visualisation of novel transcripts without the prerequisites for a gene to be represented in a cDNA library and selected for inclusion in an array.

Differential Hybridization

The various strategies that are considered under the differential hybridization procedures include subtractive hybridization and use of cDNA/oligonucleotide arrays.

Subtractive Hybridization

The method of subtractive hybridization has been the first application developed for the identification of differentially expressed genes. Hybridization of cDNA from one sample (tester) to an excess of mRNA from the second sample (driver) isolates differentially expressed genes, as transcripts expressed in both samples form a hybrid, while cDNA unique to the tester remains single-stranded. Single and double stranded molecules can be further separated using hydroxylapatite chromatography [72, 73].

Several modifications of the original protocol have been developed to increase its efficiency. The most widely used modification, [74,75], described as suppressive subtractive hybridization (SSH) is designed to selectively amplify differentially expressed transcripts while suppressing the amplification of abundantly expressed ones. This effectively eliminates the need to separate single and double stranded molecules, and also normalizes target transcripts to approximately equal abundance [76]. Other advantages of SSH include working with smaller amount of mRNA and elimination of bias towards abundant genes. However, limitations of this technology remain; only pair-wise treatment comparisons can be done, quantitative measure of differences in expression cannot be made and the procedure identifies genes that are absent, rather than weakly expressed in the driver sample.

cDNA/Oligonucleotide Arrays

This methodology, in principle, extends the hybridization-based methods such as Northern and Southern blots to identify and quantify gene expression on a genomic scale. Recent advances in materials, optics, electronics, robotics, chemistry, genetic engineering and informatics have permitted the development of integrated platforms allowing the parallel study of tens of thousands of transcripts in a variety of normal and pathological conditions. cDNA arrays (variably called microarrays, gene arrays and gene expression arrays) provide a medium for matching known and unknown DNA samples based on base-pairing rules and automating the process of identifying the unknowns. The use of cDNA microarray system requires the following basic steps:

1. Preparation of the large set of probes–which may be synthesizing short oligos or PCR amplified inserts of cDNA library clones;
2. Preparation of DNA microarrays/chips consisting of these spotted on glass slides/nylon membrane by a robotic printing device;
3. Preparation of two related mRNAs derived from cells that differ in the trait that is to be investigated;
4. Reverse transcription of these mRNAs into cDNA tagged with a fluorescent/radiolabel–fluorescent labels (Cye3 and Cye5-dUTP) are used generally with glass slides or radioactive [^{33}P]dCTP being used with nylon membranes;
5. Hybridization of the labeled cDNAs to the microarray chip–two experimental samples labeled with Cye3 and Cye5 (which are detected at different emission spectra), would undergo a single hybridization reaction and can be detected separately. Radiolabeled samples have to be hybridized individually in separate reactions;
6. Quantitative analysis of the relative abundance of mRNAs from the degree of hybridization–the intensity of the resulting hybridization signal for any sequence on the microarray is proportional to the quantity of mRNA corresponding to the sequence in the original mRNA sample;
7. Data visualization and clustering–involves the sorting and presenting of data in a logical manner.

Microarray data analysis can be divided into two tasks: grouping of genes to discover broad patterns of biological behaviour, and filtering of genes to identify specific genes of interest. The gene-grouping task is largely addressed by cluster analysis [77] to identify group(s) of genes expressing similar patterns, while the gene-filtering task relies primarily on hypothesis testing. [78]. The microarray technique thus, has massive implications for the understanding of the complicated genetic alterations involved in disease processes and also for the generation of accurate prognostic information and optimization of treatment in these situations [79].

Several companies are developing their own variants of mutation and polymorphism analysis chips, as well as ones for expression profile analysis and broader screening and diagnostics applications. In addition, further developments of the technical aspects of chip design, structure and function itself, such as microfluidic applications that are essential for the miniaturization of the large number of assays are in progress. Microarrays will thus, evolve both in terms of their physical characteristics and also in terms of the assays they carry out. For example, an interesting development is the coupling of so-called representational difference analysis (RDA) with high throughput DNA array analysis. RDA looks at cDNA from two tissues to be compared, for example a metastatic form of cancer vs. a non-metastatic one. The cDNA's from the tissues are compared in successive rounds and the different sequences are cloned, resulting in a subtracted cDNA library, which consists of the cDNA from one tissue minus that from the other. DNA inserts from the library are then used to build a high density DNA array, which is probed with fluorescently-labelled probes. Detailed explanation of the differences between these various array formats and other aspects of current DNA array technology can be found in the review by Rockett and Dix [80].

Presently, microarray applications seem to be limited in three main areas: their prohibitive cost; the restricted number of species for which they are available; the scope of genes available. Thus, although the number of genes in a microarray system seems very large the success of any study would rely on selection of the most appropriate cDNA library used for constructing the microarray. Most commercially available off-the-shelf arrays contain a cosmopolitan series of well characterized genes expressed across many cell types. As most researchers focus on one or two tissues only, this global approach to array production is somewhat wasteful. In many cases, it would be far more useful to report on the expression of only those genes that are part of the transcriptome of the tissue of interest, *e.g.* a group interested in the ovary may be better served by a chip containing only those genes expressed in that particular tissue.

Mining of Differential Protein Expression and Protein Structure Homology Data

Overshadowed by the exuberance for genomics are the observations that most disease processes and treatments are manifested at the protein level and that there may not be a direct correlation between gene expression and protein expression. Structurally, proteins are much complex than nucleic acids. They undergo several types of post-translational modifications such as acetylation, phosphorylation, glycosylation, farnesylation, etc. A single gene can code for several proteins that are produced by alternative splicing of the mRNA transcript, by varying translation start/stop signals or by frameshifting. Protein levels often do not reflect mRNA levels; even the presence of an open reading frame (ORF) need not indicate the presence of the corresponding protein. To complicate the situation, a single protein could perform several functions in a cell; conversely, several proteins can carry out a single function. A shift in localization of a protein to different subcellular locations, modifications and interactions such as cleaving, changes in binding activities, etc. are also known to ensue in response to changing environmental situations [81]. Thereby, the next logical approach in drug discovery is a study of the proteome of the experimental system. Two strategies are generally followed in proteomics: *protein profiling* (determination of the abundance, modification, and subcellular localization of proteins in a give cell or tissue) and *high–throughput function determination*.

Proteome databases integrate the expression levels and properties a large number of proteins with the genes identified on genetic maps and offer a global approach to the study of gene expression. Proteome database mining is used to identify intrinsic patterns and relationships in proteomics data. The identification of patterns in complex proteomic datasets provides two benefits *viz.* generation of insight into gene translation and post-translational modification conditions and characterization of multiple protein expression profiles in complex biological processes. Proteome databases have been extensively reviewed [82–84].

Screening libraries with a proteome mining approach accelerates the hit-rate over a conventional screen. The first step in protein data mining begin with protein profiling of a mixture, that incorporates three procedures: resolution, identification/sequencing and quantification. Protein resolution may be achieved using 2-D gel electrophoresis; identification through Edman degradation, mass spectrometry and western immunoblotting; and quantitation by using scanners and phosphorimagers.

Resolution

2-D PAGE

2-D PAGE can achieve a resolution of up to 10,000 protein spots per gel. This classical separation system has been found to be useful to isolate proteins for further identification by mass spectrometry sequencing. Stains such as Coomassie blue, silver, and SYPRO Ruby Red can be employed to visualize the proteins.

Identification/Sequencing

For identification of proteins and obtaining sequence data for probe design, mass spectrometry sequencing of a protein from a gel is the more efficient procedure because of the sensitivity and relatively simple sample preparation. For some analyses, Edman sequencing is the preferred technique.

Mass Spectrometry

Mass spectrometry is used for mass determination and can be adapted for protein identification. A mass spectrometer separates proteins (and other analytes) according to their mass-to-charge (m/z) ratio. The molecule is ionized by one of several techniques, and the ion is propelled into a mass analyzer by an electric field that resolves each ion according to its m/z ratio. The detector passes the information to the computer for analysis. Ionization methods frequently used are matrix-assisted laser desorption/ionization (MALDI) and electrospray ionization (ESI) because they cause negligible fragmentation of the molecule during the ionization and desorption process. MALDI is able to analyze proteins down to femtomole quantities and can tolerate small amounts of contaminants; therefore sample preparation is not as tedious as with ESI mass spectrometry. The information obtained from MALDI analysis can be automatically submitted to a data base search for further examination.

Proteolytic digestion of gel-separated proteins into peptides and mass analysis of the peptides provides a peptide mass fingerprint, which can be searched against theoretical fingerprints of sequences in the databases. Alternatively, the peptides can be sprayed into a tandem mass spectrometer that has the ability to resolve peptides, isolate one peptide and dissociate it into amino or carboxy terminal containing fragments. Although more complex than the peptide mass approach, the sequence information is more specific for the identification of a protein than a list of peptide masses. The data can be used to search protein sequence databases and also nucleotide databases.

Electrospray ionization (ESI) involves the production of gaseous ions by application of a potential to a flowing liquid resulting in the formation of a spray of small droplets with solvent-containing analyte. Solvent is removed from the droplet by heat or another form of energy such as collision with a gas, and multiply charged ions are formed. As the droplet size further decreases they reach a point at which become unstable and explode into even finer droplets. Finally electrostatic repulsion is sufficiently high to cause desorption of the analyte ions which are then passed to the mass spectrometer. Ions generated by ESI usually bear multiple charges. This can be transformed into a simple mass spectrum that reveals the molecular weights of the fragments. A quadropole analyzer is frequently used with ESI. Only ions of a certain m/z have the correct oscillation path that enables them to reach the detector in the presence of an electric field. Protein identification using ESI-MS/MS can be easily automated by using a liquid chromatography column may be used as the source for the proteins entering the ESI mass spectrometer.

Edman Degradation

The technique was first described for the sequential degradation of proteins and peptides in which, the amino(N)-terminal amino acid of a protein is specifically reacted with phenylisothiocyanate (PITC). This derivatized amino acid is then selectively removed, leaving the rest of the peptide chain intact. Each cycle of the degradation removes an amino acid from the N-terminal end of the protein or peptide; the resulting PTH-amino acids are analyzed sequentially to determine the amino acid sequence of the protein or peptide. The cycle is repeated which slowly reveals the primary structure of the protein

Several modifications of the method have been developed. One method is optimized for samples bound to polyvinylidene diflouride (PVDF) membranes. Statistical algorithms and software are also being developed to automatically determine the amino acid sequence of a peptide from the experimental output of Edman degradation. In the recent years, Edman sequencing has been useful in revealing the target antigen of anti-inner ear antibodies in autoimmune inner ear diseases [85], purification of proteins from human epidermal keratinocyte conditioned medium [86], purification of antimicrobial peptides [87], etc.

From the above information presented on bioinformatics, genomics and proteomics, it is apparent that the integration of experimentally derived peptide sequences, EST databases and genomic sequence database offers a unique view of genome organization and gene fine structure. These approaches enable the identification of the protein isoform most relevant to the system/disease under study, facilitate the identification of protein-coding genes not detected by gene prediction, and aid accurate and definitive identification of protein family members, including pharmaceutically-relevant target classes.

Molecular Target Identification

Target identification refers to the process of identifying molecules that play a definitive role in a disease process. This step in drug development provides a greater degree of certainty regarding identified targets, but it is still possible that these may not be the best approach to interfere with a disease process. At any given point of time, several completely new or poorly understood targets could be under analysis, which makes the choice regarding follow-up even more difficult. On the other hand, a glut of poorly defined targets creates a very costly and time-consuming bottleneck, slows the pace of drug development by introducing a large number of improbable targets in the pipeline -thereby increasing the rate of drug failure; failure costs are a big drain on profits.

Although database mining and transcriptional profiling clearly increases the number of putative targets, the next focus is to assign functions to new gene targets and the role of gene products in the cell *i.e.* to translate the data mining efforts into useful biological knowledge. Several approaches are being developed that make this process non-linear and high-throughput; some of which include: functional studies of specific genes in cells, genetic network or protein pathway studies, protein-protein interactions and protein functional studies.

Functional Studies of Specific Genes in Cells

In developing gene targets, it is necessary to understand the mechanisms that allow genes to produce a multitude of proteins with diverse functions, in order to devise high throughput strategies for drug development. Critical analytical tools are continuously being developed towards meeting this purpose. Gene sequence patterns generated by screening techniques are often clustered; which helps in identifying groups of genes that behave in a similar way. In the case of microarray data, gene clusters tend to change expression levels in a certain way in response to a variety of treatments. The theory is that if two genes are in a cluster, and the function of one is known, it is possible to infer the function of the other gene (perhaps by being in the same pathway). Specific rigorous selection criteria have to be used for identifying clusters that include a certain number of genes. For this, clustering algorithms are developed, based on discretising the time-series of the expression measurement space and including genes with expression profiles mapping to the same pattern in separate clusters. Some of the genes sharing similar expression profiles may also share similar transcription regulation mechanisms. Such studies are also applicable for correlating expression-profiling (transcriptome) data with protein-protein interaction data.

Selective expression of a gene product (mRNA or protein) is a pattern in which the expression is markedly high, or markedly low, in one particular tissue compared with its level in other tissues or sources [88]. However, mRNA content does not correlate faithfully with corresponding protein levels. An approach, in which only those mRNAs that are being translated could be isolated/concentrated, thus would be very useful. This is achieved by fractionation of cytoplasmic extracts in sucrose gradients, leading to the separation of free mRNPs (ribonucleoprotein particles) from mRNA complexes in ribosomal pre-initiation complexes and from mRNAs loaded with ribosomes/polysomes [89]. Pooled fractions can further be used to generate probes for array hybridization and changes in polysomal populations are then compared with a reference sample.

Several other systematic approaches to characterization of gene products have also been developed *e.g.* creation of living microarrays of cell clusters expressing the encoded gene products [90], custom-engineered biosensors for intracellular biochemistry, specifically localized targets and protein-protein interactions [91], etc. A combination of such methods can be used to parse the genome and identify targets worthy of inhibitor screens, *e.g.* two types of antibacterial targets have emerged from such an exercise–one *in vitro* target effective on synthetic media; the other, an *in vivo* target effective in the host organism [92]. Another example of successful functional genomics is the identification of 4 new neuropeptides (ligands) of orphan G protein-coupled receptors–GPCRs [93]. For this, reverse molecular pharmacological methodology was applied which comprises of cloning and expression of orphan GPCRs in mammalian cells and screening these cells for a functional response to cognate or surrogate agonists present in biological extract preparations, peptide libraries, and complex compound collections. Alternatively, 'humanized yeast cells' can be used, in which the yeast GPCR transduction system is engineered to permit functional expression and coupling of human GPCRs to the endogenous signalling machinery. Both systems provide an excellent platform for identifying novel receptor ligands. Once activating ligands are identified they can be used as pharmacological tools to explore receptor function and relationship to disease [94].

The advances in molecular biological, analytical and computational technologies have thus made possible investigations of the output of the gene regulatory network and unraveling the functional relationships implicit in these datasets. However, continuing development of experimental approaches and mathematical and computational resources are needed to maximize the contributions of functional genomics. Among the priorities are systems for studying physiology on a microscale, theoretical tools for understanding biological control systems, and metabolic simulators "in silico" which provide reasonable predictions of stimulus-response relationships [95].

Genetic Network or Protein Pathway Studies

Clustering of co-expression profiles further allows the inference of shared regulatory inputs and functional pathways [96]. The more complex the level at which one seeks to explain a living system, the greater the need to examine the network of interactions that lie behind the genome. Any metabolic network can be described in terms of four biologicals:

1. The pathways that compose the network;
2. The reactions that compose each pathway;
3. The metabolic compounds (substrates, activators, inhibitors); and
4. The enzymes that catalyze these reactions.

The nodes or crucial points at which genes act in concert with one another and with the environment within biochemical pathways represent potentially important targets for modulation [97]. Pathway analysis of a genome can assist target identification through an examination of the network of gene annotations. Computational pathway analyses often identify *pathway gaps*, that may be due to unidentified enzymes within the genome hidden among the uncharacterized ORFs. or, such gaps reflect pathway variation and occur at points where a step different to that in the reference is undertaken. Such analyses also identify *singleton steps*, which are single steps in a pathway where the majority of steps in the pathway are predicted to be absent. Singleton steps also could be due to incorrect functional annotations. A detailed search is undertaken to establish any similar enzymatic activities corresponding to such holes exist in the reference systems. If such biochemical evidence is found, it provides compelling evidence that corresponding genes are present within the genome.

An example in point is the fact that, a number of potential molecular targets for novel anticancer drug discovery have been identified in cell cycle control mechanisms; prominent among which are the regulatory proteins, cyclins and their effector counterparts the cyclin dependent kinases (CDKs). In fact, inhibition of CDK kinase activity has turned out to be the most productive strategy for the discovery and design novel anticancer agents specifically targeting the cell cycle, besides which, cyclin-CDK interactions, Cdc25 activation of cyclin-CDK complexes, etc. have also been targeted. The anticancer therapeutic potential of CDK inhibitors (developed through combinatorial chemistry, high throughput screening, functional genomics and bioinformatics) has been demonstrated in preclinical studies, and Phases I and II clinical trials in cancer patients are currently underway [98].

Protein kinases also play a central role in the regulation of multiple cellular responses (signal transduction pathways, control of cell growth, metabolism, differentiation and apoptosis), making them very attractive targets for the development of new therapeutic agents. Novel classes of protein kinase inhibitors have been identified using an approach based on intramolecular inhibitory mechanisms [99]. However, one of the biggest bottlenecks now facing research scientists is selectivity; approximately 550-895 protein kinases have been discovered in the human genome, but the functional interconnections of less than a fifth of these have been uncovered.

A recently developed approach *viz.* metabolic control analysis (MCA) provides a quantitative description of substrate flux in response to changes in system parameters of complex enzyme systems [100]. Proposed applications of the approach include the following: correlating the contributions of individual genes and phenotypic characteristics in metabolic disease (*e.g.*, diabetes); identifying candidate enzymes in pathways suitable as targets for cancer therapy and elucidating the function of "silent genes" by identifying metabolic features shared with genes of known pathways. MCA complements current studies of genomics and proteomics, providing a link between biochemistry and functional genomics that relates the expression of genes and gene products to cellular biochemical and physiological events. It is expected that MCA will have an increasing impact on the choice of targets for intervention in drug discovery.

Protein-Protein Interactions

Traditionally, enzymatic proteins have been preferred as therapeutic drug targets due to the ease of developing rapid, sensitive assays for screening, and to find small molecule inhibitors that block enzyme activity. However, a large number of proteins identified as potentially valuable drug targets exert their function through direct interactions with other proteins. Protein interactions such as those between antibodies and antigens, receptors and peptide- or protein-hormones, enzymes and substrates or inhibitors, etc. play pivotal roles in various aspects of the structural and functional organization of the cell, and their elucidation sheds light on the molecular mechanisms of biological processes.

Development of 3D modeling tools for elucidating the specific interactions at the atomic level has led to the reconstructional analysis of the inter-protein networks in genomes. Protein-protein interaction data strengths have been used in structure-based peptidomimetic techniques, rational drug design approaches and *in silico* and library screening. Peptidomimetics involves the identification of critical contact residues involved in protein:protein interactions through structural modelling, followed by the development of peptide mimetic agents, Pimetics™. These agents can be used for validation of therapeutic target(s) and as a basis for rational drug design and screening against compound libraries. Such techniques are useful for narrowing down the broad range of potential therapeutic targets. Protein microarrays are also increasingly being used in the elucidation of protein interactions, specifically in signal transduction, profiling activities of kinases (frequently involved in malignant cellular developments), etc. By integrating protein expression and purification technologies with protein array-based characterization assisted by automated image analysis, it becomes possible to analyze a broad set of different molecules involved in signaling processes.

One of the important developments related to protein-protein interaction has been the large-scale screening by yeast two-hybrid assays. Recently, this system has been used to detect 957 potential interactions involving 1,004 yeast proteins [101]. Software platforms and algorithms further transform raw experimental results from a two-hybrid platform into large-scale, reliable Protein Interaction Maps (PIMs) with interaction domain information. Strategies to integrate PIMs with other categories of functional information and the use of the PIM of a reference organism to predict a PIM in a target organism require to be developed in a case-specific manner and would comprise of clustering, visualization, and data integration components.

The free energy surface of proteins has been well studied in order to predict structures and folding mechanisms and has now been extended to the analysis of protein-protein interactions such as docking. Docking is a process by which two molecules fit together in 3D space. Some approaches to protein docking involve machine vision tools and genetic algorithms implemented in programs such as SurfDock and AutoDock. Several rapid new algorithms have been introduced in the original docking methodologies for discrimination against false-positive predictions [102]. Approaches are also being described based on blind predictions by homology modeling, threading, or *ab initio* methods of protein structure determined by X-ray crystallography or NMR; each being case specific and relevant to the system being investigated [103,104]. Identification of potential domain and ligand sequences through information acquisition of structure, sub-classification of domains and ligands, pattern development; sequence data-mining and analysis; protein data mining from identified sequences; and further processing on ligands such as biological priority setting have been developed. Such sequences are then supplied to the chemistry and biology laboratories for peptide synthesis and domain cloning, which in turn are screened for binding affinity. The delivered protein information is used in affinity data analysis. This sort of automated data analysis being deployed as a decision support system is critical to successful HTS in drug discovery as increasing screening speeds increases the amount of data that must be analyzed.

Even though value-additions are continuously being done, docking procedures are too slow for applications to large sets of proteins, and their capacity to identify near-native structures among a large number of docked complexes, is also low. Virtual screening (VLS) and faster docking methods through automation have been developed [105]. VLS can sample a virtually infinite chemical diversity of drug-like molecules without synthesizing and experimentally testing every screened molecule. Also, the virtual approach, as opposed to an assay, can be easily designed to select for a particular binding site or receptor specificity.

Protein Functional Studies

The status of gene expression profiling arrays is relatively advanced as compared to that of protein chips. This is primarily because, whereas the immobilization of DNA to a solid support requires a simple charge interaction, proteins must be immobilized in such a way that their folded (functional) conformation remains unchanged. Applications of protein arrays are emerging, with a promise of many more to come, corresponding to the improvements in array production methodologies. Some of these include the screening of antibody libraries, detection of the nature of protein-protein interactions, rapid protein profiling, biomarker discovery, signal transduction profiling, etc. [106, 107].

Several types of protein arrays are being produced including antibodies, receptors, ligands, nucleic acids, carbohydrates, or chromatographic surfaces (*i.e.* cationic, anionic, hydrophobic, hydrophilic). Some surfaces have broad specificity and bind whole classes of proteins, while others are highly specific and bind only a few proteins from a complex sample. After the capture step, the array is washed to reduce nonspecific binding. When subjected to short bursts of laser light, the retained proteins are uncoupled from the array surface and analyzed by radioactivity, chemiluminescence, fluorescence or laser desorption/ionization time-of-flight mass spectrometry. An example is of commercially available protein ligands that mimic antibodies in a variety of biological and functional applications (Affibodies™) that withstand a broad range of physical conditions and have very high specificity and the desired affinity for a corresponding target protein.

The microarray format offers two powerful features: the ability to screen many samples simultaneously and to screen the same set of samples repeatedly. As mentioned earlier, protein immobilization is critical, and has to be achieved in a way that preserves its folded conformations, by cross-linking in glutaraldehyde vapor or covalent immobilization of proteins via dextran spacers, etc. In addition, it is critical to minimize nonspecific binding of other proteins to the surface. These features are achieved by using of slides/membranes that are chemically treated *e.g.* with aldehyde-containing silane reagent. Classic solid phase substrates such as microtitre plates, membrane filters and microscopic slides, have also been turned into high-density, chip-like structures.

High-throughput protein chip tools could also provide an important means of evaluating the toxicity profiles of lead compounds in human cell lines prior to costly clinical evaluation. The process of chip design and implementation is highly involved, requiring the identification of toxicity-related proteins from cells of interest, the anchoring of selected toxicity marker antibodies to a chip-based platform, and optimization of reaction conditions for the detection of toxicity-related proteins from cellular samples. Positive leads emerging from this approach could find applications as diagnostic arrays for toxicity marker identification.

A derivative of protein arrays is the PISA (protein *in situ* array) method, in which, tagged functional proteins are made by cell-free expression using DNA encoding individual proteins or domains as the template (PCR, using primers designed from information in DNA databases), and immobilized *in situ*. Coupled *in vitro* transcription and translation are carried out on a surface to which the tagged protein adheres as soon as it is synthesized. This enables a protein array to be made in a single step from DNA, avoiding cloning, cell-based expression and protein purification. This technique will have advantages for proteins or individual domains where cloned genes are not available or where the proteins cannot be functionally expressed in cell-based expression systems. The PISA method has been used to functionally array human single-chain antibody fragments [108].

A derivative of the technique involves phage display in which peptide libraries are fused with the carboxy terminal domain of the coat protein, gene III protein fragment, on the surface of a filamentous phage. Relevant molecule is directly detected, screened using target molecules and amplified in the bacterial host cell. Specific clones are selected, and sequencing of the phage genome reveals the amino acid sequence of the peptide displayed on the phage [89]. Antibody fragments with affinity for specific molecules have been isolated by screening libraries. In addition, large-array approaches can be used to isolate clones directly from libraries created by chain shuffling.

Signal amplification is often necessary to achieve biologically sensitivities for most proteins on arrays. Rolling Circle Amplification (RCA) is emerging as the method of choice, since RCA uniquely allows target recognition, amplification, and detection *in situ* on the microarray. The 5′ end of an oligonucleotide primer is attached to an antibody, and in the presence of a DNA circle, DNA polymerase and nucleotides, rolling circle replication generates a concatamer of cicrcular DNA sequence copies that remain attached to the antibody. These are detected through the binding of fluorescent tagged, complementary oligonucleotide probes [109]. A modification of RCA, isothermal immuno-RCA, enables protein expression profiling on microarrays. Applications of immuno-RCA for simultaneous measurement of differential expression of many proteins in parallel on microarrays will be presented.

A remaining challenge for the full implementation of protein arrays in the acquisition of large sets of high-affinity and highly specific protein capture reagents which would revolutionize the analysis of entire proteomes [109]. A new generation of arrays has been recently described [110]. By optically encoding micron-sized polymer particles, suspension microarrays were created to enable highly multiplexed analysis of complex samples. Each element in the array comprised of a subpopulation of particles with distinct optical properties and each array element had different surface receptor; nucleic acids, proteins, lipids or carbohydrates can serve as receptors to support the analysis of a wide range of biomolecular assemblies. Coupled with such innovations for rapid serial analysis of samples, molecular analysis with microsphere arrays holds significant potential as a general analysis platform for both research and clinical applications.

Target Validation

A really tough selection process is mandatory for picking out "druggable targets" from the results of the research phase and a quick kill process for those that do not measure up in advanced trials. Before transitioning from target identification to lead development, it is necessary to validate the identified targets, define success criteria for application and focus efforts to optimize resource requirements versus time to ensure effective flow of information from drug discovery to development. The following set of technologies are usually considered during drug validation, which indicate, to a reasonable degree of probability, that a given target is critically involved in a disease process, and that a drug that modulates this target would have a therapeutic effect:

1. Cellular Assays
2. Model-Organism-Based Systems
3. Downregulation of Gene Expression

Cellular Assays

The field of cell-based screening is expanding rapidly as innovations instrumentation increase the number of targets that can be efficiently screened in cellular formats. New, noninvasive approaches involving radiometric proximity assays, arrays of fluorescence-based technologies, and reporter gene constructs in cell systems are providing more options for the selection of specific therapeutic targets. Cell-based screens can be configured to provide a broad range of data on putative drug activity. Such systems thus, serve as a springboard to better lead validation in drug discovery initiatives.

Cytotoxicity Assays

Traditionally, *in vitro* toxicity assays have relied on end-point measurement of chemical stains that either stain specific cell components, or indicate cellular enzyme activity, or mitochondrial activity (reduction of tetrazolium salts) or fluorescent dyes with flow cytometry, ATP bioluminescence [111], etc. An important aspect in cellular assays is the differentiation between the two main modes of cell death *viz.* apoptosis (cell programmed cell death) and necrosis (loss of cell integrity- usually due to cell injury). The high cost of obtaining cells for HTS has induced the performance of multiple analyses on same samples with a combination of techniques; *e.g.* labeling individual cell constituents with green fluorescent protein (GFP) emitting at different wavelengths. Imaging systems are then used for studying the translocation of the labeled targets in the cell [112]. A novel recent example is the development of a stable transformed fish (EPC) cell line containing a reporter gene for GFP-luciferase fusion protein, under the control of the tilapia HSP70 promoter and provides a rapid, cheap and easy biological test to determine cellular stress induced by environmental pollutants, alone or in conjunction with other, more specific assays [113].

Cellular Assays for Functional Studies

Several cell array technologies are commercially available that use combinatorial biology handled by sophisticated algorithms that can measure several different parameters (endpoints) of cellular behavior simultaneously. *e.g.* Tissue array technologies for genome scale expressional and clinical cancer research have been reported [114] that enable molecular analysis of large number of specimens. Tissue arrays are constructed by arranging the cylindrical punch-outs from individual tumor tissue biopsies into a tissue array block, which is then sliced onto 200 or more identical slides for probing RNA or protein targets. A single immunohistochemistry or in situ hybridization experiment provides information on specimens on the slides, while subsequent sections can be analyzed with other probes or antibodies. Thus, tissue array technology expands the scope of high-throughput molecular analysis of archival tissue specimens with multiple probes for specific genes or proteins for functional or clinical application.

Several types of assays can be carried out for providing greater detail about gene functions underlying a disease state *e.g.* activities of specific enzymes, transport of signaling molecules into the nucleus or to the plasma membrane, gene expression, cell spreading and motility, mitotic index, neurite outgrowth, and the presence, amounts, and patterns of various cellular components. One example is the application of CFU-GM assay for detection of myelosuppressive agents that can also be used as a model for predicting neutropenia [115].

Use of Cellular Assays in Functional Genomics

Expression profiling based on functional genomics can be used to investigate changes in gene transcription induced by drug treatment, which may lead to discovery of new potential drug targets. Fluorescence-based methods can be used to monitor changes within living cells that occur as a result of knocking out or modulating (*e.g.*, via an antisense reagent, a protein knockout or zinc finger protein) or overexpression of a target of interest. One can study the effects of modulating a target of interest on a range of parameters. A case in point is a strategy for drug screening in which a yellow fluorescent protein (YFP) expression vector was introduced into the colon cancer cell line DLD-1, and a blue fluorescent protein (BFP) expression vector was introduced into an isogenic derivative in which the mutant K-Ras allele had been deleted. Co-culture of both cell lines allowed facile screening of 30,000 compounds of which, a novel cytidine nucleoside analog was identified that displayed selective activity *in vitro* and inhibited tumor xenografts containing mutant Ras [116]. Another striking example is Thymidylate synthase (TS), which is a critical cellular target for cancer chemotherapeutics. Overexpression of Thymidylate synthase (TS) in a cell-based system that has been reported to be useful for the development of therapeutics for those cancers that overexpress TS *de novo* [117] particularly the fluoropyrimidine and antifolate classes of antineoplastic agents.

Model-Organism-Based Systems

Model systems are used to validate target compounds with unknown mechanisms of action. Drug targets in model organisms can also be used to probe biochemical pathways, revealing new targets and mechanisms within these pathways [118]. The rationale behind the use of prokaryotes, invertebrates and lower mammals as models in target validation is that several functionally significant regions of the genome, genetic pathways and biological functions are conserved during evolution; hence extrapolation to humans is relevant. Target drugs correlative with and predictive of drug toxicity and efficacy in suitable models is anticipated to streamline drug development by triaging lead compounds and clinical candidates that maximize efficacy while minimizing safety risks [119].

Yeast Models

Yeasts provide a useful system for drug testing in several ways, due to their use in growth-based assays for various screens, *e.g.* for investigation of complex eukaryotic cellular processes and as a recombinant system for analysis and validation of therapeutic target proteins from higher eukaryotes. When a human gene that is mutated in a disease process is identified, sequence similarity with one or more genes of known function in yeast can suggest the function performed by the human gene, *e.g.* genes causing ataxia telangiectasia [120] and Bloom syndrome [121] exhibit significant sequence similarity with *S. cerevisiae MEC1* and *SGS1* genes respectively. Sometimes, a human protein may not have a true orthologue in yeast, but may instead be related to a yeast protein as a member of a protein family with analogous but not identical functions. Recent progress in human genetic diseases provide a preview of the impact which the complete sequencing of the *S. cerevisiae* genome would have on understanding the biology of multicellular eukaryotes.

Non-Mammalian Models

Caenorhabditis Elegans

Among model organisms the nematode *Caenorhabditis elegans* offers several advantages, including well-established techniques for genetic and experimental manipulation and the first completed genome sequence for a multicellular organism (a proprietary database is available that integrates physiological and pharmacological data from in vivo chemical and/or genetic interventions on C. elegans). Additionally, the ease of culture allows adaptation of the nematode for use in high-throughput chemical screens for the identification of lead compounds in drug development [122]. Recently, using the gene knockout technology in this model system, a novel pathway essential for associative learning and memory has been defined [123].

Drosophila

Drosophila melanogaster has a wealth of genetic and bioinformatic tools available for the analyses demanded be target validation. Assays have also been developed to screen against extensive libraries of chemical compounds for the evaluation of lead structures *in vivo*. The proteome predicted from the recently sequenced *Drosophila* genome indicates that humans have more genes in common with *Drosophila* than with any other invertebrate model organism characterised to date [124]. The proprietary PathFinder' technology with FlyTag' Drosophila expressed sequencing tag (EST) database and bioinformatics tools can be used to identify and validate novel targets and to develop assays for high-throughput screening.

Zebrafish (Danio rerio)

The zebrafish has been widely utilized for the study of developmental biology with the evolution of sophisticated cellular and molecular approaches. More recently, the rapid progress of various zebrafish genomic infrastructure initiatives has facilitated the development of zebrafish models of human disease [125]. Over 90% of the zebrafish genes have human homologs. Examples include genes for insulin, hemoglobin, VEGF, myostatin, thyroid hormone receptors, leptin and erythropoietin. The transparent nature of zebrafish assists in whole body tracking of fluorescent tagged peptides. Morpholino-based zebrafish models of human disease have also been developed that would greatly assist in target evaluation. The sequencing of the zebrafish genome is expected to be complete by the end of 2002 [126], which could extend its further applications in target validation.

Nonhuman Mammalian Disease Models

Classically, lower mammals such as mice, rabbits, guinea pigs etc. have been routinely used for drug validation to answer fundamental questions of drug pharmacology and end organ toxicity, of which, the mouse has been the premier mammalian model for the identification of the genetic basis of human disorders and complex diseases involving gene-gene and gene-environment interactions. The ability to genetically manipulate the mouse through transgenesis and gene targeting allows for the hypotheses testing of specific gene functions and their role in disease. One approach is to ***knock out*** (delete) a gene of interest in an animal model of disease, study the resulting phenotype to gain clues about the function of the gene and determine whether this gene might be a useful target. Following extensive phenotypic analysis, including pathological, immunological, cardiovascular, metabolic, and behavioral tests, collected data is made available as a database of functional information, that provides direct information about mammalian gene families most relevant to small molecule drug discovery.

Modification of the basic technology now provides additional choices for a more specific and variety of manipulations of the mouse genome. This includes conditional cell-type specific gene targeting, ***knockin*** technique and the induction of the specific balanced chromosomal translocations. Knockin technology involves replacing the mouse version of the gene with its human counterpart, so that the same human target is made available for screening drugs. Over the past decade, knockin and knockout techniques have not only generated a wealth of knowledge concerning the roles of growth factors, oncogenes and hormone receptors but also helped to produce animal models for several human genetic disorders [127].

The significance of mutations in the pathogenesis of breast cancer and development of new therapeutics have been addressed through transgenic mouse systems. These include transgenic knockin mice expressing high levels of genes (such as c-myc, cyclin D1, ErbB-2 and various growth factors) in target tissues and knockout mice in which specific genes (such as p53) have been ablated [128]. Another striking example of the use of mouse genetics in target validation is the analysis of the ACC2 function. ACC2 knockout mice eat more the wild-type litter, yet burn more fat and store less fat in their adipocytes, making this enzyme, a putative target in obesity management [129]. It is hoped that through the integration of other methods with the use of these models, a deeper understanding of this disease and the generation of new therapeutic agents will result.

Methods such as gene targeting by homologous recombination, gene trapping and chemical mutagenesis are being used in a large way in mouse genetic analyses. Novel strategies include the generation of large number of mouse embryonic stem (ES) mutant clones, each carrying a single gene trap mutation [OmniBank™], and which develop into specific knockouts. For a more detailed description of mouse transgenic methodologies refer to Abuin A *et al.*, , 2002–[130].

Thus, the potential to knockout/knockin a putative drug target gene allows us to model the action of the drug, even before its formulation; making the transgenic technology a very attractive one in target validation. The potential of phenotypic information from transgenics to contribute to a high-throughput target selection/validation strategy has hitherto been limited only by the resources required to rapidly generate and characterise a large number of knock-out transgenics in a timely fashion [131].

Downregulation of Gene Expression

Although a majority of diseases are polygenic in nature, certain genes may be more actively involved and hence may be potential drug targets. Once gene function and the associated functional networks are identified, the next step is to test the hypothesized gene activity by selectively masking/eliminating the expression of that gene. Gene knockout is one approach towards this type of testing. Two other strategies based on this principle *viz.* antisense and zinc finger proteins can also be used in target validation.

Antisense Technology in Drug Validation

In principle, the best tool for target validation is an antisense oligo complementary to the mRNA coding for the protein one wishes to assess as a prospective drug target. Such an antisense oligo, specifically blocks the synthesis of the selected protein thereby providing a definitive assessment of the biological effect expected with a drug targeted against that protein. In practice, such oligos have largely failed to live up to that promise; largely due to the multiple flaws plaguing phosphorothioate antisense oligos (S-DNAs; the structural type that has been used in most antisense-based target validations). These limitations include: lack of marginal sequence specificity, multiple non-antisense effects, instability and unpredictable targeting. Efforts to develop better antisense types led to the recent introduction of three improved antisense types: chimeras, peptide nucleic acids, and morpholinos.

Chimeras typically contain a short segment (8 to 12 subunits) of S-DNA adequate to mediate RNaseH cleavage of the targeted RNA sequence. Both sides of this S-DNA are segments with 2'O-alkyl ribose backbones that serve to increase RNA binding affinity and reduce nuclease sensitivity. These chimeras show improved sequence specificity and lower non-antisense effects. Chimeras are recognized as the best tools for certain applications such as: targeting pre-messenger RNAs sites within introns, targeting point mutations, and targeting sites in the amino acid coding regions of messenger RNAs.

Peptide Nucleic Acids (PNAs) are completely resistant to nucleases and they appear to be free of most or all of the non-antisense effects which plague S-DNAs. PNAs also bind RNA with exceptionally high affinity–giving them high efficacy and predictable targeting (Nielsen PE *et al.*, , 1991). The strength of PNAs is evident in certain situations. When targeted against rare all-purine sequences, PNAs can block translation anywhere in an mRNA by forming a double-clamp structure. With RNA sequences one segment of the PNA binds to the target sequence and the other segment of the PNA binds to major-groove sites of the resulting PNA/RNA duplex. PNAs also readily form triple helix structures with rare duplex DNA sequences comprising mostly purines in one strand and mostly pyrimidines in the other strand. Lastly, under low salt conditions in cell-free systems PNAs have been shown to achieve sequence-specific invasion of duplex DNA sequences, resulting in inhibition of transcription of the invaded duplex.

Morpholinos, constitute a radical redesign of natural nucleic acids [132]. Like PNAs, morpholinos are resistant to nucleases and they appear to be free of most or all of the non-antisense effects which plague S-DNAs. In contrast to PNAs, most morpholinos exhibit excellent aqueous solubility. Morpholinos also have much higher RNA binding affinities than do S-DNAs, though not as high as PNAs. Long Morpholinos (25-mers) exhibit predictable targeting and very high efficacy. Most notable, morpholinos provide by far the best sequence specificity of all antisense types. However, the latter property also renders them unsuitable for targeting point mutations. Morpholino oligos were first introduced as a means to inhibit gene function in embryos; since then, they have been tested in a range of model organisms, including sea urchin, zebrafish, frog, chick, and mouse.

Effective delivery of antisense oligos into cultured cells can now be achieved by a variety of methods. However, the maximum value of antisense oligos for target validation can be realized only when such tools can also be routinely applied *in vivo*. Recently, gene suppression through a single intravenous injection of antisense nucleotides targeting vascular endothelial growth factor (VEGF) into pregnant mice between E7.5-8.5 enabled the definition of the critical window required to inhibit VEGF expression and recapitulate the primary loss of function phenotype observed in VEGF (-/-) embryos. This phenotype was sequence-specific and time and dose-dependent. Injection of an antisense oligo targeting a second gene, E-cadherin, into pregnant mice at E10 confirmed a hypothesized secondary phenotype. This is the first report of antisense inhibition of gene expression in utero and provides a new strategy for target validation in functional genomics [133].

Another report of the application of antisense technology in animals [134] shows reduction of liver Fas mRNA and protein expression by systemic antisense oligonucleotide dosing that was found to protect mice from some forms of fulminant hepatitis. Effective delivery of EPI-2010 (a respirable antisense oligonucleotide) as an aerosol, has also been reported in an animal model of asthma [[135]. Absorption, tissue distribution, metabolism and excretion (ADME) and safety studies of the same suggest that antisense oligos can be delivered to target respiratory tissues in low, safe, efficacious and long-acting doses.

Zinc Finger DNA-binding Proteins (ZFPs)

Zinc Finger Proteins–called so, because fingers from these proteins have a simple β-β-α structure (BBA motif) that folds around a central zinc ion, and tandem sets of fingers can contact neighboring subsites of 3-4 base pairs along the major groove of DNA [136]. ZFPs thus, can be used as modular building blocks for the construction of larger proteins that recognize and bind to specific DNA sequences. ZFPs have two distinct domains: a DNA–recognition domain that directs the transcription factor to the proper chromosomal location by recognizing a specific DNA sequence, and a functional domain that causes the gene to be expressed or repressed. This conditional regulation allows gene expression to be controlled in a reversible manner.

Phage display has been used to create a large library of different zinc fingers from which selections were made for binding to a given DNA sequence [137]. From this database there have been elucidated recognition rules that relate the amino acid sequence of a finger to its preferred DNA binding site. Control of gene expression using designed zinc finger peptides has been demonstrated by the specific inhibition in an oncogene mouse cell line and also by switching on genes in expression plasmids. These experiments demonstrate that zinc finger DNA-binding domains can be engineered *de novo* to recognise given DNA sequences. Five to six individual zinc fingers linked together would recognise a DNA sequence 15-18 bp in length, sufficiently long to constitute a rare address in the human genome. By adding functional groups to the engineered DNA-binding domains, *e.g.* silencing domains, novel transcription factors can be generated to up or down regulate expression of a target gene.

To date, thousands of ZFPs have been specifically engineered and characterized with respect to affinity of binding to their DNA target and their specificity for their intended DNA target. Such novel or "designer" zinc finger proteins with desired DNA binding specificities can serve as efficient transcription factors in various mammalian cell lines. It has been demonstrated that the expression of a target gene can be controlled by changing the *in vivo* concentration of designer ZFPs in a dose-dependent manner [138]. Naturally occurring human wig-1 (a ZFP) has been suggested to play a role in the p53-dependent growth regulatory pathway [139]; several other ZFPs have been implicated in HIV therapy strategies. Such properties of ZFPs in blocking/enhancing the function of target genes will increasingly contribute to gene expression and target validation studies.

Conclusion

Biological screening has undergone a total metamorphoses. Early efforts of the pharmaceutical community in screening libraries of compounds involved low-throughput that had limited potential of yielding and validating biological targets. It was however, realized that, most of the important diseases involve complex targets. Pharmacogenomic–based drug discovery stepped in at this stage, to identify new target molecules (genes and proteins) that underlie disease processes themselves and not the symptoms. The sheer numbers of potential targets uncovered through the bioinformatic-genomics-proteomics approach today, coupled with advanced chemical technologies like combinatorial chemistry technologies and chemical genomics widened horizons in drug development to address large unmet medical needs.

It must however, be very carefully assessed further, whether developing a molecule that blocks the action of a certain gene results in severe/unwanted side-effects that negate the therapeutic effect of the drug. On these grounds, besides others, it is very likely that many of the candidates identified through the initial screening will later prove to be either not critically involved in a disease process or not druggable for a variety of reasons. This requires additional efforts in lead optimization and prioritization that starts with detailed analyses of the number of different targets for a given indication and the variability of targets within the patient population. To meet this end, obtaining structure–activity relationships (SAR) and using combinatorial chemistry technologies and chemical genomics to find leads that have a better chance of making it through the process of discovery, is important. Pharmacokinetic modelling and simulation of preclinical and clinical trials, determination of ADME profiles, selecting and evaluating biomarkers for safety and efficacy; involvement of the target population as soon as possible; using information technologies to analyze the compiled data, are all involved at this stage. Thereby, compounds with no downstream value can be triaged as early as possible. Further aspects that remain to be considered are of a commercial nature, including market demand and potential position for the specific drug/target, existing players, competitive pharmacogenomics, intellectual property rights, marketing strategies, etc.

The genomes of any two humans exhibit 99.9% nucleotide homology, which translates into one difference (single nucleotide polymorphism–SNP), every 1000 bases. The enormous size of the genome means that a 0.1% rate of divergence is still equivalent to over 3 million SNPs between any two people. SNPs within the coding regions of genes (cSNPs) are either ***synonomous*** (cause no change in the coding sequence) or ***nonsynonomous*** (result in amino acid substitution); SNPs in the non-coding genes are ***neutral***, resulting in no impact upon health. Nonsynonomous cSNPs may sometimes be neutral, but effect protein expression and/or function, resulting in a range of phenotypes from benign to serious. Neutral SNPs, while not directly causing disease, can be extremely valuable tools in the search for elusive disease genes *e.g.* positional cloning has been used to find candidate genes in the vicinity of SNPs, and in some cases (cystic fibrosis as an instance), these have emerged as the real disease genes. However, with the advent of the complete human genome sequence, it is to be expected that most attention will now be focused upon the role of nonsynonomous cSNPs in the disease phenotype.

In the near future, surveys of SNPs known to affect patient health could be a part of medical tests, helping clinicians determine which patients are at risk for disease and what medications might be beneficial or toxic to them. Pharmacogenomics will also outline the 'best-fit' treatments for each subject according to his/her specific genomic profile. Every year, hundreds of promising drugs fail to make it to market because they cause serious side effects in a very small number of test patients. Consequently, many patients are denied access to what would be an appropriate and safe therapy for them, and the pharmaceutical industry loses millions of dollars in R and D investment. However, pharmacogenomic studies based on SNP analysis promise to alleviate this problem to a certain extent. Through the use of SNPs and microarrays, it should be possible to derive a "molecular fingerprint" diagnosis in a case-by-case manner, allowing targeted therapy with all the drugs that are needed. A similar type of analysis (*e.g.* using SNPs within/close to the genes in question) should also allow the risk of disease to be assessed by screening, enabling streamlined selection of individuals for more careful monitoring and/or more aggressive preventative approaches.

Pharmacogenomics is concerned not only with the genes and proteins involved in disease progression, but also with the enzymes involved in drug activation and/or metabolism. Individuals demonstrate a wide range of activities of such enzymes leading to variations in the active dose and duration. To clinical pharmacologists, the value of a SNP chip capable of diagnosing these various alleles would be invaluable in determining drug and dose regimes.

The advantages of pharmacogenomics have been widely touted: improving the quality of drug targets and compounds advanced into development, streamlining clinical trials, protection of market share by reducing adverse events, and laying the foundation of personalized medicine. The "other side of the coin" involves issues such as ethics, intellectual property, an overburdened/poor healthcare system, uncertain economic models of pricing and segmented markets, etc. Some of the controversies relate to genetic determinism, informed consent, privacy and confidentiality, access to genetic healthcare and patents for genetic information (SNPs, ESTs, etc.), primarily as a means to protect gene discovery and drug development programmes. However, a majority of the basic DNA code, some polymorphism and expression information databases, that currently available and more of which will emerge from the human genome project, will continue to be in the public domain. It is however unlikely that all medically important genomic information (SNP's and EST's) will be available as such.

The thrust in this article has been on the discussion of the various emerging pharmacogenomic technologies and their promise to revolutionize the drug discovery process. To cut a long story short, we expect significant solutions from these strategies for the pharmaceutical industry. There will be a drastic consolidation such that each technology achieves maximal efficiency in its respective space. Over the next few years, key participants will emerge that will corner milestone areas of the screening ensemble, rather than the hit-or-miss approaches of the past.

List of Abbreviations

Single nucleotide polymorphisms (SNPs); High-throughput screening (HTS); Haplotype Pattern Mining (HPM); Expressed sequence tags (ESTs); G-protein coupled receptor (GPCR); Age-related macular degeneration (AMD); Serial analysis of gene expression (SAGE); generation of longer cDNA fragments (GLGI); (RT-PCR); open reading frame (ORF); RNA arbitrary primed PCR (RAP-PCR); suppressive subtractive hybridization (SSH); representational difference analysis (RDA); matrix-assisted laser desorption/ionization (MALDI); electrospray ionization (ESI); Protein Interaction Maps (PIMs); structure-to-activity relationships (SARs); Rolling Cirle Amplification (RCA); phenylisothiocyanate (PITC); polyvinylidene diflouride (PVDF); cyclin dependent kinases (CDKs); metabolic control analysis (MCA); Protein Interaction Maps (PIMs); Virtual screening (VLS); PISA (protein *in situ* array); green fluorescent protein (GFP); Thymidylate synthase (TS); absorption, distribution, metabolism, excretion (ADME); embryonic stem cells (ES cells); Peptide Nucleic Acids (PNAs); vascular endothelial growth factor (VEGF); Zinc Finger Proteins (ZFPs).

REFERENCES

1. International Human Genome Sequencing Consortium: Initial sequencing and analysis of the human genome. **Nature** 2001, **409:** 860-921.
2. Dean NM–Functional genomics and target validation approaches using antisense oligonucleotide technology. **Curr. Opin. Biotech**. 2001**12**: 622–625
3. Ozaki H, Imaizumi Y, Oishi K, Kohama K–High throughput pharmacology for drug discovery. **Nippon Yakurigaku Zasshi.** 2001;**118**: 187-96.
4. Hall IP–Pharmacogenetics, pharmacogenomics and airway disease. **Respir. Res.** 2002. **3:** 10-16.
5. Roos D S; Bioinformatics–Trying to swim in a sea of data, **Science** (2001), **291:** 1260-1261

6. Landro JA, Taylor IC, Stirtan WG, Osterman DG, Kristie J, Hunnicutt EJ, Rae PM, Sweetnam PM; HTS in the new millennium: the role of pharmacology and flexibility. **J Pharmacol Toxicol Methods** (2000) **44**: 273-89.

7. Toivonen HT, Onkamo P, Vasko K, Ollikainen V, Sevon P, Mannila H, Herr M, Kere J; Data mining applied to linkage disequilibrium mapping. **Am J Hum Genet**. 2000; **67**: 133-45.

8. Sevon P, Ollikainen V, Onkamo P, Toivonen HT, Mannila H, Kere J; Mining associations between genetic markers, phenotypes, and covariates. **Genet Epidemiol**. 2001;**21 Suppl 1**:S588-93.

9. Wade Walke D, Han C, Shaw J, Wann E, Zambrowicz B and Sands A; In vivo drug target discovery: identifying the best targets from the genome. **Curr.Opin. Biotech.** (2001) **12**: 626–631

10. Scheurle D, DeYoung MP, Binninger DM, Page H, Jahanzeb M, Narayanan R.; Cancer gene discovery using digital differential display. **Cancer Res**. 2000; **60**: 4037-43.

11. Olsson P, Bera TK, Essand M, Kumar V, Duray P, Vincent J, Lee B, Pastan I; GDEP, a new gene differentially expressed in normal prostate and prostate cancer. **Prostate** 2001; **48:** 231-41

12. Wittenberger T, Schaller HC, Hellebrand; An expressed sequence tag (EST) data mining strategy succeeding in the discovery of new G-protein coupled receptors. **J Mol Biol.** 2001; 307: 799-813.

13. Conklin D, Yee DP, Millar R, Engelbrecht J, Vissing H; Mining of assembled expressed sequence tag (EST) data for protein families: application to the G protein-coupled receptor superfamily. **Brief Bioinform**. 2000; 1:93-9.

14. Stohr H, Mah N, Schulz HL, Gehrig A, Frohlich S, Weber BH; EST mining of the Uni Gene dataset to identify retina-specific genes; **Cytogenet Cell Genet**. 2000; **91:** 267-77.

15. Bloem LJ, Pickard TR, Acton S, Donoghue M, Beavis RC, Knierman MD, Wang X; Tissue distribution and functional expression of a cDNA encoding a novel mixed lineage kinase. **J Mol Cell Cardiol**. 2001; **33:** 1739-50.

16. Conklin D, Jonassen I, Aasland R, Taylor WR; Association of nucleotide patterns with gene function classes: application to human 3′ untranslated sequences; **Bioinformatics.** 2002; **18**: 182-9.

17. Jongeneel CV; Searching the expressed sequence tag (EST) databases: panning for genes; **Brief Bioinform**. 2000; **1**: 76-92

18. Velculescu, V. E., Zhang, L., Vogelstein, B., and Kinzler, K. W. (1995). Serial Analysis Of Gene Expression. **Science** 270, 484-487.

19. Velculescu VE, Zhang L, Zhou W, Vogelstein J, Basrai MA, Bassett DE Jr, Hieter P, Vogelstein B, Kinzler KW; Characterization of the yeast transcriptome; **Cell** 1997; **88**: 243-51.

20. Hermeking H, Lengauer C, Polyak K, He TC, Zhang L, Thiagalingam S, Kinzler KW, Vogelstein B; 14-3-3 sigma is a p53-regulated inhibitor of G2/M progression **Molecular Cell** 1997 **1**: 3-11

21. Jian Yu, Lin Zhang, Paul M. Hwang, Carlo Rago, Kenneth W. Kinzler, and Bert Vogelstein; Identification and classification of p53-regulated genes; **Proc Natl Acad Sci** 1999; 96: 14517-14522.

22. Velculescu VE, Madden SL, Zhang L, Lash AE, Yu J, Rago C, Lal A, Wang CJ, Beaudry GA, Ciriello KM, Cook BP, Dufault MR, Ferguson AT, Gao Y, He TC, Hermeking H, Hiraldo SK, Hwang PM, Lopez MA, Luderer HF, Mathews B, Petroziello JM, Polyak K, Zawel L, Zhang W, Zhang X, Zhou W, Haluska FG, Jen J, Sukumar S, Landes GM, Riggins GJ, Vogelstein B, Kinzler KW; Analyses of Human Transcriptomes; **Nat Genet** 1999; 23: 387-388.

23. Virlon B, Cheval L, Buhler J-M, Billon E, Doucet A, and Elalouf J-M; Serial microanalysis of renal transcriptomes; **Proc Natl Acad Sci** 1999, **96:** 15286-15291.

24. St Croix B, Rago C, Velculescu V, Traverso G, Romans KE, Montgomery E, Lal A, Riggins GJ, Lengauer C, Vogelstein B, Kinzler KW; Genes Expressed in Human Tumor Endothelium; **Science** 2000; **289**: 1197-202.

25. Paul M. Hwang, Fred Bunz, Jian Yu, Carlo Rago, Timothy A. Chan, Michael P. Murphy, Geoffry F. Kelso, Robin A. J. Smith, Kenneth W. Kinzler KW, Bert Vogelstein; Ferredoxin reductase affects p53-dependent, 5-fluorouracil-induced apoptosis in colorectal cancer cells; **Nature Medicine**. 2001: 1111-1117.

26. Saurabh Saha, Alberto Bardelli, Phillip Buckhaults, Victor E. Velculescu, Carlo Rago, Brad St. Croix, Katharine E. Romans, Michael A. Choti, Christoph Lengauer, Kenneth W. Kinzler, Bert Vogelstein; A Phosphatase Associated with Metastasis of Colorectal Cancer; **Science 294**: 1343–1346.

27. Lee S, Zhou G, Clark T, Chen J, Rowley JD, Wang SM; The pattern of gene expression in human CD15+ myeloid progenitor cells; **Proc Natl Acad Sci** 2001; **98**: 3340-5

28. Nacht M, Dracheva T, Gao Y, Fujii T, Chen Y, Player A, Akmaev V, Cook B, Dufault M, Zhang M, Zhang W, Guo M, Curran J, Han S, Sidransky D, Buetow K, Madden SL, Jen J; Molecular characteristics of non-small cell lung cancer **Proc Natl Acad Sci** 2001; **98**: 15203-8

29. Argani P, Iacobuzio-Donahue C, Ryu B, Rosty C, Goggins M, Wilentz RE, Murugesan SR, Leach SD, Jaffee E, Yeo CJ, Cameron JL, Kern SE, Hruban RH; Mesothelin is overexpressed in the vast majority of ductal adenocarcinomas of the pancreas: identification of a new pancreatic cancer marker by serial analysis of gene expression (SAGE); 2001; **7**: 3862-8

30. Aronica E, van Vliet EA, Hendriksen E, Troost D, Lopes da Silva FH, Gorter JA; Cystatin C, a cysteine protease inhibitor, is persistently up-regulated in neurons and glia in a rat model for mesial temporal lobe epilepsy; **Eur J Neurosci** 2001;**14:** 1485-91

31. Hendriksen H, Datson NA, Ghijsen WE, van Vliet EA, da Silva FH, Gorter JA, Vreugdenhil E; Altered hippocampal gene expression prior to the onset of spontaneous seizures in the rat post-status epilepticus model; **Eur J Neurosci** 2001; **14**: 1475-84

32. Seth P, Krop I, Porter D, Polyak K; Novel estrogen and tamoxifen induced genes identified by SAGE (Serial Analysis of Gene Expression); **Oncogene** 2002; **21**: 836-43.

33. Caldwell MC, Hough C, Furer S, Linehan WM, Morin PJ, Gorospe M; Serial analysis of gene expression in renal carcinoma cells reveals VHL-dependent sensitivity to TNFalpha cytotoxicity; **Oncogene** 2002; **21**: 929-36.

34. Welle S; Gene transcript profiling in aging research. Exp Gerontol 2002; **37**: 583-90.

35. Anisimov SV, Tarasov KV, Tweedie D, Stern MD, Wobus AM, Boheler KR; SAGE identification of gene transcripts with profiles unique to pluripotent mouse R1 embryonic stem cells; **Genomics** 2002; **79**: 169-76

36. Enerback C, Porter DA, Seth P, Sgroi D, Gaudet J, Weremowicz S, Morton CC, Schnitt S, Pitts RL, Stampl J, Barnhart K, Polyak K; Psoriasin expression in mammary epithelial cells in vitro and in vivo; **Cancer Res** 2002; **62**: 43-7.

37. van Ruissen F, Jansen BJ, de Jongh GJ, van Vlijmen-Willems IM, Schalkwijk J; Differential gene expression in premalignant human epidermis revealed by cluster analysis of serial analysis of gene expression (SAGE) libraries; **FASEB J** 2002; **16**:246-8

38. Sharon D, Blackshaw S, Cepko CL, Dryja TP; Profile of the genes expressed in the human peripheral retina, macula, and retinal pigment epithelium determined through serial analysis of gene expression (SAGE); **Proc Natl Acad Sci** 2002; **99:** 315-20

39. Yamamoto M. 2002; Gene expression profiling using improved SAGE; **Rinsho Byori; 50**: 52-60

40. Chen JJ, Rowley JD, Wang SM; Generation of longer cDNA fragments from serial analysis of gene expression tags for gene identification; **Proc Natl Acad Sci** 2000; **97**: 349-53

41. Chen J, Lee S, Zhou G, Wang SM; High-throughput GLGI procedure for converting a large number of serial analysis of gene expression tag sequences into 3′ complementary DNAs; **Genes Chromosomes Cancer** 2002; **33**: 252-61

42. Polyak K and Riggins GJ; Gene discovery using the serial analysis of gene expression technique: implications for cancer research. **J Clin Oncol** 2001;**19**: 2948-58.

43. Liang P and Pardee AB; Differential display of eukaryotic messenger RNA by means of the polymerase chain reaction, **Science** 1992, **257**: 967–971.

44. Zhang J, Liu WL, Tang DC, Chen L, Wang M, Pack SD, Zhuang Z, Rodgers GP; Identification and characterization of a novel member of olfactomedin-related protein family, hGC-1, expressed during myeloid lineage development; **Gene** 2002; **283**: 83-93

45. Wagnerova J, Cervenakova L, Balabanov R, Zitron I, Dore-Duffy P; Cytokine regulation of E-selectin in rat CNS microvascular endothelial cells: differential response of CNS and non-CNS vessels; **J Neurol Sci** 2002; **195**: 51-62

46. Wang Zx ZX, Hu Gf GF, Wang Hy HY, Wu Mc MC; Clone of a novel liver cancer associated gene and analysis of the second structures of the predict protein; **Zhonghua Gan Zang Bing Za Zhi** 2002;**10**: 25-7

47. Lu KP, Alejandro NF, Taylor KM, Joyce MM, Spencer TE, Ramos KS; Differential expression of ribosomal L31, Zis, gas-5 and mitochondrial mRNAs following oxidant induction of proliferative vascular smooth muscle cell phenotypes; **Atherosclerosis** 2002;**160**:273-80

48. Villablanca AC, Lewis KA, Rutledge JC; Time- and dose-dependent differential upregulation of three genes by 17[beta]-estradiol in endothelial cells; **J Appl Physiol** 2002; **92**: 1064-1073

49. McMorrow IM, Narayan K, Sachs DH, Leguern C; New intra-renal graft genes associated with tolerance or rejection; Kidney Int 2002; **61 Suppl 1**:85-93.

50. Stoll M, Hahn AW, Jonas U, Zhao Y, Schieffer B, Fischer JW, Unger T; Identification of a zinc finger homoeodomain enhancer protein after AT(2) receptor stimulation by differential mRNA display; **Arterioscler Thromb Vasc Biol** 2002; **22**: 231-7

51. Wang X, Lan M, Shi YQ, Lu J, Zhong YX, Wu HP, Zai HH, Ding J, Wu KC, Pan BR, Jin JP, Fan DM; Differential display of vincristine-resistance-related genes in gastric cancer SGC7901 cell; **World J Gastroenterol** 2002; **8**: 54-9

52. Joseph P, Lei YX, Whong WZ, Ong Tm TM; Molecular Cloning and Functional Analysis of a Novel Cadmium-responsive Proto-oncogene; **Cancer Res** 2002; **62**: 703-7

53. Bahn S, Mimmack M, Ryan M, Caldwell MA, Jauniaux E, Starkey M, Svendsen CN, Emson P; Neuronal target genes of the neuron-restrictive silencer factor in neurospheres derived from fetuses with Down's syndrome: a gene expression study; **Lancet** 2002;359:310-5

54. Morimoto I, Sasaki Y, Ishida S, Imai K, Tokino T; Identification of the osteopontin gene as a direct target of TP53; **Genes Chromosomes Cancer** 2002;**33**: 270-8

55. Hursting SD, Shen JC, Sun XY, Wang TT, Phang JM, Perkins SN; Modulation of cyclophilin gene expression by N-4-(hydroxyphenyl)retinamide: association with reactive oxygen species generation and apoptosis; **Mol Carcinog** 2002;**33**:16-24

56. Kim CD, Choe Y, Shim C, Kim K; Interferon Beta secreted from human hair dermal papilla cells inhibits the growth of outer root sheath cells cultured in vitro; **Biochem Biophys Res Commun** 2002;290:1133-8

57. Sommerfeldt DW, Zhi J, Rubin CT, Hadjiargyrou M; Proline-rich transcript of the brain (prtb) is a serum-responsive gene in osteoblasts and upregulated during adhesion; **J Cell Biochem** 2002; 84: 301-8

58. Le Poole IC, Sarangarajan R, Zhao Y, Stennett LS, Brown TL, Sheth P, Miki T, Boissy RE; 'VIT1', a novel gene associated with vitiligo; **Pigment Cell Res** 2001;14:475-84

59. Liu QR, Zhang PW, Zheng Q, Walther D, Wang XB, Uhl GR; KEPI: A PKC-dependent protein phosphatase 1 inhibitor regulated by morphine; **J Biol Chem** 2002; **25**;

60. Aoki K, Sun YJ, Aoki S, Wada K, Wada E; Cloning, expression, and mapping of a gene that is upregulated in adipose tissue of mice deficient in bombesin receptor subtype-3; **Biochem Biophys Res Commun** 2002;**290**: 1282-8

61. Welsh J, Pretzman C, Postic D, Saint Girons I, Baranton G, McClelland M–Genomic fingerprinting by arbitrarily primed polymerase chain reaction resolves Borrelia burgdorferi into three distinct phyletic groups. **Int J Syst Bacteriol**. 1992; 42: 370-7.

62. Paulino LC, de Mello MP, Ottoboni LM; Differential gene expression in response to copper in Acidithiobacillus ferrooxidans analyzed by RNA arbitrarily primed polymerase chain reaction; **Electrophoresis** 2002; 23: 520-7

63. Neumann E, Kullmann F, Judex M, Justen HP, Wessinghage D, Gay S, Scholmerich J, Muller-Ladner U; Identification of differentially expressed genes in rheumatoid arthritis by a combination of complementary DNA array and RNA arbitrarily primed-polymerase chain reaction; **Arthritis Rheum** 2002; **46**: 52-63

64. Kullmann F, Judex M, Ballhorn W, Justen HP, Wessinghage D, Welsh J, Yen TJ, Lang B, Hittle JC, McClelland M, Gay S, Scholmerich J, Muller-Ladner U; Kinesin-like protein CENP-E is upregulated in rheumatoid synovial fibroblasts; **Arthritis Res** 1999;**1**:71-80

65. Sato M, Yoshida S, Nagao K, Nishizono S, Kusaba M, Hung MC, Ikeda I, Imaizumi K; Application of the random arbitrary primed polymerase chain reaction differential display method to isolate genes of cholesterol metabolism-related proteins from rat liver; **Biosci Biotechnol Biochem** 2000; **64:** 1058-60

66. Sakamoto H, Kinoshita K, Nakayama Y, Takami T, Ohtani K, Izuka S, Satoh K; Gene expression in primary, metastatic and recurrent lesions of endometrial cancer; **Gynecol Obstet Invest** 1999; **48** 215-20

67. Neto ED, Harrop R, Correa-Oliveira R, Wilson RA, Pena SD, Simpson AJ; Minilibraries constructed from cDNA generated by arbitrarily primed RT-PCR: an alternative to normalized libraries for the generation of ESTs from nanogram quantities of mRNA; **Gene** 1997; **186**: 135-42

68. Diachenko LB, Ledesma J, Chenchik AA, Siebert PD; Combining the technique of RNA fingerprinting and differential display to obtain differentially expressed mRNA; **Biochem Biophys Res Commun** 1996; **219**: 824-8

69. Yamashita R, Matsubara K, Kato K; A comprehensive collection of mouse zinc finger motifs compiled by molecular indexing; **Gene** 2001; **274**: 101-10

70. Doyu M, Sawada K, Mitsuma N, Niwa J, Yoshimoto M, Fujii Y, Sobue G, Kato K; Gene expression profile in Alzheimer's brain screened by molecular indexing; **Brain Res Mol Brain Res** 2001; **87**: 1-11

71. Shaw-Smith CJ, Coffey AJ, Leversha M, Freeman TC, Bentley DR, Walters JR Characterisation of a novel murine intestinal serine protease, DISP; **Biochim Biophys Acta** 2000; **1490:** 131-6

72. Sargent TD and Dawid IB (1983)–Differential gene expression in the gastrula of Xenopus laevis. **Science 222**: 135–139.

73. Davis MM, Cohen DI, Nielsen EA, Steinmetz M, Paul WE and Hood L (1984)–Cell-type-specific CDNA probes and the murine 1 region: the localization and orientation of Ad. **Proc Natl Acad Sci 81:** 2194–2198

74. Gurskaya NG, Diatchenko L, Chenchik A, Siebert PD, Khaspekov GL, Lukyanov KA, Vagner LL, Enmolaeva OD, Lukyanov SA and Sverdlov ED (1996)–Equalising cDNA subtraction on selective suppression of polymerase chain reaction: cloning of Jurkat cell transcriptsinduced by phytochemagglutinin and phorbol 12 myristate 13-acetate. **Anal. Biochem 240:** 90–97.

75. Diatchenko L, Lau Y.-F.C, Campbell AP, Chenchik A, Moqadam F, Huang B, Lukyanov S, Pukyanov K, Gurskaya N, Sverdlov E and Siebert PD (1996)–Supression subtractive hybridization: A method for generating differentially regulated or tissue-specific cDNA probes and libraries. **Proc Natl Acad Sci 93**: 6025–6030.

76. Moody DE (2001)–Genomics techniques: An overview of methods for the study of gene expression. **J. Anim. Sci. 79** (E.Suppl.): E128–E135.

77. Eisen, MB, Spellman PT, Brown PO and Botstein D; Cluster analysis and display of genome-wide expression pattern; **Proc Natl Acad Sci** 1998; **95**:14863–14868.

78. Thomas D. Wu; Analysing gene expression data from DNA microarrays to identify candidate genes; **The Journal of Pathology**; 2001; **195**: 53-65

79. Nicola J. Maughan, Fraser A. Lewis, Victoria Smith; An introduction to arrays; **The Journal of Pathology**; 2001; **195**: 3-6

80. Rockett, JC and Dix, DJ: DNA arrays: technology, options and toxicological applications. **Xenobiotica** 2000, **30**:155–177.

81. Celis, J.E., Ostergaard, M., Jensen, N.A., Gromova, I., Rasmussen, H.H. and Gromov, P. (1998). Human and mouse proteomic databases: novel resources in the protein universe. **FEBS Lett**. 430, **1-2**, 64-72.

82. Celis, J.E., Gromov, P., Ostergaard, M., Madsen, P., Honore, B., Dejgaard, K., Olsen, E., Vorum, H., Kristensen, D.B., Gromova, I., Haunso, A., Van Damme, J., Puype, M., Vandekerckhove, J. and

Rasmussen, H.H. (1996). Human 2-D PAGE databases for proteome analysis in health and disease: http://biophase.dk/cgi-bin/celis. **FEBS Lett.** 398, **2-3**, 129-134.

83. Bairoch, A. (1997). Proteome databases. In M.R. Wilkins, K.L. Williams, R.D. Appel and D.F. Hochstrasser (Eds.), Proteome Research: New Frontiers in Functional Genomics. Springer, New York, pp. 93-132.

84. Anderson, N.L. and Anderson, N.G. (1998). Proteome and proteomics: new technologies, new concepts and new words. **Electrophoresis** 19, 11, 1853-1861.

85. Boulassel MR, Tomasi JP, Deggouj N, Gersdorff M–COCH5B2 is a target antigen of anti-inner ear antibodies in autoimmune inner ear diseases. Otol **Neurotol** 2001; **22**:614-8

86. Ahmed A, Kandola P, Ziada G, Parenteau N- Purification and partial amino acid sequence of proteins from human epidermal keratinocyte conditioned medium. **J Protein Chem** 2001;**20:** 273-8

87. Ferchichi M, Fathallah M, Mansuelle P, Rochat H, Sabatier JM, Manai M, Mabrouk K Chemical synthesis, molecular modeling, and antimicrobial activity of a novel bacteriocin, MMFII **Biochem Biophys Res Commun** 2001; **289**: 13-8

88. Greller LD and Tobin FL- Detecting selective expression of genes and proteins.**Genome Res** 1999; **9**: 282-96.

89. Haberkorn U, Altmann A and Eisenhut M–Functional genomics and proteomics–the role of nuclear medicine. **Eur. J. Nuc. Med.** 2002; 29: 115–132.

90. Ziauddin J, Sabatini DM. Microarrays of cells expressing defined cDNAs. **Nature** 2001; 411: 107-10

91. Gonzalez JE and Negulescu PA Intracellular detection assays for high-throughput screening **Curr Opin Biotechnol** 1998; **9**: 624-31

92. Buysse JM The role of genomics in antibacterial target discovery. **Curr Med Chem** 2001; **8**: 1713-26

93. Civelli O and Nothacker HP. Functional genomics and the discovery of new drug targets. **Diabetes Technol Ther** 1999; **1**: 71-6

94. Wilson S, Bergsma DJ, Chambers JK, Muir AI, Fantom KG, Ellis C, Murdock PR, Herrity NC, Stadel JM. Orphan G-protein-coupled receptors: the next generation of drug targets? **Br J Pharmacol** 1998; **125**: 1387-92

95. Bailey JE–Reflections on the scope and the future of metabolic engineering and its connections to functional genomics and drug discovery. **Metab Eng** 2001; **3**: 111-4

96. Patrik D'haeseleer, Shoudan Liang and Roland Somogyi Genetic network inference: from co-expression clustering to reverse engineering. **Bioinformatics** 2000; **16**:707-726

97. del Rio G, Bartley TF, del-Rio H, Rao R, Jin KL, Greenberg DA, Eshoo M, Bredesen DE. Mining DNA microarray data using a novel approach based on graph theory.**FEBS Lett**. 2001; **509**: 230-4.

98. Buolamwini JK. Cell cycle molecular targets in novel anticancer drug discovery. **Curr Pharm Des** 2000; **6**: 379-92

99. Lukas TJ, Mirzoeva S, Slomczynska U, Watterson DM.Identification of novel classes of protein kinase inhibitors using combinatorial peptide chemistry based on functional genomics knowledge. **J Med Chem** 1999;**42**: 910-9

100. Cascante M, Boros LG, Comin-Anduix B, de Atauri P, Centelles JJ, Lee PW. Metabolic control analysis in drug discovery and disease. **Nat Biotechnol** 2002; **20:** 243-9

101. Peter Uetz, Loic Giot, Gerard Cagney, Traci A. Mansfield, Richard S. Judson, James R. Knight, Daniel Lockshon, Vaibhav Narayan, Maithreyan Srinivasan, Pascale Pochart, Alia Qureshi-Emili, Ying Li, Brian Godwin, Diana Conover,Theodore Kalbeisch, Govindan Vijayadamodar, Meijia Yang, Mark Johnston, Stanley Fields and Jonathan M. Rothberg–A comprehensive analysis of protein-protein interactions in Saccharomyces cerevisiae **Nature** 2000; **403**: 623-627.

102. Smith GR, Sternberg MJ. Prediction of protein-protein interactions by docking methods. **Curr Opin Struct Biol** 2002; **12**: 28-35

103. Lind KE, Du Z, Fujinaga K, Peterlin BM, James TL. Structure-Based Computational Database Screening, In Vitro Assay, and NMR Assessment of Compounds that Target TAR RNA. **Chem Biol** 2002; 9: 185-193

104. Zabell AP, Post CB. Docking multiple conformations of a flexible ligand into a protein binding site using NMR restraints **Proteins** 2002; **46**: 295-307

105. Schneider G, Bohm HJ. Virtual screening and fast automated docking methods. **Drug Discov Today** 2002; **7**: 64-70

106. Emili AQ and Cagney G–Large scale functional analysis using peptide or protein arrays, **Nat Biotech** 2000, **18**: 393–397.

107. Zhu H and Snyder M–Protein arrays and microarrays, **Curr opin Chem** Biol2001; **5**:40– 45.

108. He M, Taussig MJ.-Single step generation of protein arrays from DNA by cell-free expression and in situ immobilisation (PISA method); **Nucleic Acids Res.** 2001; **29**:E73-3.

109. Haab BB. Advances in protein microarray technology for protein expression and interaction profiling. **Curr Opin Drug Discov Devel** 2001; **4**: 116-23

110. Nolan JP, Sklar LA. Suspension array technology: evolution of the flat-array paradigm **Trends Biotechnol** 2002; **20**: 9-12

111. Petty R, Sutherland L, Hunter E and Cree I–Comparison of MTT and ATP-based assays for the measurement of viable cell number. **J Biolumin Chemilumin** 1995; **10**: 29–34.

112. Kain S–Green fluorescent protein (GFP) applications in cell based assays for drug discovery, **Drug Disc** Today; 1999; **4**: 304–312

113. Molina A, Carpeaux R, Martial JA, Muller M–A transformed fish cell line expressing a green fluorescent protein-luciferase fusion gene responding to cellular stress **Toxicol In Vitro** 2002;**16**: 201-7

114. Kim WH; Tissue array technology for translational research. From gene discovery to application **Exp Mol Med** 2001; **33(1 Suppl)**: 135-48

115. Pessina A, Albella B, Bueren J, Brantom P, Casati S, Gribaldo L, Croera C, Gagliardi G, Foti P, Parchment R, Parent-Massin D, Sibiril Y, Van Den Heuvel R–Prevalidation of a model for predicting acute neutropenia by colony forming unit granulocyte/macrophage (CFU-GM) assay; **Toxicol In Vitro** 2001;**15**: 729-40

116. Torrance CJ, Agrawal V, Vogelstein B, Kinzler KW Use of isogenic human cancer cells for high-throughput screening and drug discovery **Nat Biotechnol** 200; **19:** 940-5

117. Parr AL, Myers TG, Holbeck SL, Loh YJ, Allegra CJ–Thymidylate synthase as a molecular target for drug discovery using the National Cancer Institute's Anticancer Drug Screen; **Anticancer Drugs** 2000;**12**: 569-74

118. Matthews DJ, Kopczynski J. Using model-system genetics for drug-based target discovery. **Drug Discov Today** 2001; **6**: 141-149

119. Bonnie E and Rothberg G -The Use of Animal Models In Expression Pharmacogenomic Analyses **The Pharmacogenomics Journal** 2001 **1**:48-58

120. Vanagaite L, James MR, Rotman G, Savitsky K, Bar-Shira A, Gilad S, Ziv Y, Uchenik V, Sartiel A, Collins FS, **et al.,** A high-density microsatellite map of the ataxia-telangiectasia locus **Hum Genet** 1995; **95**: 451-4

121. Ellis NA, Groden J, Ye TZ, Straughen J, Lennon DJ, Ciocci S, Proytcheva M, German J. The Bloom's syndrome gene product is homologous to RecQ helicases. **Cell** 1995; **83**: 655-66

122. Link EM, Hardiman G, Sluder AE, Johnson CD, Liu LX, Therapeutic target discovery using Caenorhabditis elegans **Pharmacogenomics** 2000; **1**: 203-17

123. Gomez M, De Castro E, Guarin E, Sasakura H, Kuhara A, Mori I, Bartfai T, Bargmann CI, Nef P–Ca2+ signaling via the neuronal calcium sensor-1 regulates associative learning and memory in C. elegans **Neuron** 2001; **30**:241-8

124. Stanyon CA, Finley RL Jr. Progress and potential of Drosophila protein interaction maps **Pharmacogenomics** 2000; **1**: 417-31

125. Ward AC and Lieschke GJ The zebrafish as a model system for human disease. **Front Biosci** 2002; **7**: D827-33

126. Nasevicius A, Ekker SC.–Effective targeted gene 'knockdown' in zebrafish. **Nat Genet.** 2000;**26**:216-20.

127. Shastry BS–Gene disruption in mice: models of development and disease, **Mol Cell Biochem** 1998, **181:** 163-79

128. Hutchinson JN, Muller WJ. Transgenic mouse models of human breast cancer, **Oncogene** 2000;**19:** 6130-7

129. Abu-Elheiga L **et al.,** 2001–Continuous fatty acid oxidation and reduced fat storage in mice lacking Acetyl CoA carboxylase 2. **Science 291**: 2613–2616.

130. Abun A, Holt KH, Platt KA, Sands AT and Zambrowicz BP–Full-sped mammalian genetics: **in vivo** target validation in the drug discovery process. **Trends Biotechnol** 2002; **20**: 36–42.

131. Harris S and Foord SM. Transgenic gene knock-outs: functional genomics and therapeutic target selection **Pharmacogenomics** 2000;**1**: 433-43.

132. Summerton J, Stein D, Huang SB, Matthews P, Weller D, Partridge M. Morpholino and phosphorothioate antisense oligomers compared in cell-free and in-cell systems **Antisense Nucleic Acid Drug Dev** 1997;7:63-70.

133. Driver SE, Robinson GS, Flanagan J, Shen W, Smith LE, Thomas DW, Roberts PC–Oligonucleotide-based inhibition of embryonic gene expression. **Nat Biotechnol** 1999;**17**: 1184-7.

134. Zhang H, Cook J, Nickel J, Yu R, Stecker K, Myers K and Dean NM–Reduction of liver fas expression by an antisense oligonucleotide protects mice from fulminant hepatitis. **Nat Biotech** 2000; **18**: 862–867.

135. Sandrasagra A, Tang L, Leonard SA, Teng K, Li Y, Mannion JC, Nyce JW. RASONs: a novel antisense oligonucleotide therapeutic approach for asthma.

136. Pabo CO, Peisach E, Grant RA Design and selection of novel Cys2His2 zinc finger proteins. **Annu Rev Biochem** 2001; **70**:313-40.

137. Klug A. Zinc finger peptides for the regulation of gene expression. **J Mol Biol** 1999 22;293(2):215-8.

138. Kang JS, Kim JS–Zinc finger proteins as designer transcription factors. **J Biol Chem** 2000 24;**275**:8742-8.

139. Hellborg F, Qian W, Mendez-Vidal C, Asker C, Kost-Alimova M, Wilhelm M, Imreh S, Wiman KG–Human wig-1, a p53 target gene that encodes a growth inhibitory zinc finger protein; **Oncogene** 2001; **20**: 5466-5474.

Index